1795451
AF807956

GUIDE TO HUMAN ANATOMY

RONALD PHILO, Ph.D.

Assistant Professor, Department of Neurobiology and Anatomy, The University of Texas Health Science Center at Houston, School of Medicine, Houston, Texas

MATTHEW S. BOSNER, M.D.

Department of Medicine, Washington University School of Medicine, St. Louis, Missouri

ANNE Le MAISTRE, M.D.

Department of Pathology, The University of Texas Health Science Center at Houston, School of Medicine, Houston, Texas

JOHN G. LINNER, Ph.S.

Research Assistant Professor, Speech and Hearing Institute, Houston, Texas

BENJAMIN H. VENGER, M.D.

Department of Neurosurgery, Baylor College of Medicine, Houston, Texas

1985

W. B. SAUNDERS COMPANY

Philadelphia, London, Toronto, Mexico City, Rio de Janeiro, Sydney, Tokyo

W. B. Saunders Company: West Washington Square
Philadelphia, PA 19103

1 St. Anne's Road
Eastbourne, East Sussex BN21 3UN, England

1 Goldthorne Avenue
Toronto, Ontario M8Z 5T9, Canada

Apartado 26370 - Cedro 512
Mexico 4, D.S., Mexico

Rua Coronel Cabrita, 8
Sao Cristovao Caixa Postal 21176
Rio de Janeiro, Brazil

9 Waltham Street
Artarmon, N.S.W. 2064, Australia

Ichibancho, Central Bldg., 22-1 Ichibancho
Chiyoda-Ku, Tokyo 102, Japan

Philo, Ronald.
Guide to human anatomy.

1. Anatomy, Human--Outlines, syllabi, etc.
I. Bosner, Matthew S. II. LeMaistre, Anne.
III. Linner, John G. IV. Venger, Benjamin H.
V. Title. [DNLM: 1. Anatomy. QS 4 P565g]
QM31.P45 1985 611 84-22166
ISBN 0-7216-1203-2

Guide to Human Anatomy ISBN 0-7216-1203-2

© 1985 by W. B. Saunders Company. Copyright under the Uniform Copyright Convention. Simultaneously published in Canada. All rights reserved. This book is protected by copyright. No part of it may be reproduced, stored in a retrieval system, or transmitted in any form or by any means, electronic, mechanical, photocopying, recording or otherwise, without written permission from the publisher. Made in the United States of America. Press of W. B. Saunders Company. Library of Congress catalog card number 84-22166

Last digit is the print number: 9 8 7 6 5 4 3 2 1

DEDICATED TO

THE

STUDENTS, FACULTY AND STAFF

OF

THE UNIVERSITY OF TEXAS
MEDICAL SCHOOL AT HOUSTON

IN HONOR OF THE CLASS OF 1983

FOREWORD

This book is the result of a four-year-long student-faculty interaction to produce a study aid for human anatomy. Included are pictorial summaries of body regions, succinct text, clinical comments and a section of computed tomograms of key body levels. We do not intend for this book to be the primary source for your education, but rather to serve as an aid in the integration of lecture, text and laboratory materials and for future quick reference.

We wish to acknowledge many people in this endeavor, but especially The University of Texas Health Science Center at Houston, Roger Bulger, M.D., President, for a grant to prepare this work. Grateful appreciation is also extended to Joe G. Wood, Ph.D., Chairman of Neurobiology and Anatomy, for his support in the collation of the guide, and to Kathleen G. Andreoli, D.S.N., Special Assistant for Educational Affairs, for her help in the project.

THE AUTHORS

CONTENTS

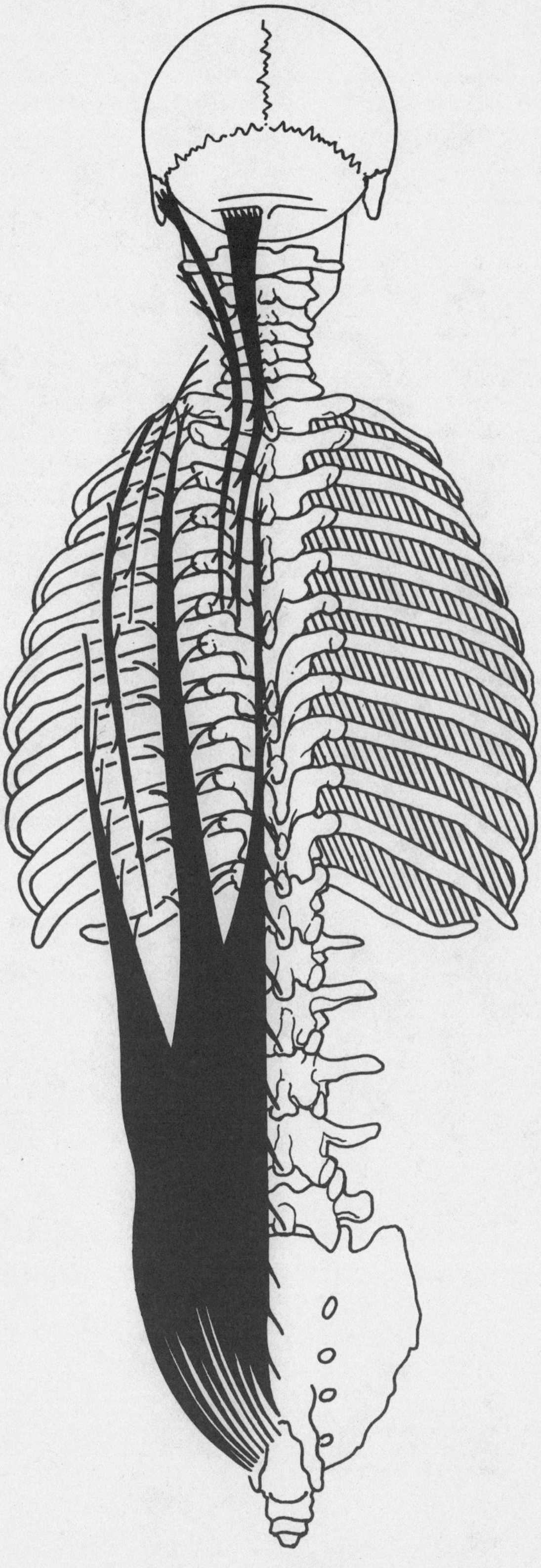

BACK

BACK, SPINAL CORD AND MENINGES

I. SKIN AND SUBCUTANEOUS TISSUES

SKIN

The skin is composed of two layers, the EPIDERMIS AND DERMIS. The epidermis is thin, but composed entirely of epithelial cells in various strata. The dermis is made of two different layers, the papillary layer and reticular layer. Blood supply and innervation to the skin travel through the subcutaneous tissues on the deep surface of the skin.

SUBCUTANEOUS TISSUES

These are the tissues deep to the skin. The subcutaneous tissues are best described as FASCIAS, or sheets of connective tissue. Often the subcutaneous fascias are infiltrated with fat. The SUBCUTANEOUS FASCIAS are often separated into superficial subcutaneous fascias--fatty--and deep subcutaneous fascias--membranous. The FATTY layers are well distributed on the trunk and limbs, while the MEMBRANOUS layers are well developed over the lower abdomen and in the upper arm and thigh. INVESTING FASCIAS are deep to the subcutaneous tissues but are often encountered in association with the subcutaneous fascias. The investing fascias invest each skeletal muscle and bone.

SKIN AND SUBCUTANEOUS TISSUES

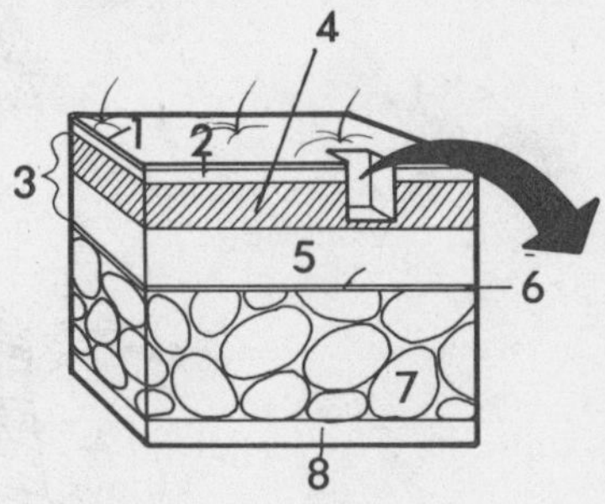

STRUCTURES:

1) Epidermis
2) Dermis
3) Hypodermis--also known as tela subcutanea or subcutaneous connective tissue
4) Superficial or Fatty subcutaneous fascia
5) Deep or Membranous subcutaneous fascia
6) Deep or Investing fascia
7) Muscle
8) Bone

II. THE BACK

OSTEOLOGY

The SPINAL COLUMN is formed by a series of 33 bones called VERTEBRAE which make a flexible shaft to support the weight of the trunk, head, and arms. The spinal column is divided into the following VERTEBRAL REGIONS:

1) Cervical--7 vertebrae
2) Thoracic--12
3) Lumbar--5
4) Sacral--5 fused
5) Coccygeal--3-5

CURVATURES:

Primary--in thoracic and sacral regions, concave anteriorly
Secondary--in cervical and lumbar regions, convex anteriorly
Kyphosis--a 'humpback' if extreme, but there is a natural kyphosis in the thoracic and sacral regions
Lordosis--a 'swayback' if extreme, but there is a natural lordosis in the cervical and lumbar regions
Scoliosis--a lateral deviation of the vertebral column

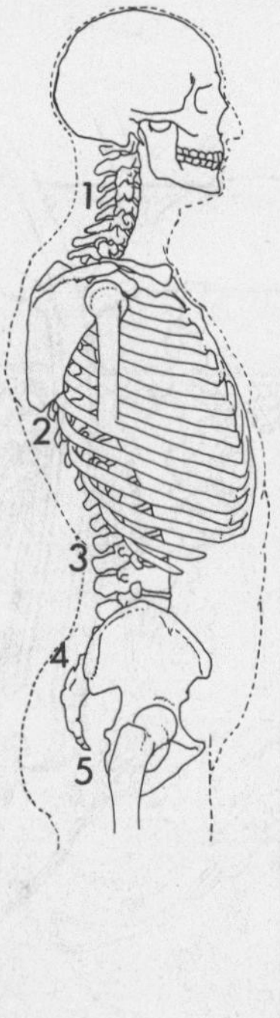

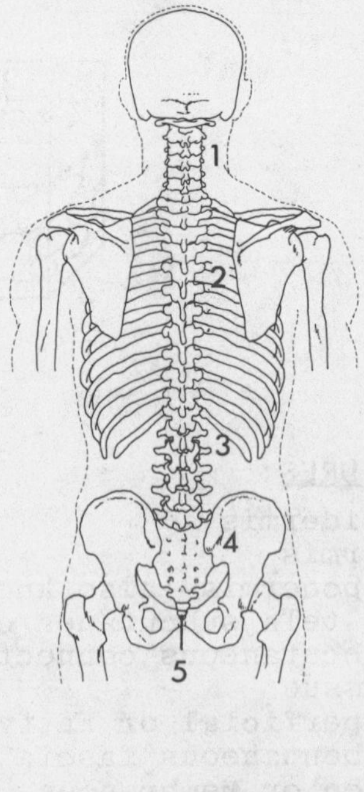

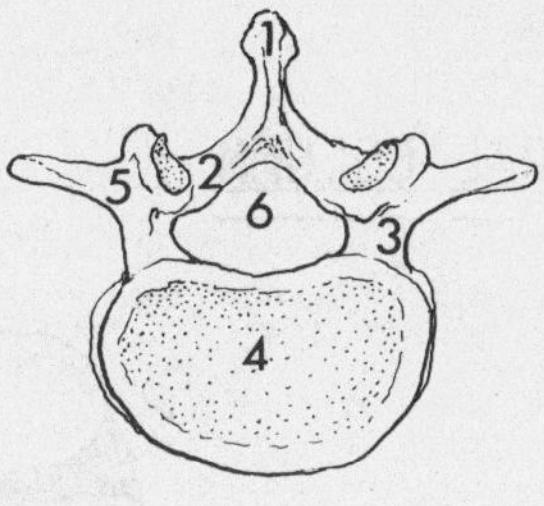

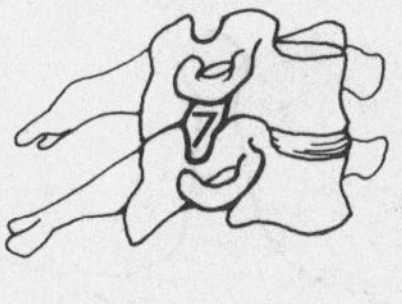

THE VERTEBRAE

PARTS OF A TYPICAL VERTEBRA:

1) Spine
2) Lamina
3) Pedicle--the pedicles are the feet or pillars which hold up the roof or lamina
4) Body
5) Transverse process
6) Vertebral foramen
7) Intervertebral foramen--shown between 2 adjacent vertebrae, the spinal nerve emerges here

CERVICAL VERTEBRAE--7

Oval body, triangular neural canal, bifid spine and (8) foramen transversarium for vertebral artery

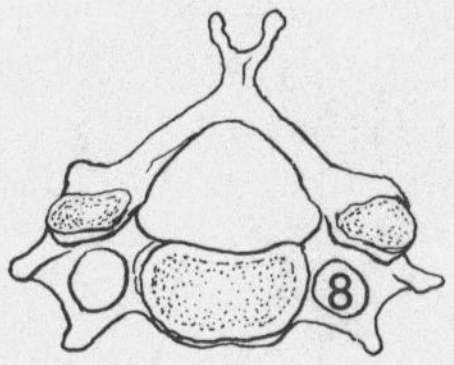

THORACIC VERTEBRAE--12

Round body and neural canal, long downward sloping spine, facet (9) on transverse process for tubercle of rib, demifacet (10) for head of rib

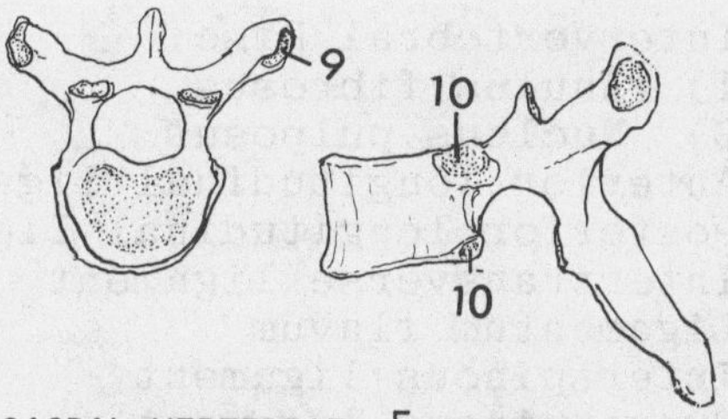

LUMBAR VERTEBRAE--5

Large oval body, triangular neural canal, thick posteriorly directed spine

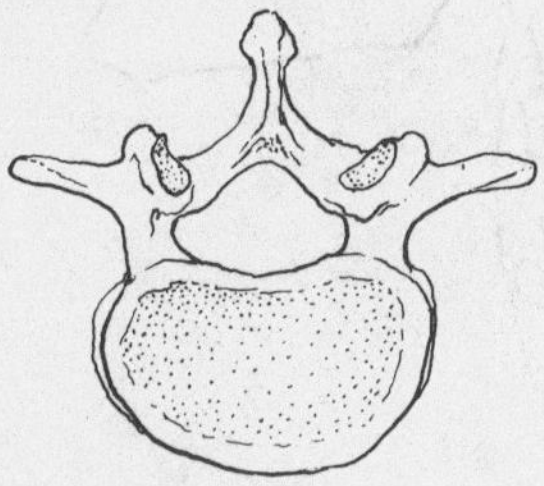

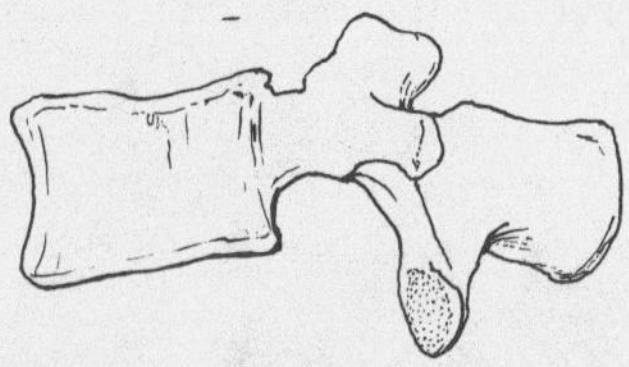

SACRAL VERTEBRAE--5 FUSED

11) Ala
12) Auricular surface for articulation with pelvis
13) Sacral canal 14) Hiatus

COCCYX--3-5 RUDIMENTARY

15) Cornu 16) Transverse process

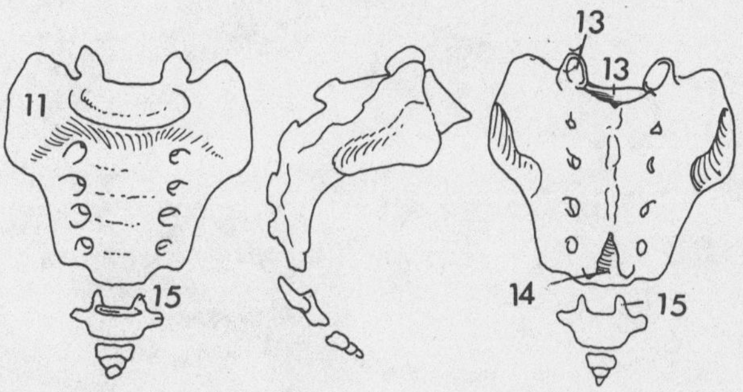

ARTICULATIONS OF THE VERTEBRAL COLUMN

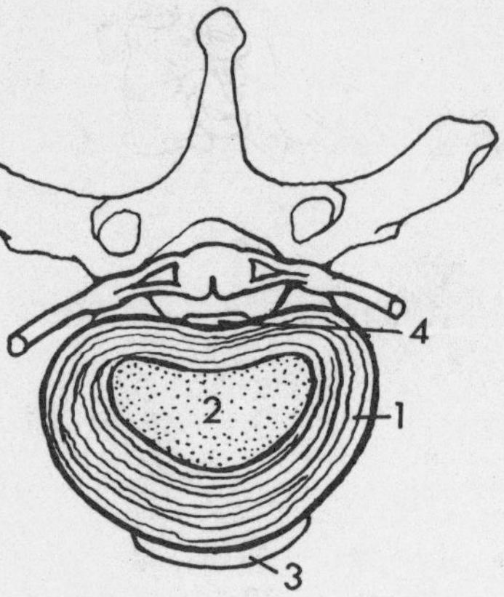

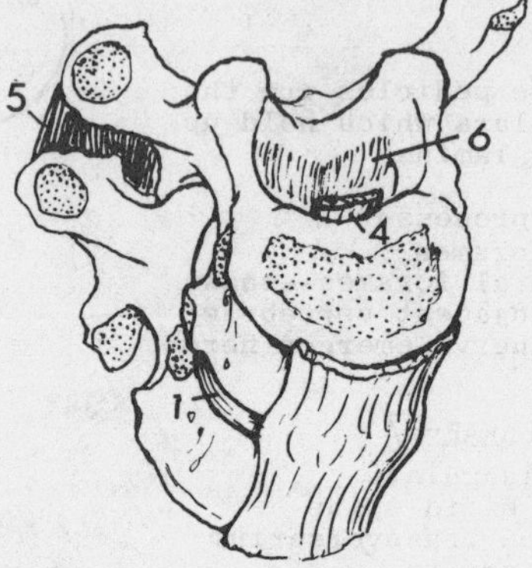

LIGAMENTOUS:

The Intervertebral Disc:
- 1) Anulus fibrosus
- 2) Nucleus pulposus

3) Anterior longitudinal ligament
4) Posterior longitudinal ligament
5) Intertransverse ligament
6) Ligamentum flavum
7) Interspinous ligament
8) Supraspinous ligament
9) Ligamentum nuchae--thickened portion of (8)

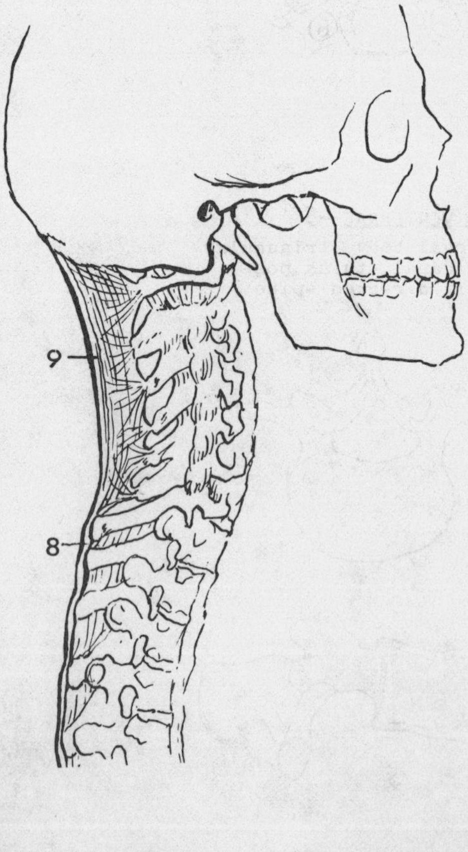

SYNOVIAL:

TRUE SYNOVIAL JOINTS occur at the articular processes. Shown here on lumbar vertebrae are the:

1) Superior articular surface
2) Inferior articular surface

An anterior and posterior surface of successive vertebrae articulate in a joint commonly called a zygapophyseal joint. Also, thoracic vertebrae have articulations with ribs and sacral vertebrae have articulations with os coxae.

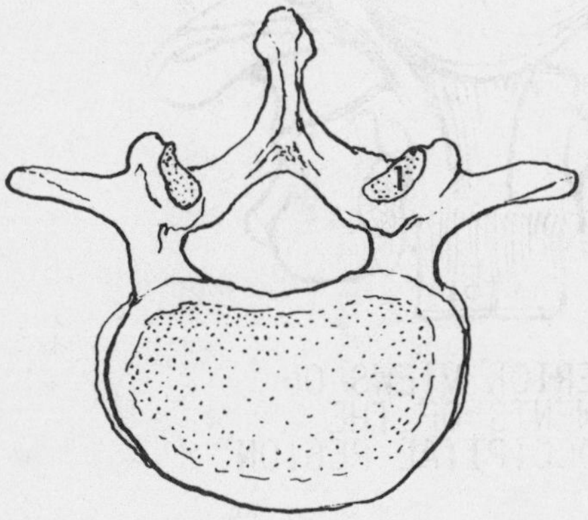

SUPERIOR VIEW

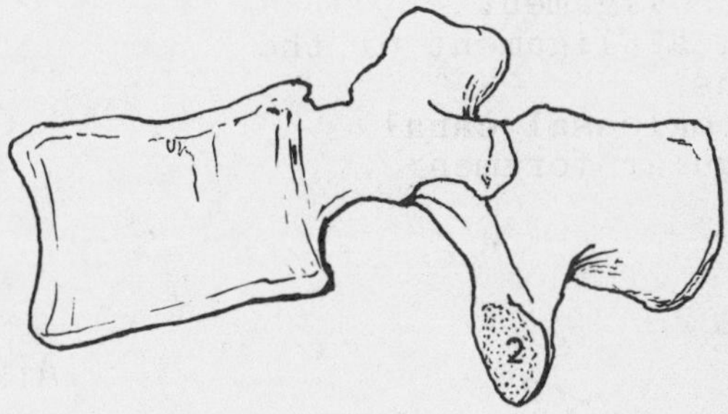

LATERAL VIEW

THE ATLANTO-OCCIPITAL REGION:

The first cervical vertebra or Atlas articulates with the skull. It does not have a spine and its body is fused to the 2nd vertebra or Axis as the Odontoid process or Dens. The Dens forms a pivot for rotation of the Atlas.

STRUCTURES:

1) Atlas
2) Dens
3) Articular surface for occipital bone
5) Tectorial membrane, divided and reflected
6) Articular capsule
7) Cruciform or Cruciate ligament--"cross" shaped
8) Alar ligament
9) Apical ligament of the Dens
10) Hypoglossal canal
11) Jugular foramen

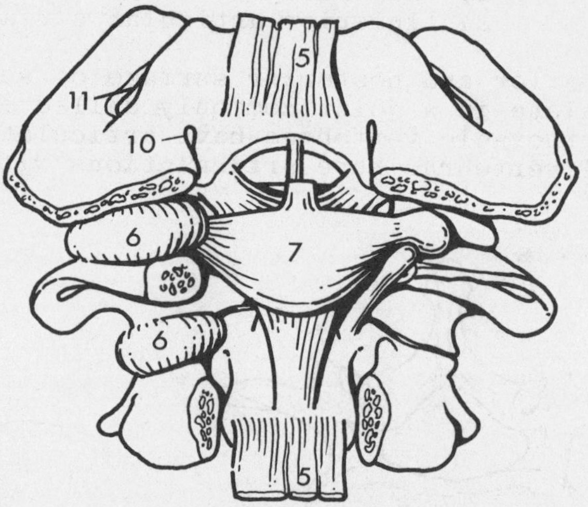

POSTERIOR VIEWS OF LIGAMENTS OF THE ATLANTO-OCCIPITAL REGION

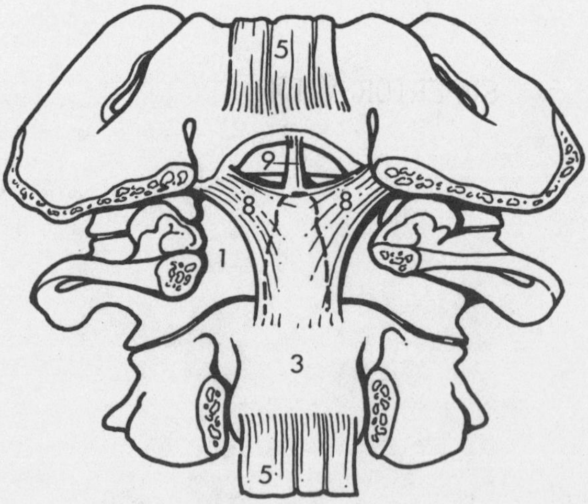

ANTERIOR VIEW OF ATLAS AND AXIS

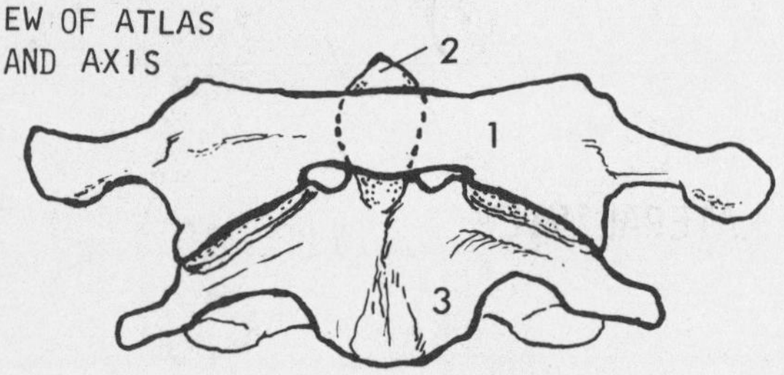

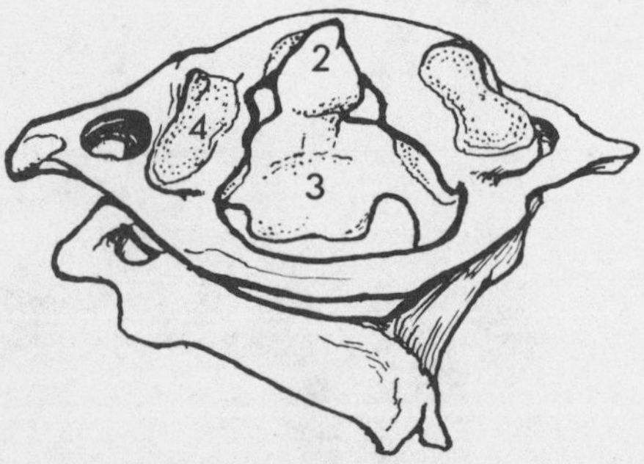

POSTERIOR VIEWS OF ATLAS AND AXIS

MUSCULATURE OF THE BACK

The musculature may be divided into those muscles originating on the back, but inserting elsewhere--EXTRINSIC MUSCLES--and those originating and inserting on the back for movement of the back--INTRINSIC MUSCLES.

EXTRINSIC MUSCLES include muscles of the upper extremity and two respiratory muscles. The trapezius, levator scapulae, latissimus dorsi, rhomboideus major and minor will be discussed with the upper extremity. The serratus posterior muscles, accessories to respiration, include:

1) SERRATUS POSTERIOR SUPERIOR
O-Ligamentum nuchae, spines C6,7-T1,2,3
I-Ribs 3,4,5--lateral to the angle
A-May elevate ribs, often the muscle is absent

2) SERRATUS POSTERIOR INFERIOR
O-Supraspinous ligament, spines L1,2,3
I-Lower 4 ribs, lateral to the angle
A-Depress ribs

BOTH MUSCLES receive their innervation and vasculature from intercostal nerves and arteries

INTRINSIC MUSCLES iclude the erector spinae, transversospinales, splenius and suboccipital muscles. The broad, flat tendon of the THORACOLUMBAR APONEUROSIS serves as an origin for several muscles, including the erector spinae and transversospinales groups.

STRUCTURES--SCHEMATIC:
1) Posterior lamina thoracolumbar aponeurosis
2) Anterior lamina thoracolumbar aponeurosis
3) Latissimus dorsi
4) External oblique
5) Internal oblique
6) Transverse abdominus
7) Quadratus lumborum
8) Psoas
9) Erector spinae group
10) Transversospinales group

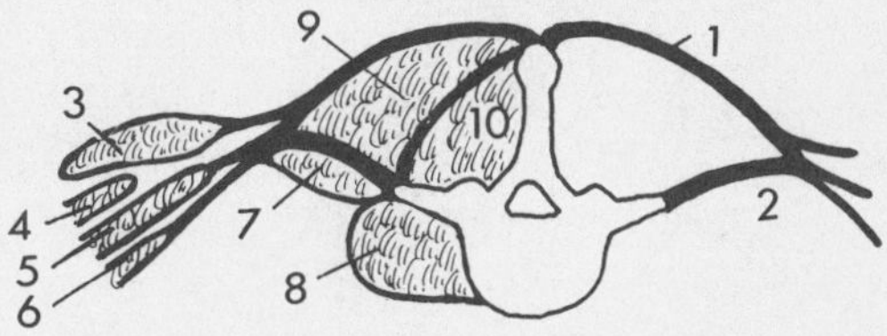

SCHEMATIC VIEW BACK MUSCLES
CROSS SECTIONAL VIEW

THE ERECTOR SPINAE AND TRANSVERSOSPINAL GROUPS:

THE ERECTOR SPINAE can be considered as one large muscle with one origin that splits into three masses--iliocostalis, longissimus, and spinalis--with intermediate tendons and multiple insertions. The VASCULATURE is derived from intercostals and lumbar segmental arteries as well as arteries to the posterior neck. INNERVATION is from posterior primary rami of spinal nerves. ACTION is to extend vertebral column and head and to assist in rotation and lateral movements. The ORIGIN of the erector spinae is from the thoracolumbar aponeurosis, lower lumbar vertebrae, sacrum, and coccyx. The INSERTIONS of the various groups will now be shown.

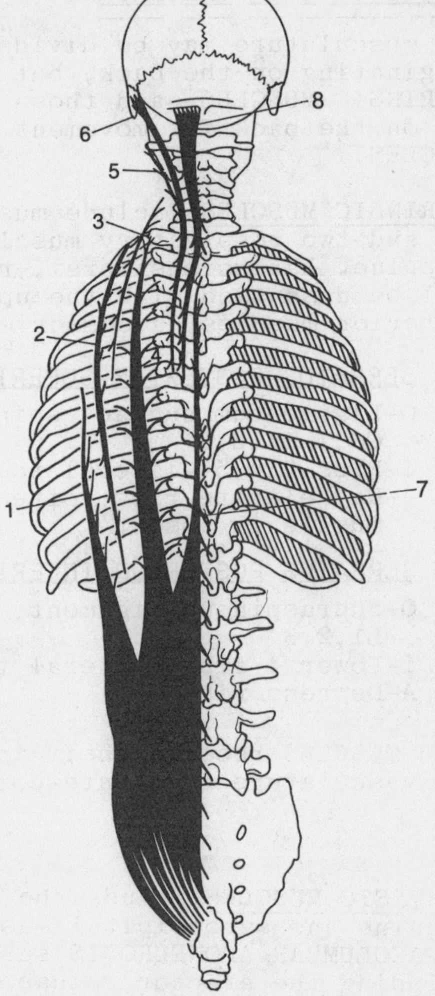

ILIOCOSTALIS

1) Lumborum--inserts at the angles of the lower 6-7 ribs
2) Thoracis--inserts at the angles of upper 6 ribs, transverse process C7
3) Cervicis--inserts transverse process C4,5,6

LONGISSIMUS

4) Thoracis--inserts transverse process all T vertebrae, and medial aspect lower 9-10 ribs
5) Cervicis--inserts transverse processes C2-6, T1-5
6) Capitus--inserts mastoid process

SPINALIS

7) Thoracis--united with semispinalis, inserts on spines upper T vertebrae
   Cervicis--often absent, blends with underlying semispinalis, inserts on spines C2-4
8) Capitis--blends with semispinalis, inserts on occipital bone

TRANSVERSOSPINALIS GROUP

These run obliquely from the transverse process of one vertebra to the spinous process of a preceding vertebra.

9) Semispinalis--blends with spinalis, found in thorax, cervical and occipital (capitis) regions, spans 5+ vertebrae
10) Multifidus--in sacral and lumbar regions, spans 2-4 vertebrae
11) Rotatores--mainly in thorax, the long rotators insert at the second preceding vertebra, the short at the 1st preceding

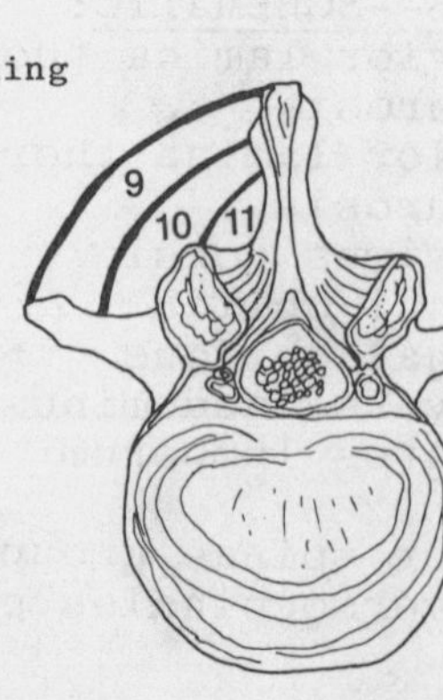

SUBOCCIPITAL TRIANGLE:

This triangle of muscles is found on the upper posterior neck and contains muscles important in extension and rotation of the head. The vertebral artery is found on the floor of the triangle.

MUSCLES:

1) RECTUS CAPITIS POSTERIOR MAJOR
   O-Spine axis
   I-Occipital bone
   A-Extend and rotate head
   N-C1, posterior ramus
   B-Vertebral, ascending and deep cervical arteries, and deep branch occipital a.
2) RECTUS CAPITIS POSTERIOR MINOR
   O-Arch atlas
   I-Occipital bone
   A-Extend head
   N-C1, posterior ramus
   B-Vertebral, ascending and deep cervical arteries, and deep branch occipital a.
3) OBLIQUE CAPITIS INFERIOR
   O-Spine axis
   I-Transverse process atlas
   A-Rotation of head
   N & B--same as above
4) OBLIQUE CAPITIS SUPERIOR
   O-Transverse process atlas
   I-Occipital bone
   A-Rotation of head
   N & B--same as above

OTHER STRUCTURES:

5) Vertebral artery
6) Occipital artery
7) Splenius capitis
8) Longissimus capitis
9) Semispinalis capitis
10) Trapezius
11) C1
12) C2--Greater occipital n.

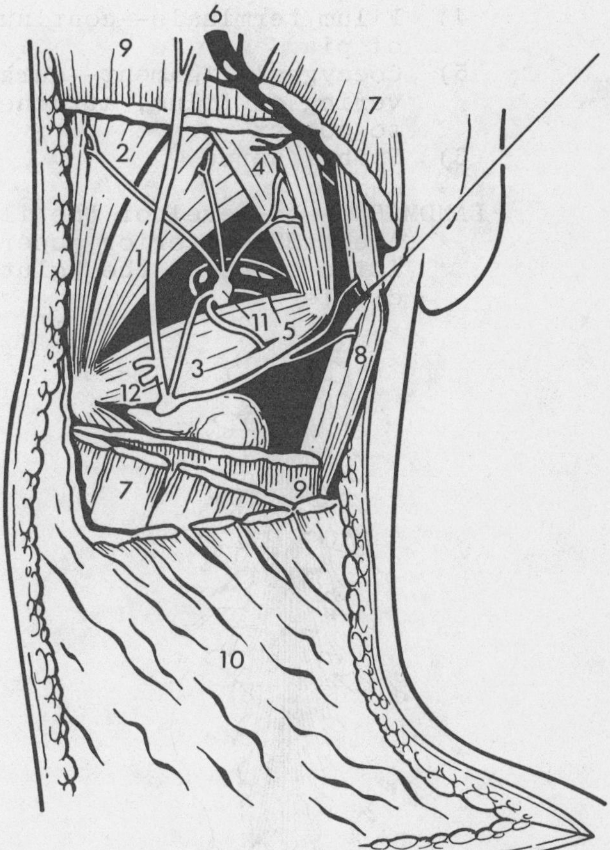

SUBOCCIPITAL TRIANGLE

III. SPINAL CORD AND MENINGES

THE SPINAL CORD

The spinal cord is a portion of the central nervous system, continuous with the medulla oblongata of the brain and occupies much of the vertebral canal, formed by successive vertebral foramina.

STRUCTURES:

1) Spinal cord
2) Conus medullaris--end of the cord, caudal border L1 or cranial border L2 in adult
3) Cauda equina--roots of the spinal nerves
4) Filum terminale--continuation of pia
5) Coccygeal ligament--dural covering of filum, attaches to coccyx
6) Spinal nerve

LANDMARKS--L4-Level of the iliac crest; S2-posterior superior iliac spine at the "gluteal dimple"

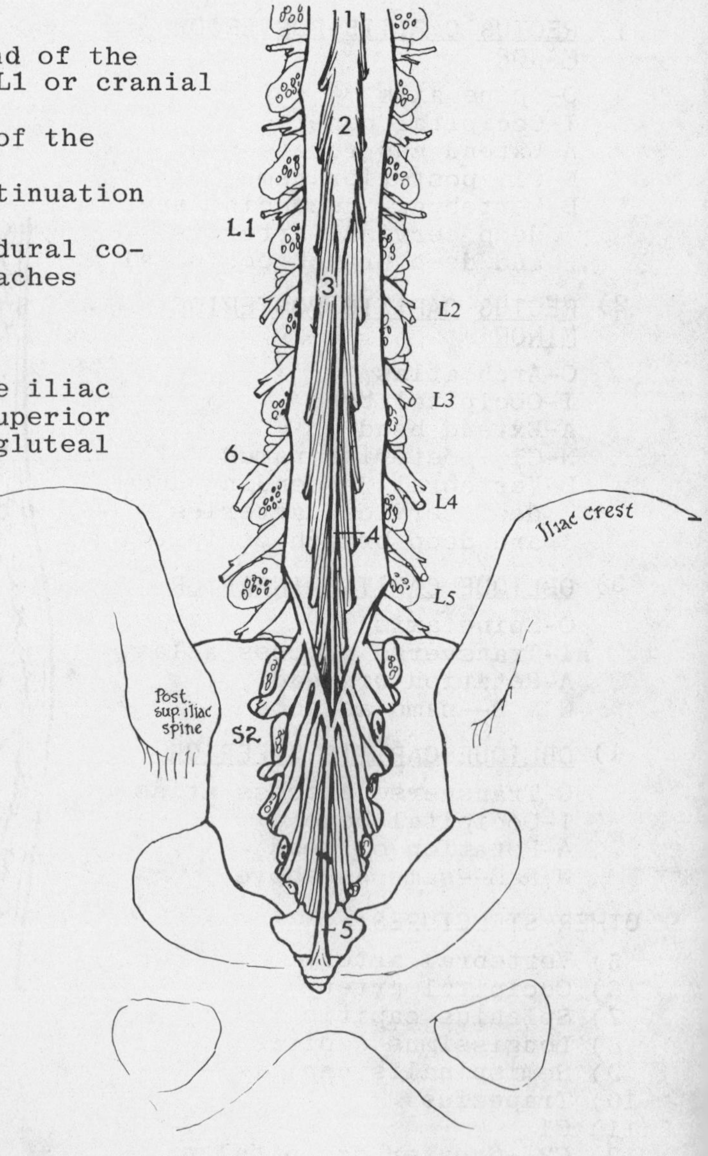

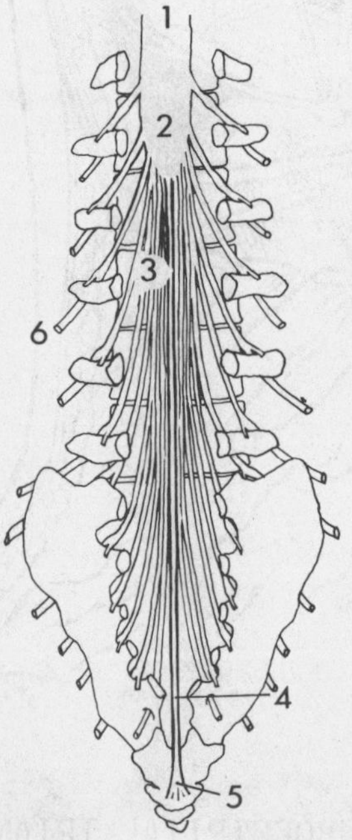

SPINAL NERVES

There are 31 pairs of spinal nerves that exit the intervertebral foramina; they are formed from anterior and posterior roots that attach to the spinal cord.

PARTS OF A SPINAL NERVE:

1) Anterior root--with rootlets attaching to cord
2) Posterior root ganglion--on posterior root
3) Spinal nerve
4) Posterior primary ramus
   5) Medial branch
   6) Lateral branch
7) Anterior primary ramus
   8) Lateral branch with anterior and posterior divisions
   9) Anterior branch with medial and lateral divisions

OTHER STRUCTURES:

10) Spinal cord
11) Sympathetic chain with ganglia and communicating rami
12) Vertebra

CUTANEOUS INNERVATION BACK:

From posterior primary rami

13) Greater occipital--C2
14) Medial branch posterior primary ramus
15) Lateral branch posterior primary ramus

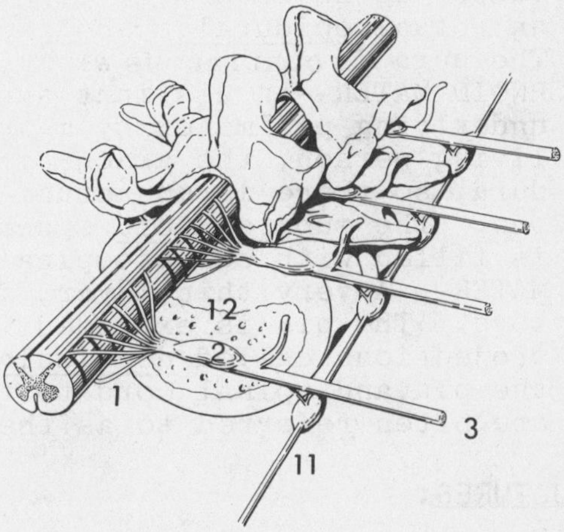

COMPOSITE PROJECTION

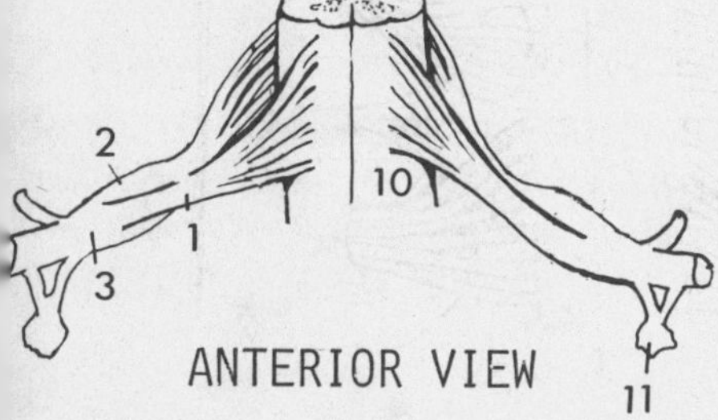

ANTERIOR VIEW

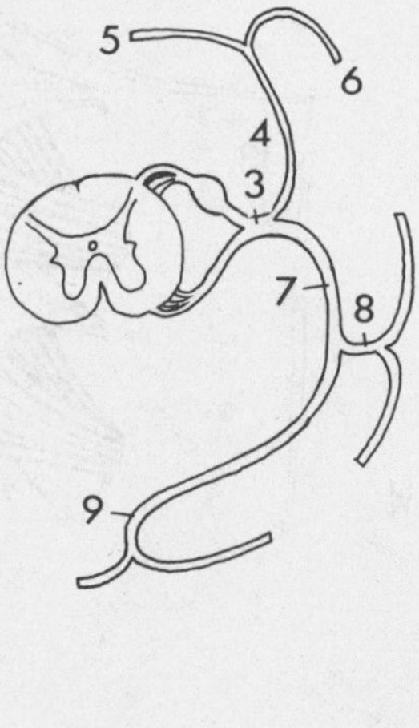

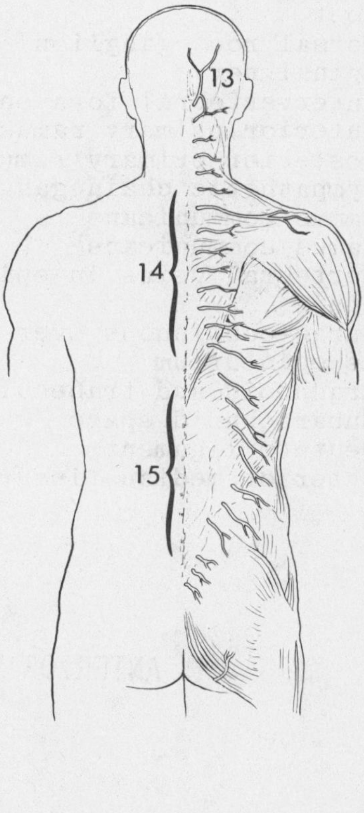

THE MENINGES

THE MENINGES are structures which line the spinal canal and ensheath the spinal cord. The meninges are three in number:

DURA MATER-- Tough and leather-like outer covering, which is not fused to the endosteum of the vertebrae. Therefore, there is an actual epidural space that contains fat and blood vessels. The dura is continuous with the epineurium of the spinal nerves.

ARACHNOID MATER-- A distinct and separate membrane connected to the underlying pia mater by means of fibrous trabeculae. In the living person, the arachnoid membrane is contiguous with the dura, so there is no actual subdural space, only a potential one. The subarachnoid space, between the arachnoid and pia, is filled with cerebrospinal fluid.

PIA MATER-- A very thin layer, intimately connected to the spinal cord. The pia is extended into distinct, triangular fibrous projections called denticulate ligaments. These serve to secure the pia and spinal cord to the dura. The pia and arachnoid are often referred to as the leptomeninx.

STRUCTURES:

1) Spinal cord
2) Anterior--motor--root
3) Posterior--sensory--root
4) Dorsal root ganglion
5) Spinal nerve
6) Intervertebral foramen
7) Anterior primary ramus
8) Posterior primary ramus
9) Sympathetic chain ganglion
10) Ramus communicans
11) Ramus communicans
12) Vertebral veins in epidural space
13) Dura--continuous over nerves as epineurium
14) Arachnoid and trabeculae
15) Subarachnoid space
16) Dentate ligament
17) Anterior median fissure

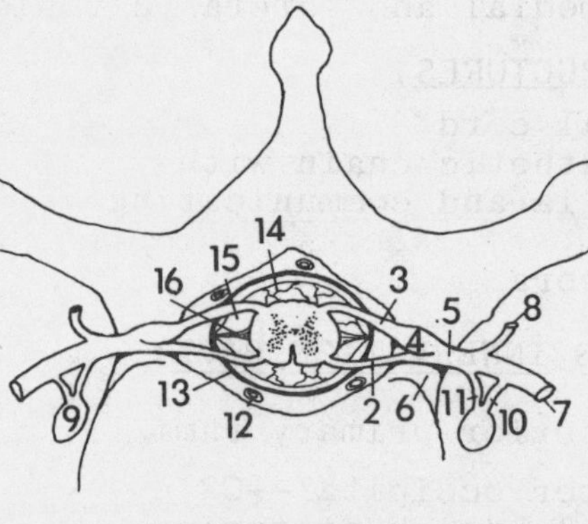

SUPERIOR VIEW

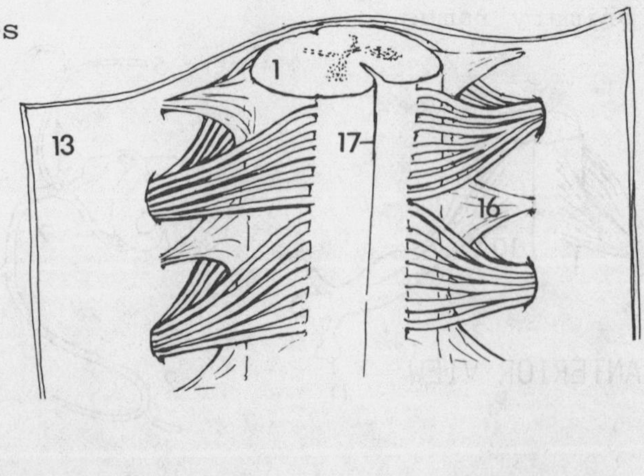

ANTERIOR VIEW

VASCULATURE OF THE SPINAL CORD

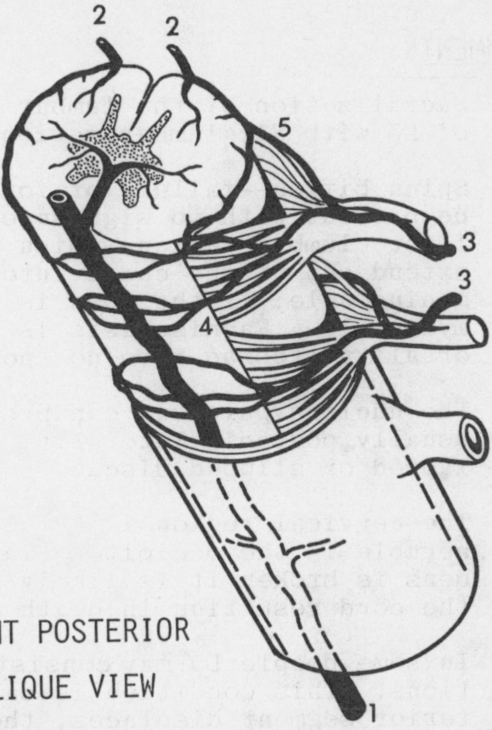

RIGHT POSTERIOR OBLIQUE VIEW

ARTERIES:

Derived from vertebral and segmental --intercostal, lumbar, etc. --arteries

1) Anterior spinal artery--from vertebral
2) Posterior spinal arteries--from vertebral
3) Segmental artery
4) Anterior radicular artery
5) Posterior radicular artery

These arteries form a vascular crown--vasocorona--around the cord. The vertebral arterial branches (1 & 2) can supply only the cervical segment in total; the rest is mainly supplied by segmentals.

VERTEBRAL VENOUS PLEXUS:

THE PLEXUS forms a valveless communication from pelvis to cranium, communicates with all major venous channels in the abdomen, chest and neck, and provides a route for spread of disease as well as a bypass for inferior vena caval circulation.

STRUCTURES:

1) Spinal cord
2) Dura
3) Epidural fat
4) Internal vertebral venous plexus
5) Posterior external plexus
6) Anterior external plexus

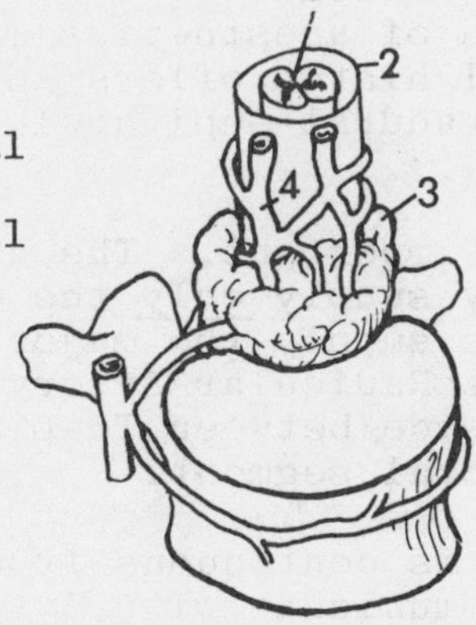

SUPERIOR

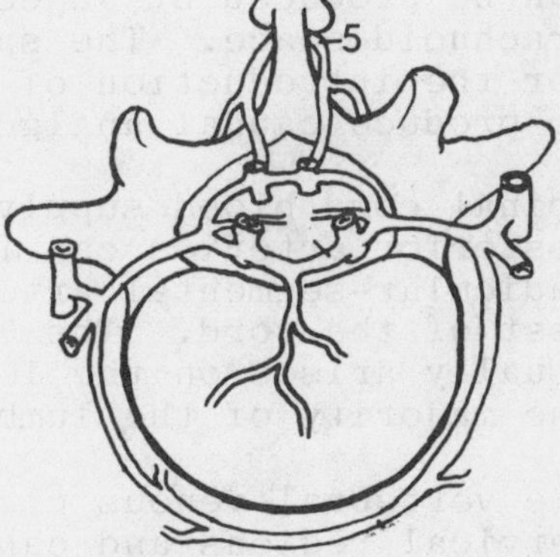

CROSS

COMMENTS.

1. Sacralization of the lumbar vertebra refers to fusions of L5 with S1; lumbarization of S1 refers to a separate S1.

2. Spina bifida--failure of formation of posterior arch--may be present with no sign save for a tuft of hair over the lower lumbar region--spina bifida occulta. If the meninges extend out of the canal into the defect, the condition is a meningocele; if the cord is involved the term is meningomyelocele. Rachischisis is a condition in which several or all vertebrae have no posterior arch.

3. The nucleus pulposus can protrude through the anulus fibrosus, usually posterio-laterally, and is called a herniated, prolapsed or slipped disc.

4. The cervical region is prone to trauma. Particularly vulnerable is the occipito-atlanto-axial joint. If the dens is broken it is likely to be driven posteriorly into the cord resulting in death or paralysis.

5. In some people L5 may consist of anterior and posterior portions. This condition is called spondylolysis. If the anterior segment displaces, the condition is spondylolisthesis.

6. A cisternal tap or puncture is the introduction of a needle into the cisterna magna percutaneously through the foramen magnum. A spinal tap is usually performed at L3-4 or L4-5 into the subarachnoid space. This procedure should not be performed if increased intracranial pressure is suspected because the lower brainstem and cerebellum may herniate through the foramen magnum. Alternately, spinal anesthesia can be produced by injection of anesthetics into the subarachnoid space. The sacral hiatus offers an easy approach for the introduction of extradural--epidural--anesthetics to produce caudal analgesia.

7. Spinal cord blood supply is important. The anterior and posterior arteries can fully supply <u>only</u> the cervical cord. Radicular-segmental arteries supply the majority of the rest of the cord. The Great Radicular Artery--of Adamkiewicz--usually arises on the left side between T6-L3; it supplies the majority of the lumbosacral segment.

8. The vertebral venous plexus is continuous from sacrum to cervical regions and can communicate with intracranial venous sinuses. It may be a route for spread of disease.

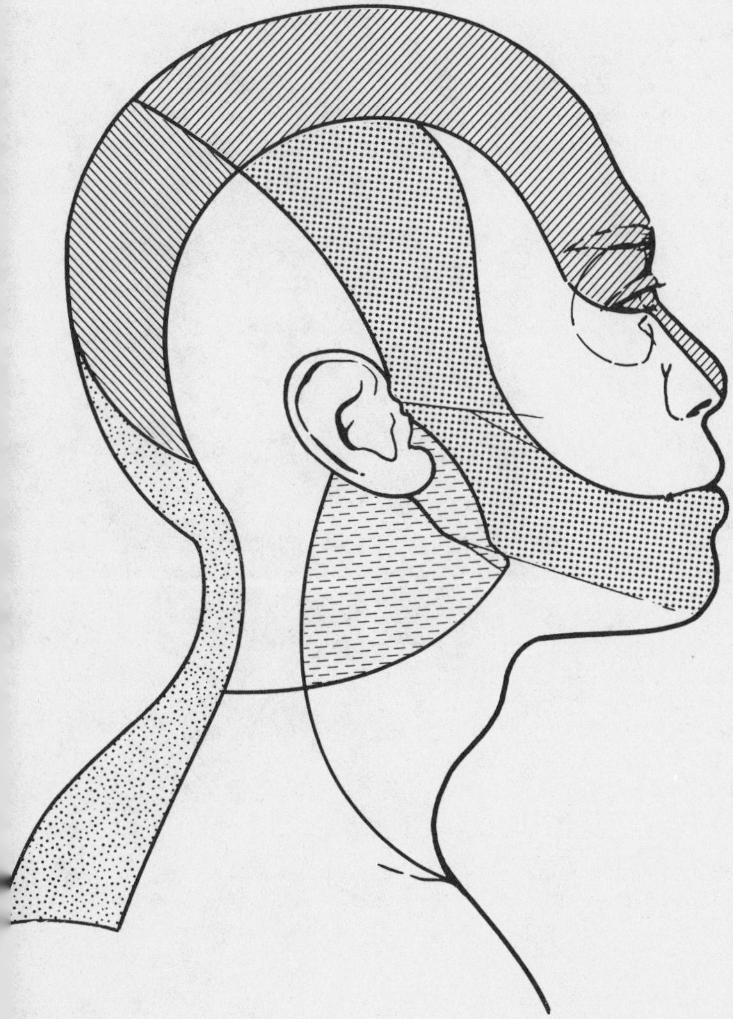

HEAD & NECK

SCALP, CALVARIA, CRANIAL CAVITY, AND BASE OF BRAIN

I. SCALP AND CALVARIA

THE SKULL is the skeleton of the head; CRANIUM is the skull without the mandible; CALVARIA is the dome or vault of the cranium; SCALP is the skin and structures covering the calvaria.

<u>STRUCTURES</u>:

SCALP--

1) <u>S</u>--skin
2) <u>C</u>--connective tissue; highly vascularized with many nerves
3) <u>A</u>--aponeurosis--galea aponeurotica; broad flat tendon of occipito-frontalis complex
4) <u>L</u>--loose connective tissue; 'DANGER ZONE' of scalp
5) <u>P</u>--Periosteum

Skin unit--layers (1,2,3)-- move as a unit

Epicranius--frontalis (6) and occipitalis (7) muscles, their aponeurosis (3) and some auricular muscles around the ear

CALVARIA--Portions frontal, parietal and occipital bones

8) Outer table--thick
9) Diploe--cancellous bone
10) Inner table--thin
11) Endocranium--also called periosteal layer of dura mater

<u>VASCULATURE</u>:

12) Supratrochlear
13) Supraorbital
14) Superficial temporal
15) Posterior auricular
16) Occipital

<u>INNERVATION</u>:

17) Supratrochlear
18) Supraorbital
19) Zygomatico-temporal
20) Auriculotem-poral
21) Lesser occipital
22) Greater occipital
23) 3rd occipital

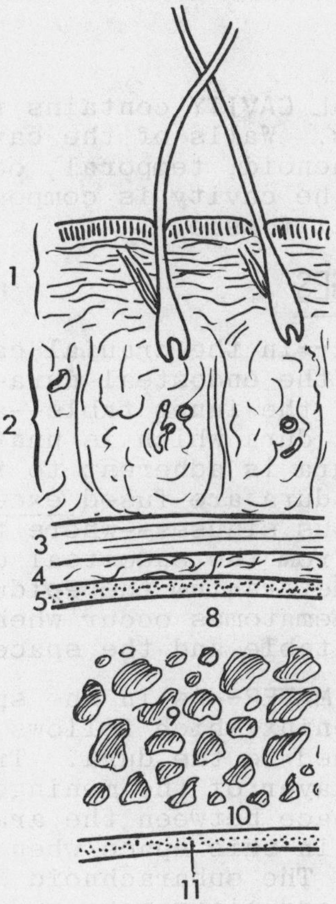

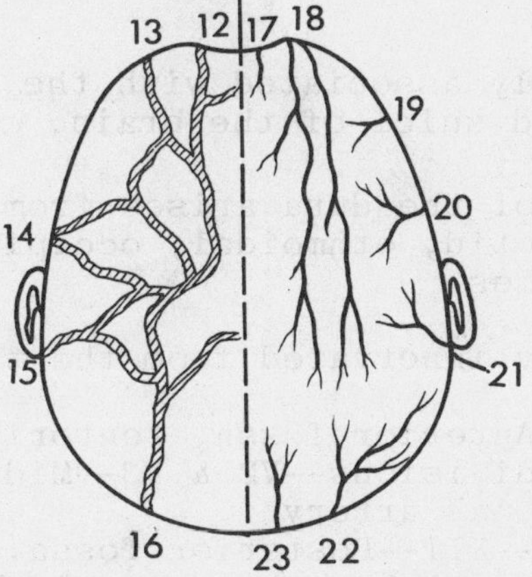

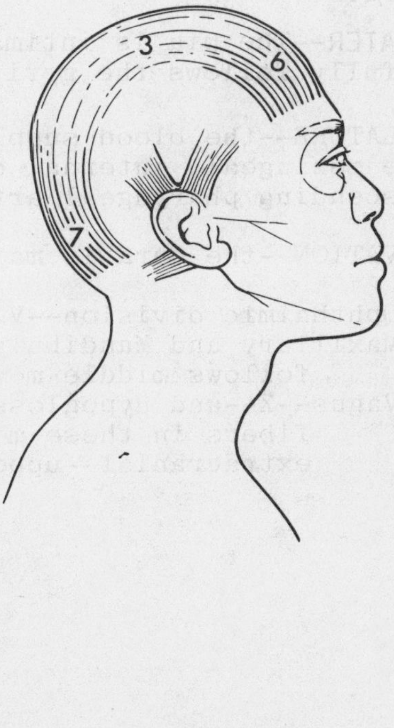

II. THE CRANIAL CAVITY

THE CRANIAL CAVITY contains the meninges, brain, blood vessels and nerves. Walls of the cavity are formed by the frontal, parietal, sphenoid, temporal, occipital and ethmoid bones. The floor of the cavity is composed of anterior, middle and posterior cranial fossae.

THE MENINGES

DURA MATER--In the cranial cavity, the dura consists of two portions, the endosteal dura--actually the periosteum or endosteum of the inner table--and the meningeal dura. Unlike the spinal dura which is unattached to the spinal canal, the cranial dura is adherent to the calvaria. The endosteal and meningeal dura are fused except at special locations--the dural venous sinuses--where the meningeal dura separates and reflects from the endosteal dura to form folds or projections. Thus in the cranium the epidural space is potential, not actual. Epidural hematomas occur when trauma separates the dura from the inner table and the space fills with blood.

ARACHNOID MATER--As in the spinal cord, the arachnoid is a spidery meninx which follows the contours of the dura but is not attached to the dura. Trabeculae connect the arachnoid to the pial layer of the meninges. The subdural space is the potential space between the arachnoid and the dura. Bleeding can occur in this space when veins entering the venous sinuses are torn. The subarachnoid space is that space between the arachnoid and pia mater, and is filled with cerebrospinal fluid. Hemorrhage in this area occurs when surface vessels of the brain rupture.

PIA MATER--The pia is intimately associated with the brain and faithfully follows the gyri and sulci of the brain.

VASCULATURE--the blood supply of the dura arises from the middle meningeal, internal carotid, ethmoidal, occipital, and ascending pharyngeal arteries.

INNERVATION--the dura is mainly innervated from the trigeminal:

- Ophthalmic division--V1--Anterior fossa, tentorium, falx
- Maxillary and Mandibular divisions--V2 & V3--Middle fossa, follows middle meningeal artery
- Vagus--X--and Hypoglossal--XII--Posterior fossa. Sensory fibers in these meningeal branches may also be of extracranial--upper cervical--origin

DURAL FOLDS

These folds compartmentalize the cranial cavity and contain the dural venous sinuses.

STRUCTURES:

1) Falx cerebri
2) Falx cerebelli
3) Tentorium cerebelli
4) Diaphragma sellae

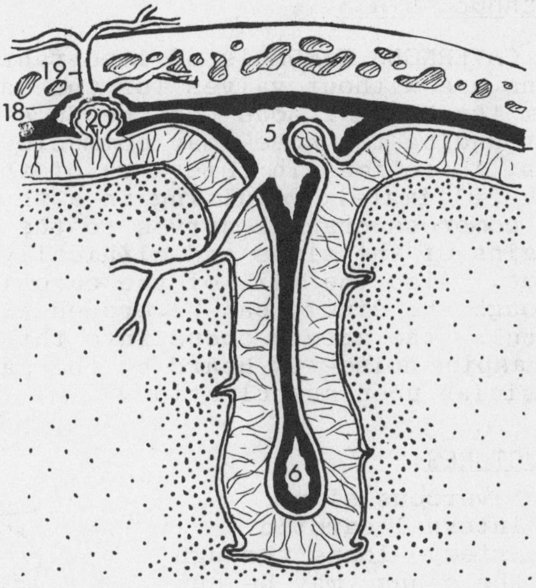

DURAL VENOUS SINUSES

The sinuses are spaces formed by reflections of the meningeal layer of the dura. These spaces are lined with endothelium of the blood vascular system. Veins draining the brain enter the sinuses which then drain, for the most, to the great veins of the neck.

STRUCTURES:

5) Superior sagittal sinus
6) Inferior sagittal sinus
7) Great cerebral vein --of Galen
8) Straight sinus
9) Confluence of the sinuses
10) Transverse sinus
11) Sigmoid sinus
12) Jugular vein
13) Superior petrosal sinus
14) Inferior petrosal sinus
15) Cavernous sinus
16) Ophthalmic vein
17) Venous channel to pterygomandibular sinus
18) Lateral lacuna
19) Emissary vein--connects sinuses to scalp veins
20) Arachnoid granulation--for reabsorption of CSF

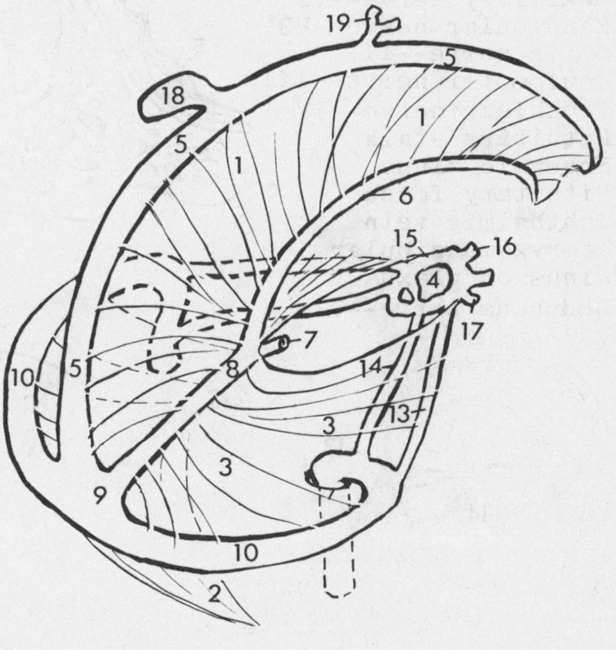

CAVERNOUS SINUS

THE CAVERNOUS SINUS is of important clinical concern. Facial veins are without valves for the sake of mobility and those from the "Danger Zone" of the face--a roughly triangular area surrounding the nose--may communicate with the cavernous sinus. Thus, infections in this area of the face may spread to the sinus, ultimately resulting in a thrombosis. The thrombosis can compress cranial nerves to the eye which passes through the margins of the sinus and clinically neurological impairment is noted. Also the wall of the carotid artery, which travels through the sinus, can be eroded with a resultant arterial-venous fistula (the artery opens into the sinus). A 'bruit'--murmur or rasping noise--is heard by the patient if conscious, and by the physician upon auscultation.

STRUCTURES:

1) Cavernous sinus
2) Internal carotid artery--loops through the sinus, may be cut twice on section
3) Cranial nerve V
4) Ophthalmic nerve--V1
5) Maxillary nerve--V2
6) Mandibular nerve--V3
7) Optic nerve--II
8) Oculomotor nerve--III
9) Trochlear nerve--IV
10) Pituitary stalk
11) Sphenoid sinus
12) Pituitary fossa
13) Ophthalmic vein
14) Pterygomandibular sinus or plexus
15) Abducens nerve--VI

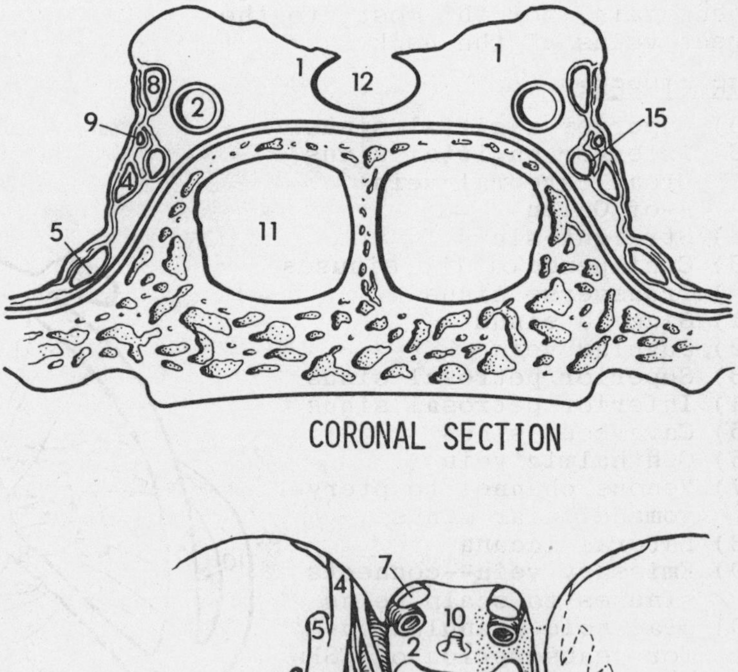

CORONAL SECTION

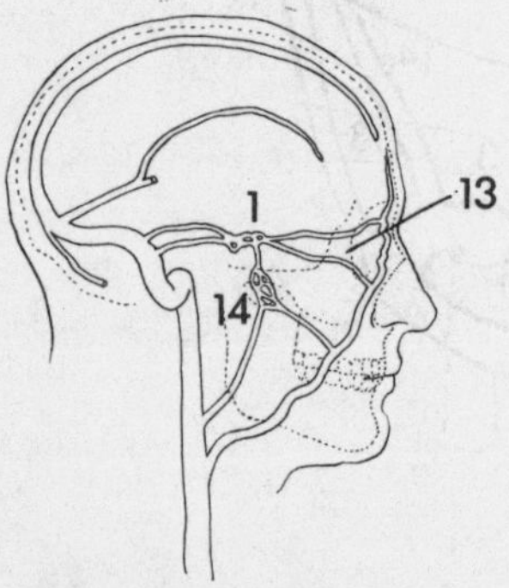

LATERAL PROJECTION

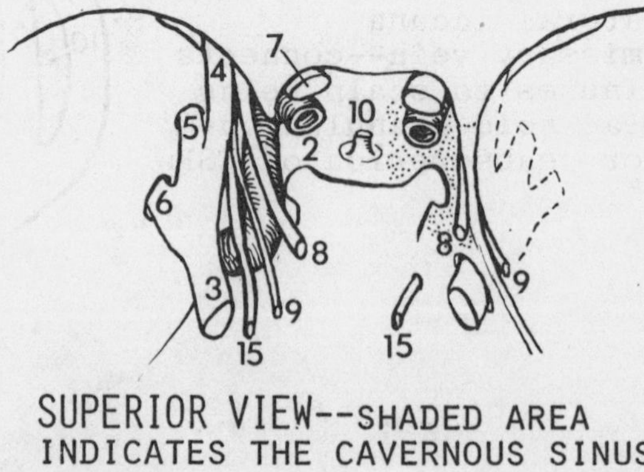

SUPERIOR VIEW--SHADED AREA INDICATES THE CAVERNOUS SINUS

III. BASE OF BRAIN AND CRANIAL NERVES

CRANIAL FOSSAE

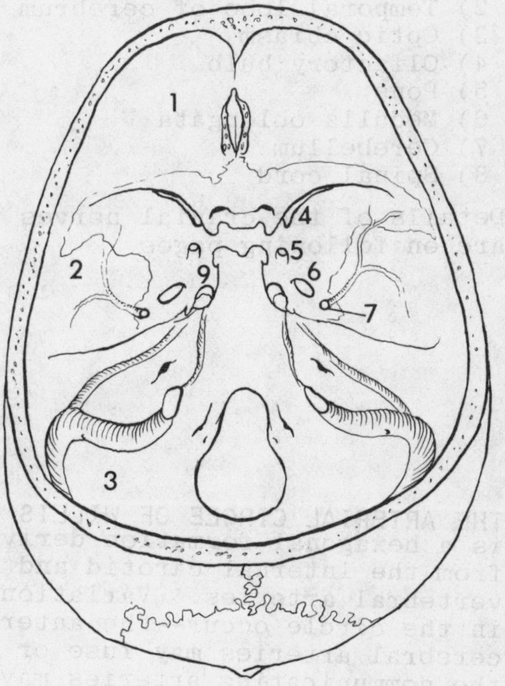

1) ANTERIOR FOSSA:

Features--crista galli, cribiform plate, ethmoid, foramen caecum
Relations--Inferior surface frontal lobes, olfactory bulb, falx cerebri

2) MIDDLE FOSSA:

Features--Hypophyseal fossa, sella turcica, superior orbital fissure, foramina rotundum, ovale, spinosum and lacerum, sulci for venous sinuses

Relations--Trigeminal ganglion, pituitary gland, cavernous sinus, divisions of the trigeminal nerve, cranial nerves II, III, IV, VI, temporal lobes, internal carotid artery, middle meningeal artery, venous sinuses, internal ear

3) POSTERIOR FOSSA:

Features--Sulci for venous sinuses, foramen magnum, internal acoustic meatus, clivus, hypoglossal canal, jugular foramen, internal occipital crest
Relations--Cranial nerves VII, VIII, IX, X, XI, XII, venous sinuses, brain stem, cerebellum , medulla, internal ear

DETAIL SPHENOID BONE MIDDLE FOSSA:

4) Superior orbital fissure
5) Foramen rotundum
6) Foramen ovale
7) Foramen spinosum
8) Foramen lacerum
9) Carotid canal

BASE OF THE BRAIN

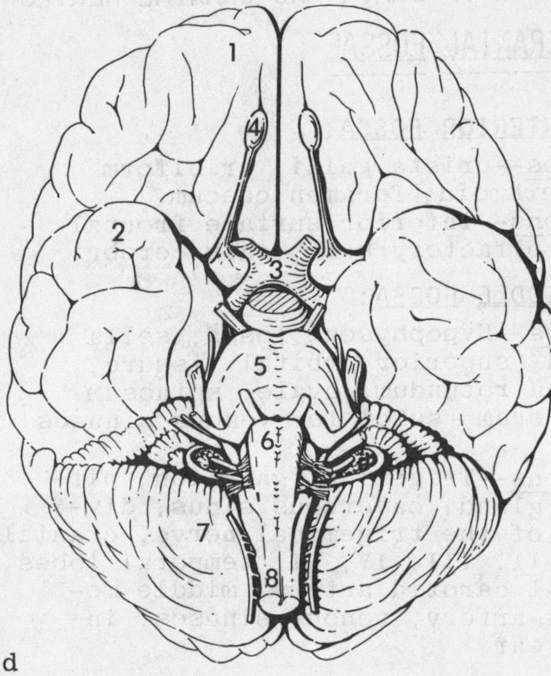

STRUCTURES:

1) Frontal lobe of cerebrum
2) Temporal lobe of cerebrum
3) Optic chiasm
4) Olfactory bulb
5) Pons
6) Medulla oblongata
7) Cerebellum
8) Spinal cord

Details of the cranial nerves are on following pages

THE ARTERIAL CIRCLE OF WILLIS is a hexagonal formation derived from the internal carotid and vertebral arteries. Variations in the circle occur--the anterior cerebral arteries may fuse or the communicating arteries may be absent.

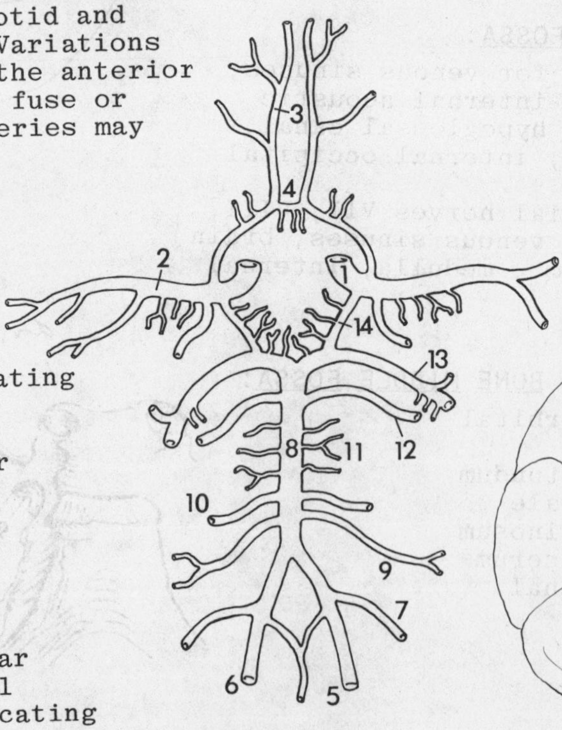

ARTERIES:

1) Internal carotid
2) Middle cerebral
3) Anterior cerebral
4) Anterior communicating
5) Vertebral
6) Anterior spinal
7) Posterior inferior cerebellar
8) Basilar
9) Anterior inferior cerebellar
10) Labyrinthine
11) Pontine
12) Anterior cerebellar
13) Posterior cerebral
14) Posterior communicating

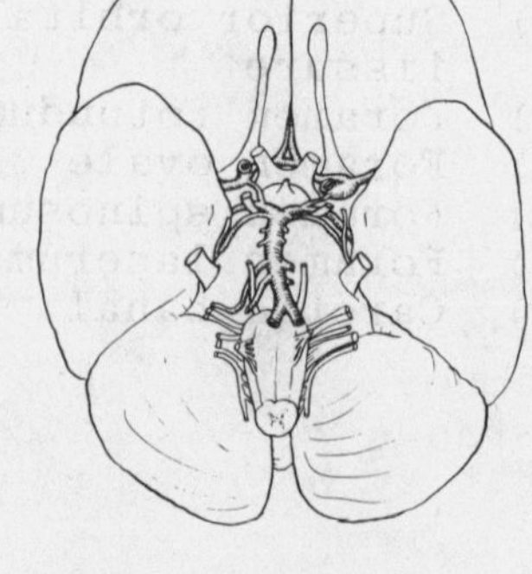

THE CRANIAL NERVES

NERVES:

I. OLFACTORY
II. OPTIC
III. OCULOMOTOR
IV. TROCHLEAR
V. TRIGEMINAL
VI. ABDUCENS
VII. FACIAL
VIII. ACOUSTIC
IX. GLOSSOPHARYNGEAL
X. VAGUS
XI. SPINAL ACCESSORY
XII. HYPOGLOSSAL

STRUCTURES:

1) Optic chiasm
2) Optic tract
3) Cerebral peduncle
4) Pons
5) Medulla

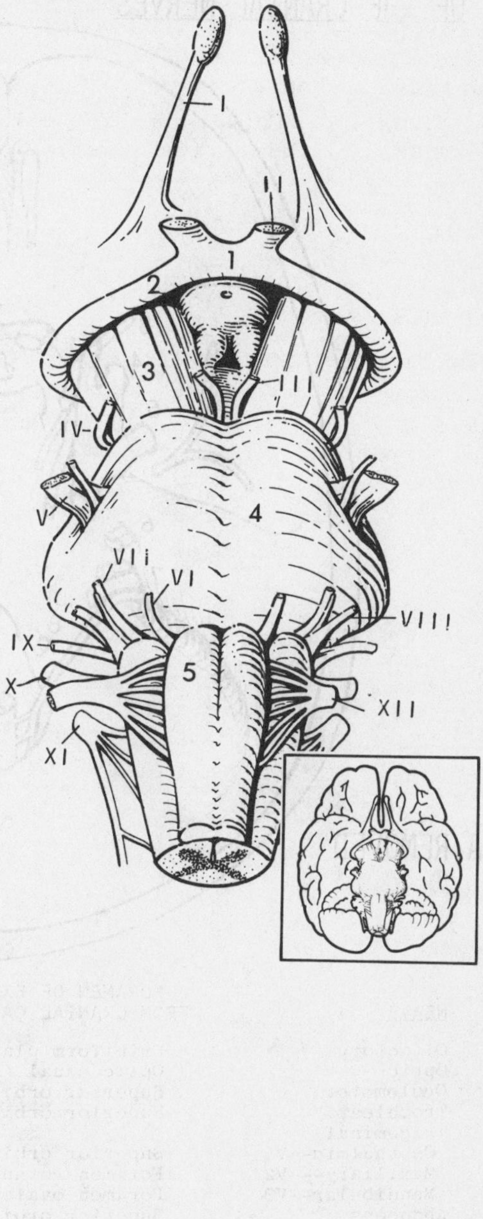

The acoustic nerve is also called the vestibulocochlear nerve; the sensory component of the facial nerve is called the nervus intermedius; note also a motor root for the trigeminal.

MNEMONICS abound for the cranial nerves; however the most familiar --and non-pornographic--is the verse "On Old Olympus' Towering Tops A Finn And German Vended Some Hops" with the first letter of each word standing for the first letter of each nerve. Another verse is helpful in remembering the function of each nerve--S stands for sensory, M stands for motor, B stands for both--"Some Saps Make Merry, But My Brother Says Bad Business Making Merry."

EXITS OF THE CRANIAL NERVES

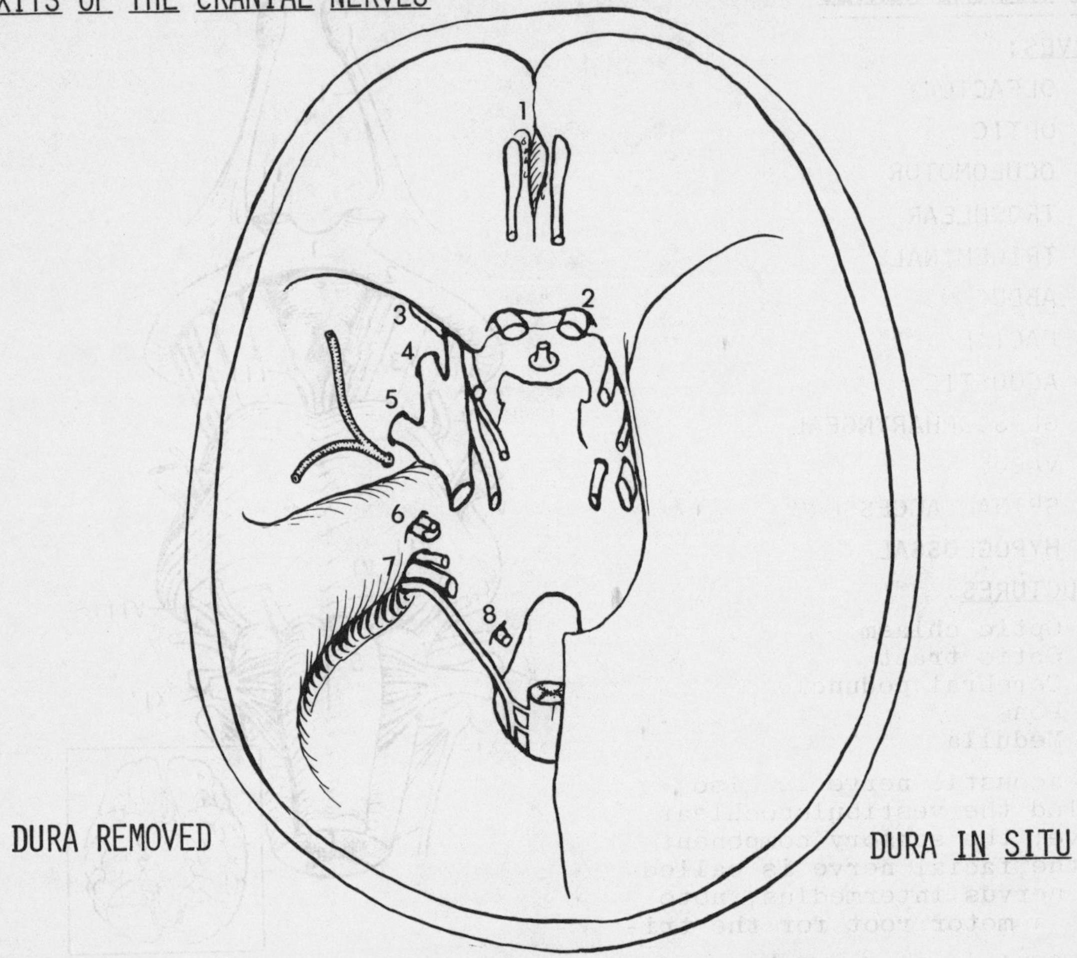

NERVE	FORAMEN OF EXIT FROM CRANIAL CAVITY	FORAMEN OF EXIT FROM THE CRANIUM
Olfactory	Cribiform plate (1)	Same
Optic	Optic canal (2)	Same
Oculomotor	Superior orbital fissure(3)	Same
Trochlear	Superior orbital fissure	Same
Trigeminal		
Ophthalmic--V1	Superior orbital fissure	Same
Maxillary---V2	Foramen rotundum (4)	Same
Mandibular--V3	Foramen ovale (5)	Same
Abducens	Superior orbital fissure	Same
Facial		
Motor	Internal acoustic meatus(6)	Stylomastoid foramen
Nervus intermedius	Internal acoustic meatus	
Chorda tympani		Petrotympanic fissure
Acoustic	Internal acoustic meatus	None
Glossopharyngeal	Jugular foramen (7)	Same
Vagus	Jugular foramen	Same
Spinal Accessory	Jugular foramen	Same
Hypoglossal	Hypoglossal canal (8)	Same

A BRIEF SUMMARY OF THE CRANIAL NERVES

OLFACTORY AND OPTIC NERVES are essentially outgrowths of the brain and are special nerves too complex to be discussed here.

OCULOMOTOR

This nerve is motor to 4 extrinsic eye muscles and autonomic

MOTOR

Motor nucleus III → (Nerve III) → Medial rectus, superior rectus, inferior oblique, inferior rectus, levator palpebrae

PARASYMPATHETIC

Nucleus of Edinger-Westphal → (Nerve III) → Synapse in ciliary ganglion → (Short ciliary nerves) → Constrictor (sphincter) of pupil and ciliary body

TROCHLEAR

This nerve is motor to 1 extrinsic eye muscle

Trochlear nucleus → (Nerve IV) → Superior oblique

TRIGEMINAL

This nerve is both sensory and motor

MOTOR

Motor nucleus V → (Motor root V) → Temporalis, masseter, pterygoids, tensor tympani, tensor veli palatini, mylohyoid, anterior belly digastric

SENSORY

V1, V2, V3 → Gasserian, semilunar or trigeminal ganglion → (Sensory root V) → Brain

The trigeminal is the great sensory nerve of the head--see the unit on FACE.

ABDUCENS

This nerve is motor to 1 extrinsic eye muscle

Abducens nucleus → (Nerve VI) → Lateral rectus

FACIAL

This nerve has sensory, motor and autonomic components. The nerve has a prominent bend or genu within the petrous temporal bone associated with the geniculate ganglion.

MOTOR

Facial nucleus → (Nerve VII) → Mimetic muscles, stapedius, posterior belly digastric, stylohyoid

SENSORY

Taste buds anterior 2/3 tongue → (Lingual nerve of V3) → Taste fibers leave lingual

↓

to Chorda Tympani

↓

Cell bodies in Geniculate Ganglion

↓ Nervus intermedius

Brain

PARASYMPATHETIC

Superior salivatory nucleus → Nervus intermedius → Chorda tympani

Nervus intermedius ↓ Greater petrosal nerve ↓ Synapse in pterygopalatine ganglion ↓ V2 to orbit, V1 to ↓ Lacrimal gland--lacrimation

Chorda tympani ↓ Lingual nerve ↓ Synapse in submandibular ganglion ↓ Sublingual and submandibular glands--salivation

ACOUSTIC OR VESTIBULOCOCHLEAR

This nerve subserves the modalities of hearing, equilibrium and movement and is too complex for this discussion.

GLOSSOPHARYNGEAL

This nerve is sensory, motor and autonomic.

MOTOR

Nucleus ambiguus → Nerve IX → Stylopharyngeus, maybe superior pharyngeal constrictor

SENSORY

Taste buds posterior 1/3 tongue pharyngeal lining → Lingual and Pharyngeal branch → Petrosal ganglion Inferior

Tympanic plexus, linings auditory tube, middle ear → Tympanic n. → Petrosal ganglion Inferior

Skin posterior to auricle → Superior ganglion

Petrosal ganglion Inferior → Nerve IX

Superior ganglion → Nerve IX

Nerve IX → Brain

PARASYMPATHETIC

Inferior salivatory nucleus → Nerve IX → Tympanic n.

Tympanic n. → Tympanic plexus → Lesser petrosal n. → Synapse in otic ganglion

Synapse in otic ganglion → V3 Auriculotemporal n. → Parotid gland salivation

VAGUS

This nerve is sensory, motor and autonomic.

MOTOR

Nucleus ambiguus → Nerve X → Levator veli palatini, pharyngeal constrictors salpingopharyngeus all laryngeal muscles via the superior and recurrent laryngeal nerves

SENSORY

Skin of the auricle, pinna, external acoustic meatus → Jugular--superior ganglion → Nerve X → Brain

Taste buds, epiglottis, larynx, some laryngopharynx → Nodosal--inferior ganglion → Nerve X → Brain

PARASYMPATHETIC

Dorsal motor nucleus → Nerve X → Synapse in terminal ganglia → Internal organs

SPINAL ACCESSORY

This is a motor nerve.

Nucleus ambiguus
Spinal nucleus of XI → Nerve XI → Trapezius, Sternocleido-mastoid

HYPOGLOSSAL

This is a motor nerve.

Hypoglossal nucleus → Nerve XII → Extrinsic and intrinsic tongue musculature

COMMENTS

1. Arteries in the subcutaneous tissue of the scalp are attached so firmly by surrounding connective tissue that they will not retract or collapse immediately if cut, but are held open. This is the reason scalp wounds bleed so profusely. In addition, the anterior-posterior pull of the occipito-frontalis muscles will cause coronally oriented cuts to gape while sagittally oriented ones gape less. Local anesthetic for the scalp should be injected into the tela, where nerves and arteries are located.

2. The loose connective tissue area offers a perfect space for fluid to collect and for infections to spread--the "Danger zone of the scalp." This spread is limited posteriorly by the occipitalis origin, but the frontalis origin is in the skin so fluids and infections may spread even into the upper eye lids. Since emissary veins go from the scalp to the dural sinuses, scalp infection may spread to the cranial cavity--or vice versa.

3. The inner table of the calvaria is thinner than the outer table and may be fractured by blows that don't fracture the outer table. Occasionally a persistent frontal suture--metopic--is mistaken on X-rays for a fracture.

4. Fontanelles are membranous portions of an infant's skull located at intersections of the major bones. As the bones continue to develop, the 'soft spots' close. The posterior fontanelle--between the parietals and occipital bone--closes at roughly 3 months. The anterior fontanelle between frontal and parietals closes at about 2-3 years. In particular the anterior fontanelle offers a readily accessible spot for venipuncture of the underlying superior sagittal sinus as well as a gauge of intracranial pressure.

5. Epidural hematomas are frequently caused when a vessel of the dura--such as middle meningeal artery--is ruptured as in a fracture. Subdural hematomas are often the result of tears of cerebral vessels as they cross from the brain surface to enter the superior sagittal sinus. Such tears could be caused by blows to the head which momentarily shift the brain and shear the vessels from the fixed sinus. Subarachnoid hematomas result from rupture of surface vessels of the brain.

6. The arachnoid granulations are major sites for absorption of CSF into the blood vascular system. If pressure in the sinuses increases above pressure in the CSF the granulations will collapse. Granulations may be so large and prominent that they pit the internal surface of the calvaria.

7. Cerebrospinal cisterns are areas where the subarachnoid space is large. The cisterna magna, just inside the foramen magnum, between the cerebellum and medulla is just one of at least 8 named cisterns. The circle of Willis, for example, is contained in the interpeduncular cistern.

8. The numerous foramina of the base of the skull make it a likely place for fractures; however fractures of the calvaria--the area of the pterion in particular--are more common.

9. An incomplete Circle of Willis--absence of posterior and/or anterior communicating arteries--is not uncommon and it should not be assumed that a blockage of the circle is of minor consequence. For collateral circulation here, as well as anywhere else, a slow occlusion--with time for anastomoses to enlarge--is far superior to a rapid occlusion. Berry aneurysms--solitary round dilatations--are common at bifurcations of arteries and particularly so around the arterial Circle of Willis.

10. Cranial nerve III is particularly vulnerable to increased intracranial pressure which compresses it against bone. Function of the nerve is compromised and is detected by varying degrees of and combinations of ptosis, pupillary dilation and paralysis of extraocular muscles innervated by III.

11. Other than conventional X-rays for fractures, radiological procedures of the skull and contents are numerous. For examination of ventricles of the brain and CSF spaces, air or radio-opaque oils may be injected in the lumbar region and the patient positioned to encourage flow of the substances to the brain. Arterial injections of compounds for cerebral angiography--with an arterial and venous phase--are common to observe patterns and alterations in flow of blood. Computed tomography with or without injected contrast media is becoming increasingly important in diagnostic procedures. Ultrasound procedures are also becoming widely used for cranial examinations.

FACE

I. CUTANEOUS INNERVATION

The trigeminal nerve is the great sensory nerve of the face. Cutaneous innervation for the scalp and neck is also indicated.

NERVES:

1) V1--Ophthalmic division of the trigeminal nerve
2) V2--Maxillary division of the trigeminal nerve
3) V3--Mandibular division of the trigeminal nerve
4) Greater occipital nerve --C2
5) Lesser occipital nerve --cervical plexus
6) Greater auricular nerve --cervical plexus
7) Transverse cervical nerve --cervical plexus
8) Posterior primary rami C3,4
9) Supraclavicular nerves --cervical plexus

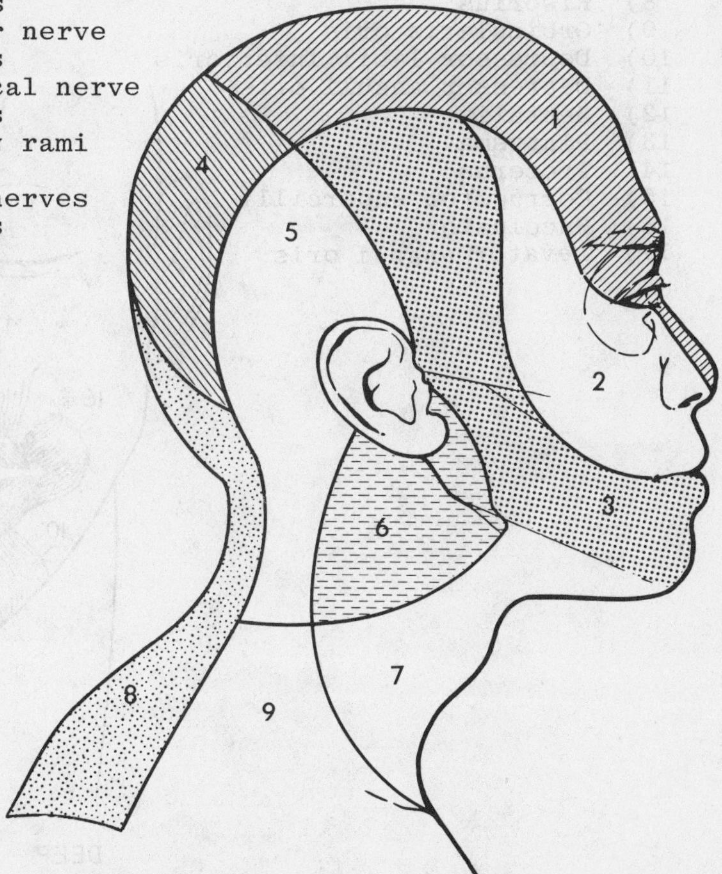

II. MIMETIC MUSCLES

THE MIMETIC MUSCLES are the muscles of facial expression. Many originate or insert in the skin. ALL are innervated by the Facial nerve. BLOOD supply is from the ophthalmic, facial, submental, transverse facial and superficial temporal arteries.

1) Frontalis
2) Orbicularis oculi
3) Nasalis
4) Levator labii superioris
5) Levator labii superioris alequae nasi
6) Zygomaticus major
7) Zygomaticus minor
8) Risorius
9) Orbicularis oris
10) Depressor labii inferioris
11) Depressor anguli oris
12) Mentalis
13) Platysma
14) Procerus
15) Corrugator supercilli
16) Buccinator
17) Levator anguli oris

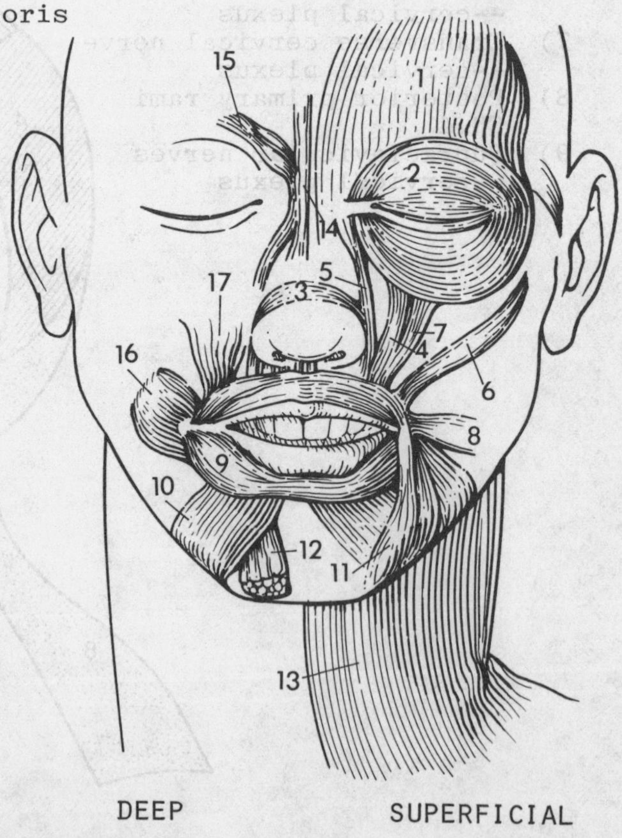

THE FACIAL NERVE--VII

THE MOTOR ROOT of the facial nerve is motor to all muscles of facial expression or mimetic muscles. Other sensory components of the nerve are not here indicated.

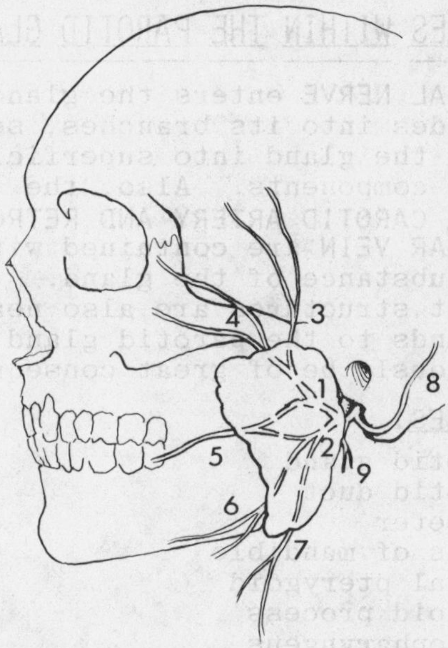

DIVISIONS:

1) Temporofacial
2) Cervicofacial

BRANCHES:

3) Temporal
4) Zygomatic
5) Buccal
6) Mandibular
7) Cervical--these first five may be approximated by placing the hand with abducted digits on the face with the palm over the ear and the index finger on the ramus of the mandible
8) Occipital
9) Branches to stylohyoid and digastric

III. PAROTID GLAND

THE PAROTID GLAND is the largest salivary gland--it extends from the zygomatic arch to the angle of the mandible and overlaps on the masseter muscle. The gland is encased in a sheet of fascia from the SUPERFICIAL LAYER OF DEEP CERVICAL FASCIA--SEE THE SECTION ON NECK.

STRUCTURES:

1) Parotid gland
2) Parotid duct--enters opposite upper 2nd molar, accessory glandular tissue along duct is normal
3) Masseter
4) Buccinator
5) Buccal fat pad

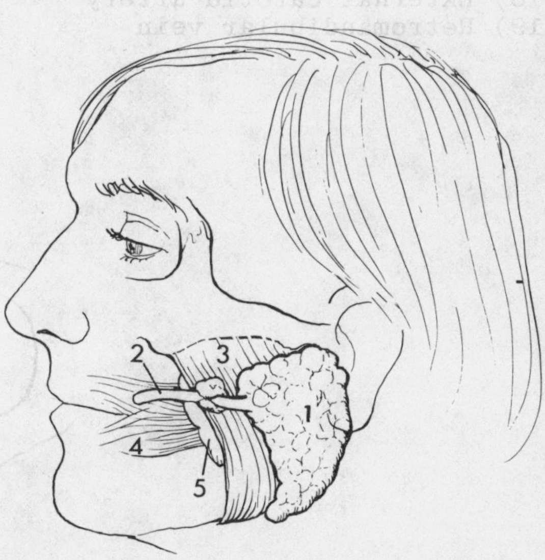

STRUCTURES WITHIN THE PAROTID GLAND

THE FACIAL NERVE enters the gland and divides into its branches, separating the gland into superficial and deep components. Also, the EXTERNAL CAROTID ARTERY AND RETROMANDIBULAR VEIN are contained within the substance of the gland. Important structures are also nearby. Wounds to the parotid gland may obviously be of great consequence.

STRUCTURES:

1) Parotid gland
2) Parotid duct
3) Masseter
4) Ramus of mandible
5) Medial pterygoid
6) Styloid process
7) Stylopharyngeus
8) Styloglossus
9) Stylohyoid
10) Mastoid process
11) Facial nerve
12) Buccinator
13) Pharyngeal constrictor
14) Atlas
15) Posterior belly digastric
16) Internal carotid
17) Internal jugular vein
18) External carotid artery
19) Retromandibular vein

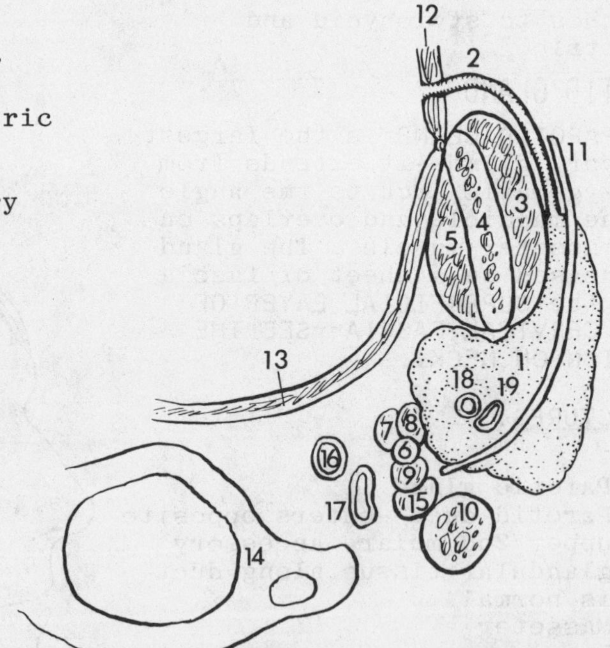

IV. MUSCLES OF MASTICATION

ALL MUSCLES OF MASTICATION ARE INNERVATED BY THE TRIGEMINAL NERVE. These muscles include the masseter, temporalis and pterygoids. The buccinator aids in mastication but is not involved in closing the jaws. These muscles are VASCULARIZED by branches of the external carotid including maxillary and superficial temporal.

20) MASSETER
O-Zygomatic arch
I-Ramus and angle of mandible
A-Elevate and protract mandible

21) TEMPORALIS
O-Temporal fossa and fascia
I-Coronoid process mandible
A-Elevate and retract mandible

22) Superior temporal line--origin of the temporal fascia
23) Buccinator
24) Buccal branch of V--not motor to buccinator, it is sensory to the mucosa of the cheek

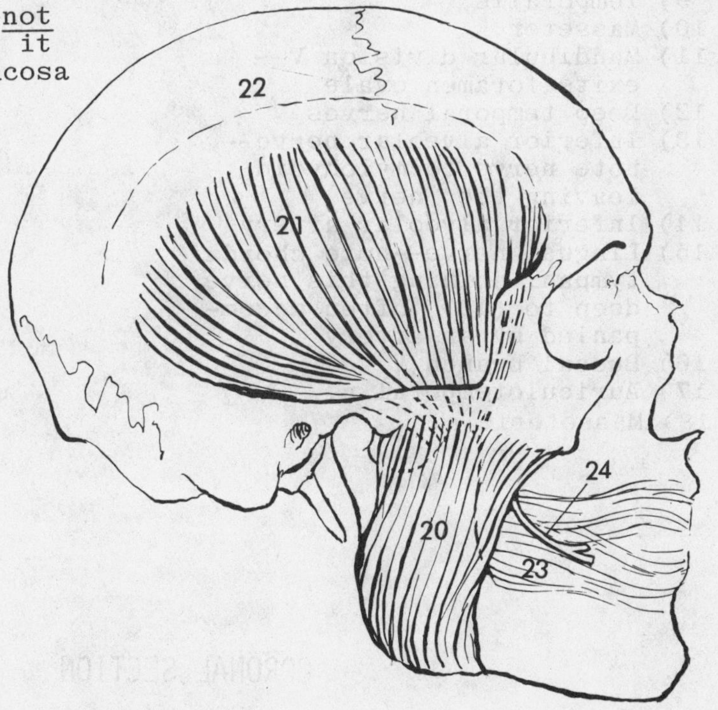

1) MEDIAL PTERYGOID

O-Medial surface lateral pterygoid plate
I-Internal surface angle of mandible
A-Elevate and protract mandible

2) LATERAL PTERYGOID

O-Upper head--infratemporal surface greater wing sphenoid
Lower head--lateral surface lateral pterygoid plate
I-Pterygoid fovea neck of mandible
A-Protract mandible, pull articular disc forward

<u>OTHER</u> <u>STRUCTURES</u>:

3) External carotid
4) Superficial temporal a.
5) Maxillary artery
6) Middle meningeal a.--enters at foramen spinosum
7) Infraorbital a.
8) Superior alveolar branch a.
9) Temporalis
10) Masseter
11) Mandibular division V--exits foramen ovale
12) Deep temporal nerves
13) Inferior alveolar nerve--note nerve to mylohyoid leaving the nerve
14) Inferior alveolar artery
15) Lingual nerve--note chorda tympani joining this nerve deep to (13), often accompanied by an artery
16) Buccal branch V
17) Auriculotemporal n.
18) Masseteric n.

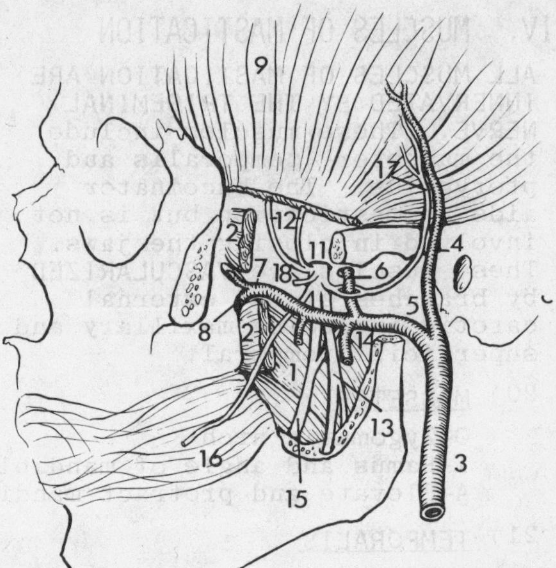

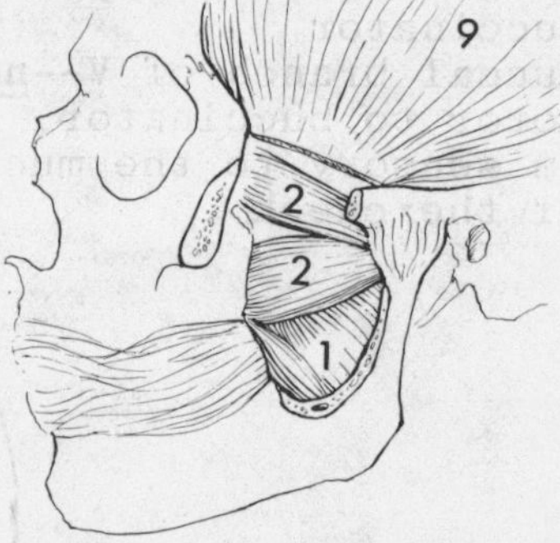

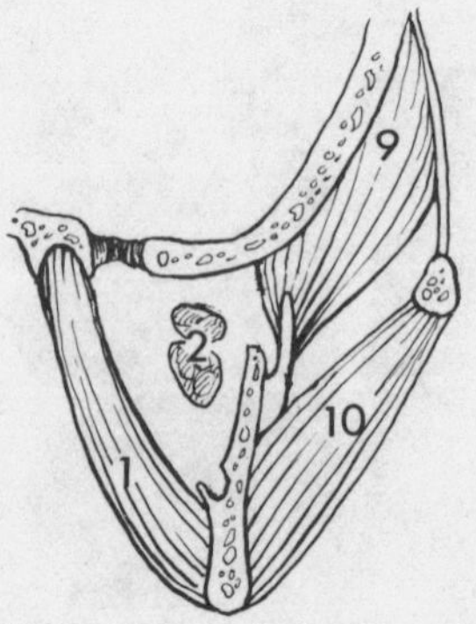

CORONAL SECTION

TEMPOROMANDIBULAR JOINT

THE TMJ is an important joint. When the mouth opens, the jaw is depressed and the condyle of the mandible moves forward; therefore the TMJ is a hinge and sliding joint--the sliding is an integral part of the action. THE ARTICULAR DISC acts as a cushion to prevent wear to the articular surfaces during action of the jaw.

THE SPHENOMANDIBULAR LIGAMENT acts as an accessory ligament to the TMJ. This strong ligament attaches to the mandible at an axis of movement of the mandible--that is to say this point is rather stationery while other portions of the mandible move in different directions. The ligament can then act as an important strut for stabilization of the jaw. It also protects the neurovascular elements that enter the jaw at the mandibular foramen--an ideal place since elements of wear such as stretch are minimal at that pivotal point.

STRUCTURES:

1) Joint capsule
2) Articular disc
3) Articular tubercle
4) Lateral pterygoid m.
5) Condyle of mandible
6) Spine of sphenoid
7) Lingula of mandible
8) Sphenomandibular ligament
9) Foramen ovale; mandibular div. V
10) Inferior alveolar nerve and artery
11) Nerve to the mylohyoid

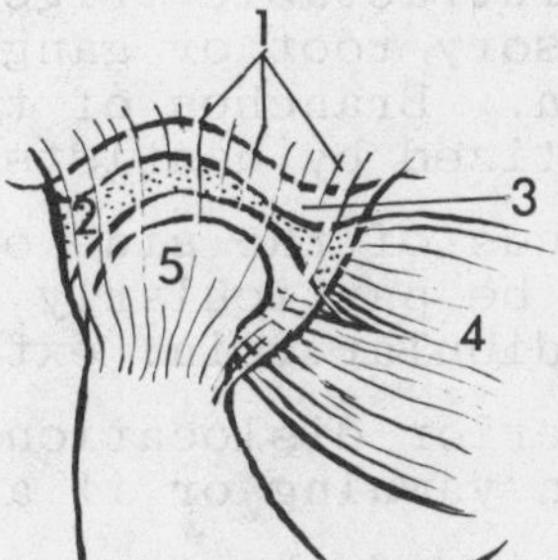

LATERAL VIEW

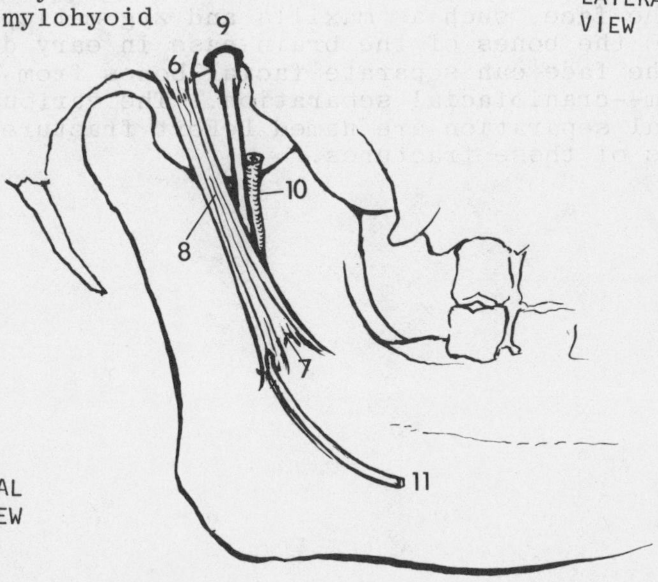

MEDIAL VIEW

COMMENTS

1. Innervation of facial musculature is important in normal facial appearance. Paralysis of the facial nerve, as in Bell's Palsy, causes asymmetry of the face. The lids droop and the cornea may dry; the mouth will not close properly on the affected side. Bell's palsy frequently shows spontaneous remission.
2. Vasculature of the face is abundant and the veins are without valves. The rich vasculature assures prompt healing of facial wounds. Collateral circulation in the face is well developed and arteries communicate across the midline. In hemorrhage of the face, it may be necessary to control bleeding from both sides of the laceration.
3. The trigeminal nerve is the great sensory nerve of the face. In intractable trigeminal neuralgia--tic douloureux--the sensory root or ganglion is often destroyed to eliminate the pain. Branches of the trigeminal nerve are routinely anesthetized by dentists and surgeons for deadening of pain.
4. Wounds of the side of the face involving the parotid gland can be particularly dangerous due to injuries of the retromandibular vein, external carotid artery and nerve VII.
5. Anterior dislocations of the TMJ are common, particularly when yawning or if a blow hits the opened mandible.
6. Fascia of the face is not highly organized and the looseness of the fascia allows extravasates to travel widely under the skin.
7. Bones of the face, such as maxilla and zygomatic, became attached to the bones of the brain case in eary development. Blows to the face can separate facial bones from bones of the cranium--craniofacial separation. The various types of craniofacial separation are named LeFort fractures; there are 3 types of these fractures.

ORBIT

I. SUPERFICIAL FEATURES

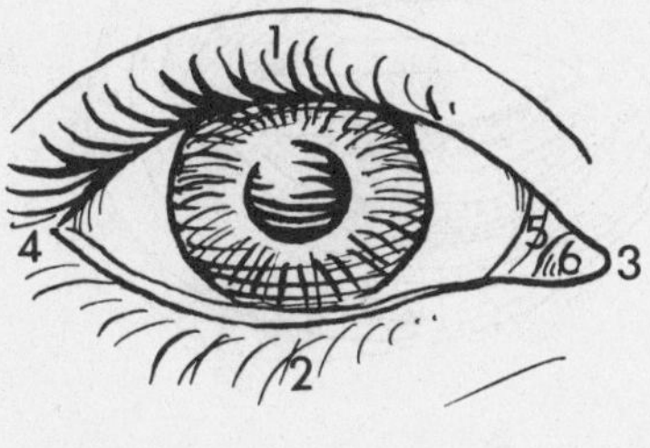

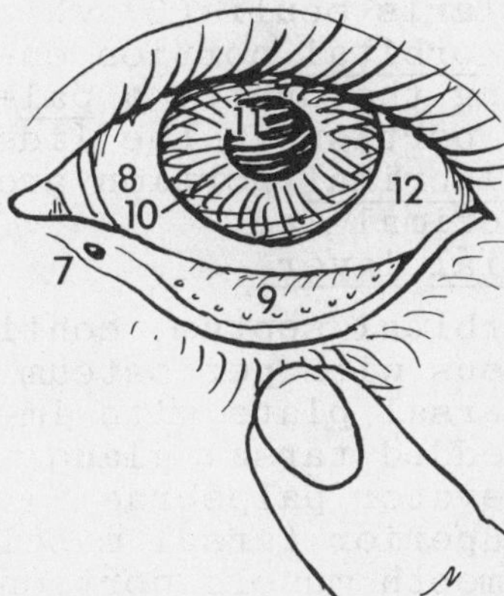

1) Superior palpebrum
2) Inferior palpebrum
3) Medial canthus--angle
4) Lateral canthus
5) Plica semilunaris
6) Lacrimal caruncle
7) Lacrimal punctum--one on each lid
8) Lake of tears--triangular area superficial to plica
9) Openings of meibomian glands
10) Iris--pigmented
11) Pupil
12) Sclera--the cornea, a transparent tissue continuous with the sclera, covers the pupil and iris--the limbus is the junction of the sclera and cornea
13) Lacrimal gland
14) Lacrimal sac
15) Nasolacrimal duct
16) Ostium into the inferior meatus of the nasal cavity

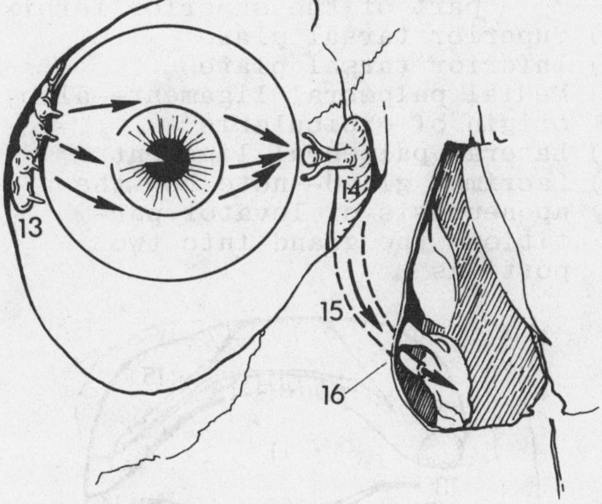

The Palpebral Fissure is simply the space between the two lids through which one sees when the lids are open.

THE EYELIDS

Structures: Layers of the lids

1) Skin--thin and delicate
2) Muscular layer--made up of orbicularis oculi (3) which has an orbital portion encircling the orbit, a palpebral portion in the lids and a lacrimal portion around the lacrimal sac.

Tarso-fascial layer

4) Orbital septum, continuous with periosteum
5) Tarsal plate with imbedded tarsal gland
6) Levator palpebrae
7) Superior tarsal muscle, smooth muscle portion of levator palpebrae

Conjunctival layer

8) Palpebral conjunctiva
9) Orbital conjunctiva
10) Conjunctival fornices--the ducts of the lacrimal gland open into the lateral part of the superior fornix

11) Superior tarsal plate
12) Inferior tarsal plate
13) Medial palpebral ligament--also origin of orbicularis
14) Lateral palpebral ligament
15) Lacrimal gland--note how the aponeurosis of levator partitions the gland into two portions

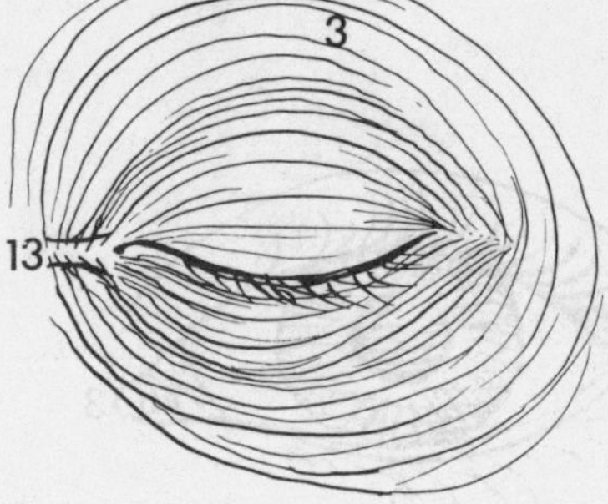

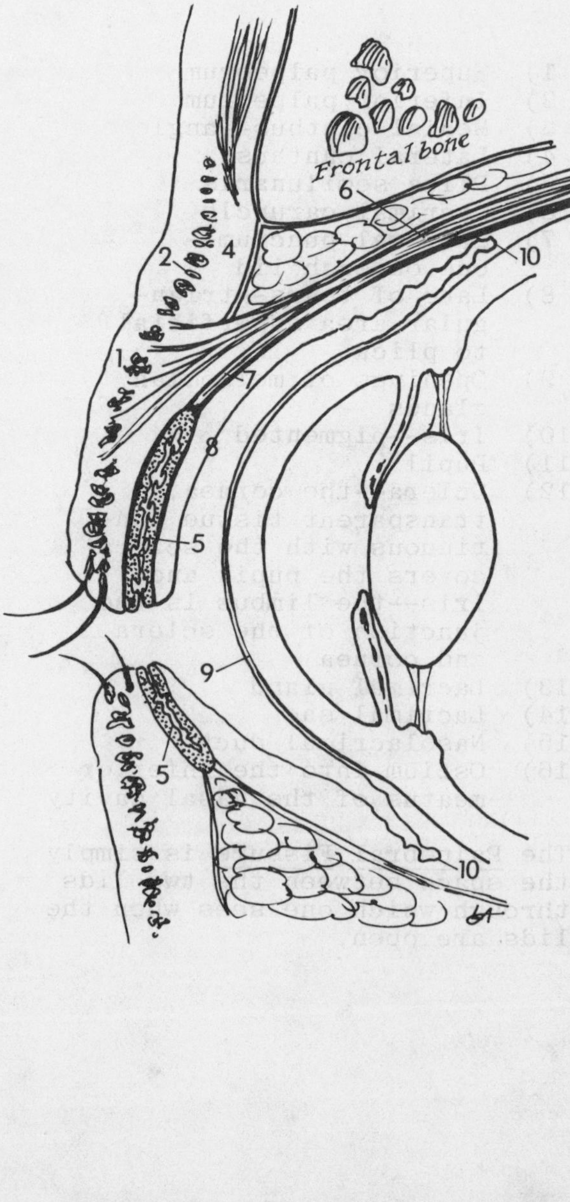

II. THE BONY ORBIT

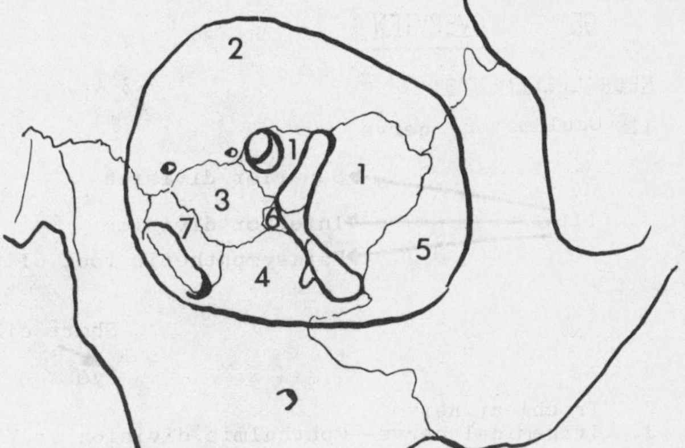

BONES:

1) Sphenoid
2) Frontal
3) Ethmoid
4) Maxilla
5) Zygomatic
6) Palatine
7) Lacrimal

FEATURES:

8) Nasolacrimal canal
9) Anterior ethmoid foramen
10) Posterior ethmoid foramen
11) Optic canal
12) Superior orbital fissure
13) Inferior orbital fissure
14) Infra-orbital groove
15) Infra-orbital foramen
16) Lacrimal fossa--depression for lacrimal gland

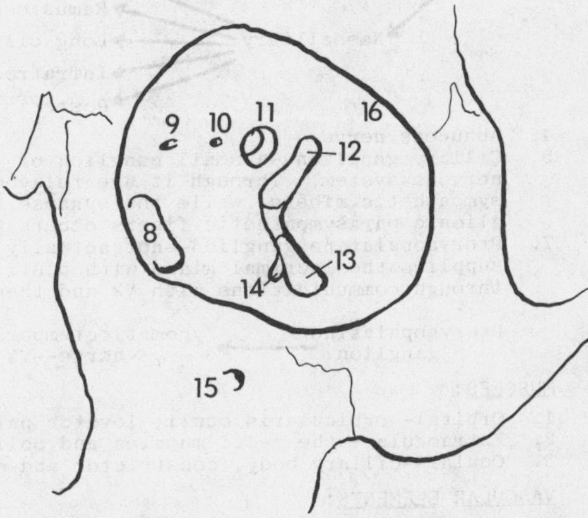

THE ORBIT is shaped like a cone, with the apex pointing postero-medially. The medial walls of the orbits (M) are parallel while the lateral walls (L) are perpendicular to one another--important in understanding action of extra-ocular muscles.

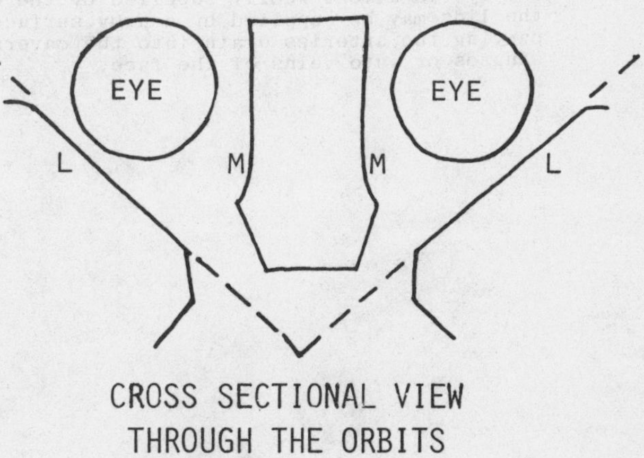

CROSS SECTIONAL VIEW
THROUGH THE ORBITS

III. CONTENTS OF THE ORBIT

GENERAL OVERVIEW

NEURAL ELEMENTS:

1. Oculomotor nerve

III → Superior division
III → Inferior division
III → Parasympathetic root of the ciliary ganglion → Short ciliary nerves → Eye

2. Trochlear nerve
3. Trigeminal nerve--Ophthalmic division or V1

V1 → Frontal → Supratrochlear
Frontal → Supraorbital
V1 → Lacrimal
V1 → Nasociliary → Anterior ethmoidal → Internal nasal
Anterior ethmoidal → External nasal
Nasociliary → Ramus communicans to ciliary ganglion
Nasociliary → Long ciliary nerves → Eye
Nasociliary → Infratrochlear
Nasociliary → Posterior ethmoidal

4. Abducens nerve
5. Ciliary ganglion--a small ganglion of the cranial parasympathetic nervous system. Through it are relayed sensory and postganglionic sympathetic fibers, while the synapse between pre and post ganglionic parasympathetic fibers occurs within the ganglion.
7. Pterygopalatine ganglion--not actually within the orbit, but supplies the lacrimal gland with postganglionic parasympathetics through communications with V2 and then V1--lacrimal nerve

Pterygopalatine ganglion → Zygomaticotemporal nerve--V2 → Lacrimal nerve--V1

MUSCLES:

1. Orbital--orbicularis oculi, levator palpebrae, superior tarsus
2. Extraocular--the recti muscles and oblique muscles
3. Ocular--ciliary body, constrictor and dilator of the pupil

VASCULAR ELEMENTS:

The eye is almost wholly supplied by the ophthalmic artery, while the lids may be supplied by nearby surface arteries. Veins accompanying the arteries drain into the cavernous sinus, other cranial sinuses or into veins of the face.

CONTENTS OF THE ORBIT

STRUCTURES:

1) Periorbita
2) Lining of ethmoidal sinus cells
3) Frontal sinus
4) Optic chiasm
5) Lacrimal gland
6) Frontal n.
7) Supraorbital n.
8) Supratrochlear n.
9) Trochlear n.
10) Lacrimal n.--observe the communicating branch of the zygomaticotemporal n. carrying parasympathetic fibers
11) Abducens n.
12) Nasociliary n.
13) Oculomotor n., superior division
14) Superior oblique m.
15) Levator palpebrae m.
16) Superior rectus
17) Lateral rectus
18) Inferior rectus
19) Inferior oblique
20) Trochlea
21) Ciliary ganglion

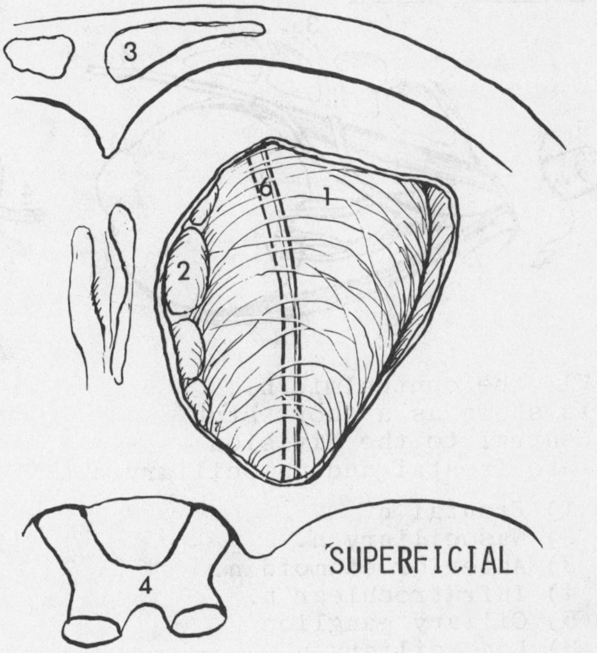

SUPERFICIAL

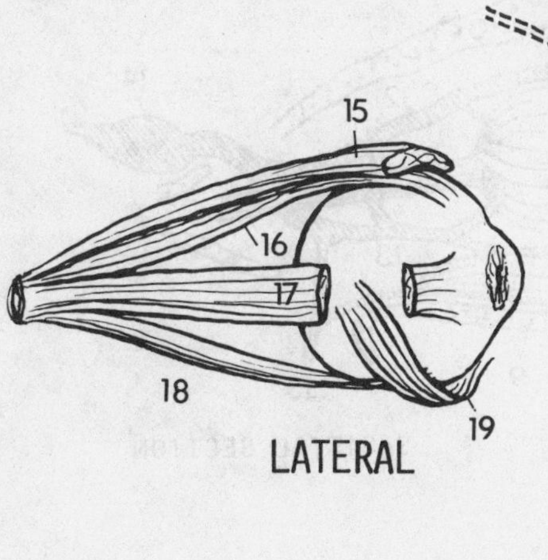

LATERAL

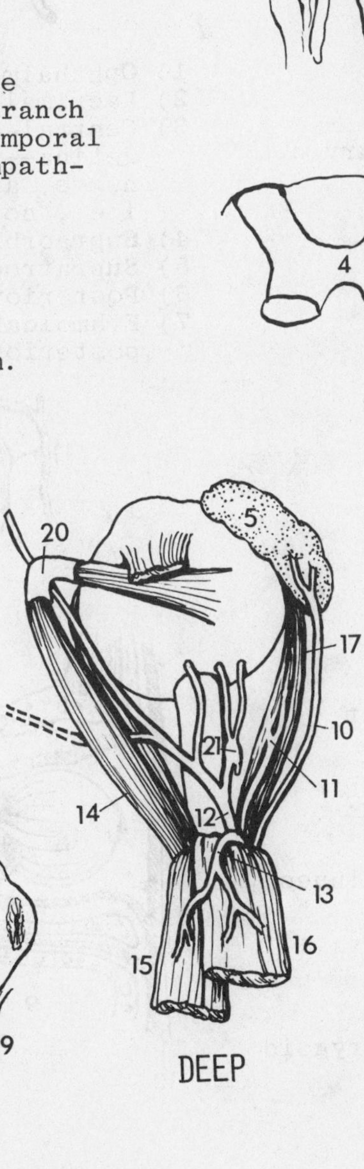

DEEP

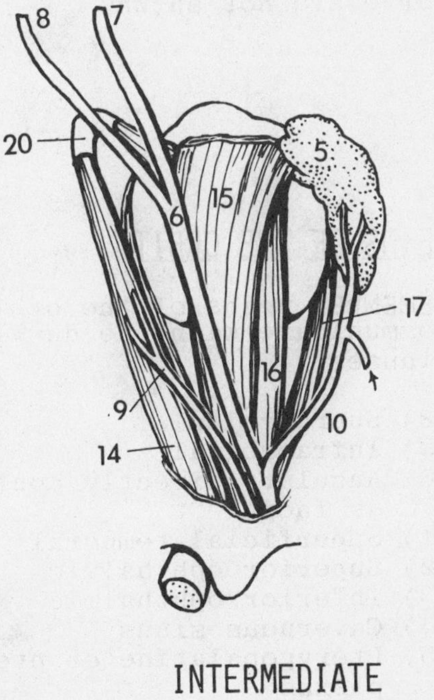

INTERMEDIATE

BRANCHES OF V1

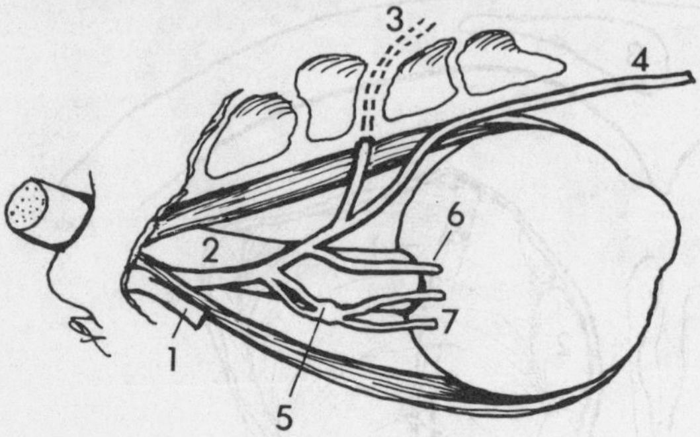

V1, the ophthalmic n., is shown as a short stub central to the division into frontal and nasociliary n.

1) Frontal n.
2) Nasociliary n.
3) Anterior ethmoid n.
4) Infratrochlear n.
5) Ciliary ganglion
6) Long ciliary n.
7) Short ciliary nerves

Branches of the frontal nerve are not shown

OPHTHALMIC ARTERY

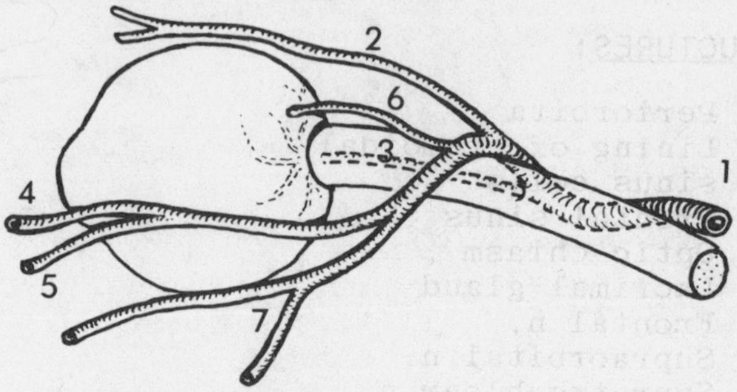

1) Ophthalmic artery
2) Lacrimal a.
3) Central artery of the retina--within the optic nerve, an end artery, i.e., no anastomoses
4) Supraorbital a.
5) Supratrochlear a.
6) Posterior ciliary
7) Ethmoidal--anterior and posterior

VEINS OF THE ORBIT

REMEMBER veins of the orbit communicate with the dural sinuses.

8) Supraorbital
9) Infraorbital
10) Angular--directly continues as facial
11) Superficial temporal
12) Superior ophthalmic
13) Inferior ophthalmic
14) Cavernous sinus
15) Pterygopalatine or pterygoid plexus

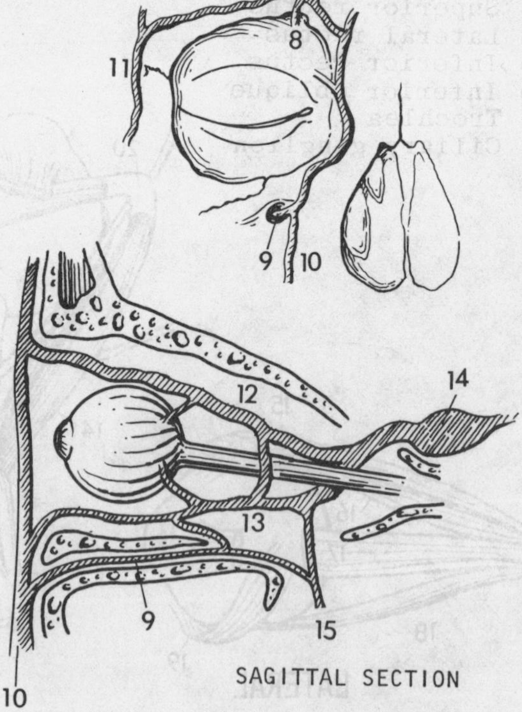

SAGITTAL SECTION

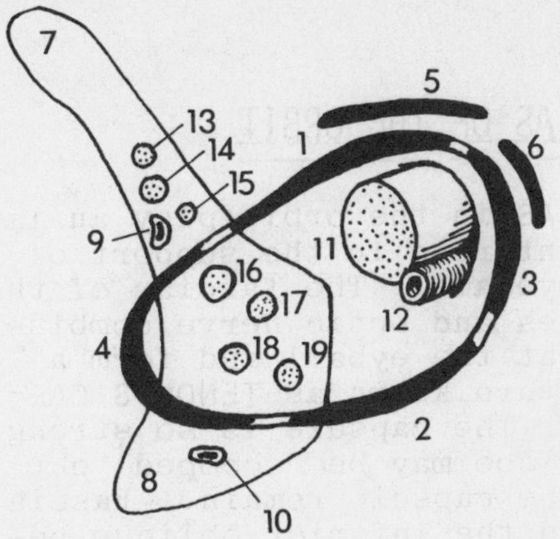

EXTRAOCULAR MUSCLES

THE EXTRAOCULAR OR EXTRINSIC eye muscles are those muscles concerned with movement of the eye.

MUSCULAR ANULUS:

1) Origin of superior rectus
2) Origin of inferior rectus
3) Origin of medial rectus
4) Origin of lateral rectus

OTHER STRUCTURES:

5) Origin of levator palpebrae
6) Origin of superior oblique--inferior oblique oribinates on the anterior inferior aspect of the bony orbit
7) Superior orbital fissure
8) Inferior orbital fissure
9) Superior ophthalmic vein
10) Inferior ophthalmic vein
11) Optic n.
12) Ophthalmic a.
13) Lacrimal n.
14) Frontal n.
15) Trochlear n.
16) Superior division III
17) Nasociliary n.
18) Inferior division III
19) Abducens n.

MENMONICS: For the nerves in order "Lovely French Tart Sitting Naked In Anticipation"; for nerves outside the ring, "Three Fat Ladies"; for innervation of the extraocular muscles, think of the chemical formula $(LR_6SO_4)_3$--lateral rectus by VI, superior oblique by IV and all others by III.

ACTIONS OF THE EXTRAOCULAR MUSCLES are indicated in the diagram to the right, which indicates the eye in the 'position of rest,' with the eyes straight ahead. Since the axis of the orbit is directed to the lateral or temporal aspect of the skull, this means all actions of the eye muscles are mixed, although each has a primary action. The actions of a muscle are pure when the A-P axis of the eye coincides with the direction of a muscle's fibers. The primary action of the muscles are: Lateral rectus -abduction; medial rectus - adduction; superior rectus and inferior oblique-elevation; inferior rectus and superior oblique-depression.

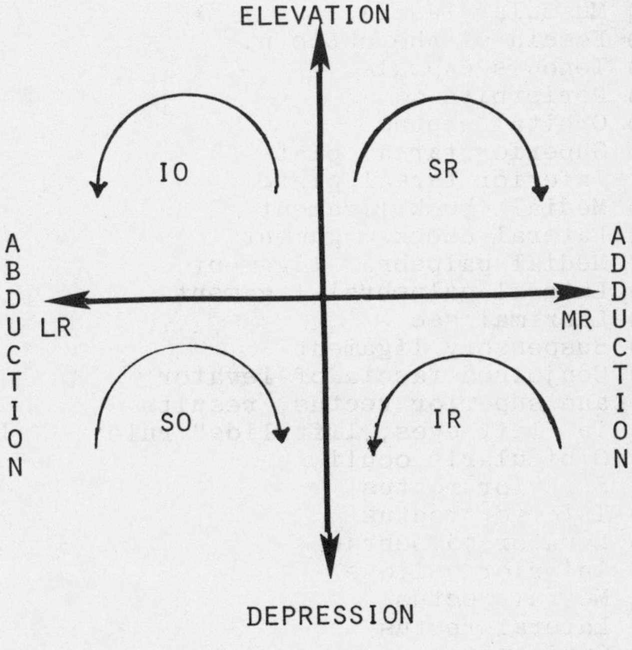

FASCIAS OF THE ORBIT

FASCIAS in the orbit play an important role in the support of the eyeball. The fascias of the muscles and optic nerve combine to coat the eyball and form a structure known as TENON'S CAPSULE. The capsule is so strong the globe may be 'scooped' out and the capsule remain. Fascia around the inferior oblique becomes thickened and attaches laterally to the wall of the orbit and is called the SUSPENSORY LIGAMENT. In addition, the fascias of the medial and lateral rectus muscles 'fan out' to form medial and lateral CHECK LIGAMENTS.

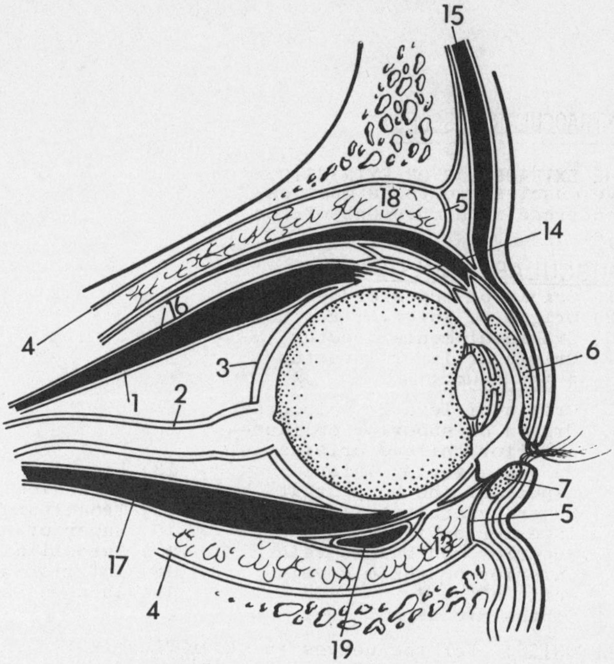

SAGITTAL SECTION

STRUCTURES:

1) Muscular fascia
2) Fascia of the optic n.
3) Tenon's capsule
4) Periorbita
5) Orbital septum
6) Superior tarsal plate
7) Inferior tarsal plate
8) Medial check ligament
9) Lateral check ligament
10) Medial palpebral ligament
11) Lateral palpebral ligament
12) Lacrimal sac
13) Suspensory ligament
14) Conjoined fascia of levator and superior rectus, results in "lift eyes, lift lids" rule
15) Orbicularis oculi
16) Superior rectus
17) Inferior rectus
18) Levator palpebrae
19) Inferior oblique
20) Medial rectus
21) Lateral rectus
22) Optic nerve

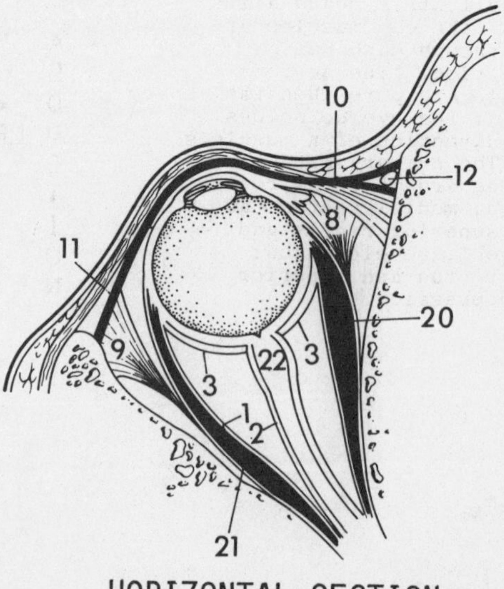

HORIZONTAL SECTION

IV. THE EYEBALL

COATS OF THE EYEBALL

FIBROUS:

1) Sclera
2) Cornea

VASCULAR OR PIGMENTED:

Also called Uveal tract, because the structures contain a deep purple pigment

3) Choroid
4) Ciliary body
5) Iris

NERVOUS:
Has three subdivisions;

6) Pars optica--light sensitive, the anterior limit is the Ora serrata (7)

Ciliary and Iridial, not pictured.

OTHER STRUCTURES:

8) Anterior chamber --filled with Aqueous humor
9) Posterior chamber --filled with Aqueous humor
10) Vitreous body
11) Lens
12) Suspensory ligament
13) Fovea centralis--surrounded by macula lutea
14) Central artery of the retina, note branches, it supplies the retina and some of the pigmented layer, but most of the vascular or pigmented layer receives its blood from other branches of the ophthalmic artery and the supplies of Uveal tract and retina do not mix.

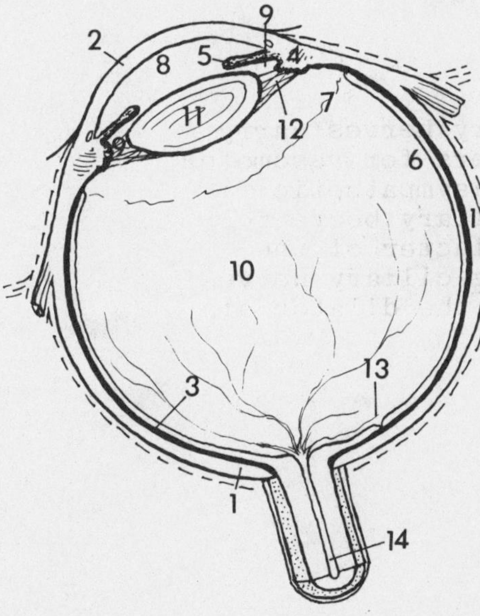

V. THE CILIARY GANGLION

This ganglion is an important relay for sensory and post-ganglionic sympathetic fibers and the site of synapses for the parasympathetic system.

SYMPATHETIC FIBERS ARE DASHED, PARASYMPATHETIC FIBERS ARE SOLID, SENSORY FIBERS ARE NOT SHOWN.

STRUCTURES:

1) Trigeminal n.
2) Ophthalmic division
3) Nasociliary n.
4) Long ciliary nerve--one of several
5) Oculomotor n.
6) Optic n.
7) Internal carotid a.

THE GANGLION--it may not be a discrete structure

8) Sensory root
9) Parasympathetic root
10) Sympathetic root
11) Short ciliary nerves

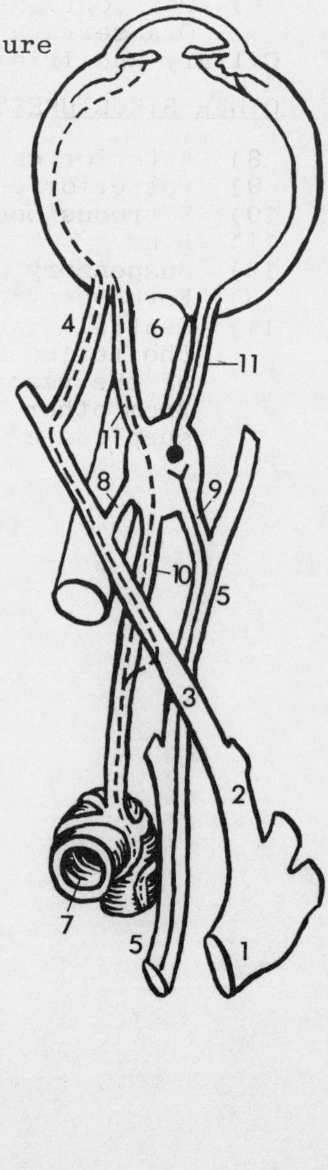

The short ciliary nerves carry sympathetic fibers for vasomotor control and parasympathetic fibers for the ciliary body--focusing--and sphincter of the pupil. The long ciliary nerves carry fibers to the dilator of the pupil.

COMMENTS

1. Fractures of the bones of the orbit are common. In a "blow-out" fracture, increased pressure in the orbit due to a blow from the front fractures a wall of the orbit--usually inferior--and the eyeball 'sinks' inferiorly. Muscles and nerves may be pinched in fragments of bone. Fracture of the roof of the orbit may result in leakage of CSF into the orbit.

2. The fascia of the globe and muscles--Tenon's Capsule--is important in the positioning of the eye. It is so dense a cup is formed when the eye is removed. The cup serves as a convenient receptacle for a prosthesis.

3. Veins in the orbit communicate with facial veins and with the cavernous sinus. Infections in the danger zone of the face can spread into the cavernous sinus and result in a thrombosis. The wall of the carotid artery may be eroded and an arterial-venous fistula result. A bruit--rasping sound--is heard by the patient and by the physician upon auscultation.

4. The sheath of the optic nerve is an extension of the cranial dura. Increases in intracranial pressure are transmitted by the CSF around the nerve and compromise venous flow. The ophthalmic veins then have trouble draining and the optic disc becomes elevated--papilledema.

5. The retinal artery is an end artery--that is there are no anastomoses to supply the retina if the ophthalmic artery is occluded. Blindness will result.

6. Diplopia is double vision; conjunctivitis is an inflammation of the conjunctiva; a sty is an inflammation of a sebaceous gland around the lashes of the eye; glaucoma is an increase in pressure in the aqueous humor and can result in blindness; cataract refers to the development of opacity in the lens; ptosis is a drooping of the lid; crying results in a runny nose because the nasolacrimal duct opens into the nose; eye color results from a combination of color of pigmentation and color of the iridial membrane.

7. An extraocular muscle is tested by positioning of the eye in such a manner as to minimize action of muscles other than the one being tested. Action of an extraocular muscle is pure when it is lined up with its visual axis--that is when its direction of muscle fibers is coincidental with an axis passing through the center of the eyeball.

8. The sclera is fenestrated--lamina cribosa--where the optic nerve enters the globe. This is the weakest point of the globe and increased intraocular pressure causes a bulging outward of the area--'cupped disc.'

9. Fascias of the superior rectus muscle and levator palpebrae are interconnected such that there is a rule: "Lift eyes, lift lids."

10. People are so conscious about their eyes and facial appearance that apparent suicides with gunshots anterior to the temple or in the orbit are considered murder unless proven otherwise.

THE EAR

I. DEFINITIONS

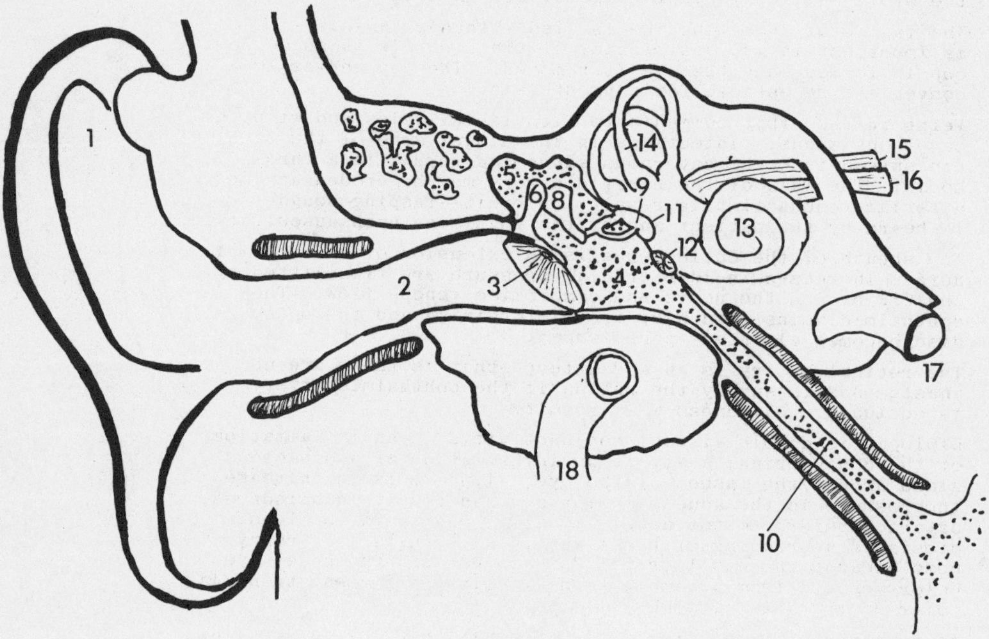

EXTERNAL EAR:

1) Auricle or Pinna
2) External auditory meatus, with bony and cartilaginous portions
3) Tympanic membrane

MIDDLE EAR:

4) Tympanic cavity
5) Attic of tympanic cavity
6) Malleus
8) Incus
9) Stapes
10) Auditory tube
11) Oval window
12) Round window

INNER EAR:

13) Cochlea
14) Semicircular canals
15) Nerve VII
16) Nerve VIII

ALSO:

17) Internal carotid artery
18) Internal jugular vein

II. EXTERNAL EAR

PINNA OR AURICLE

STRUCTURES:

1) Tragus
2) Lobule
3) Helix
4) Antitragus
5) Concha
6) Antihelix

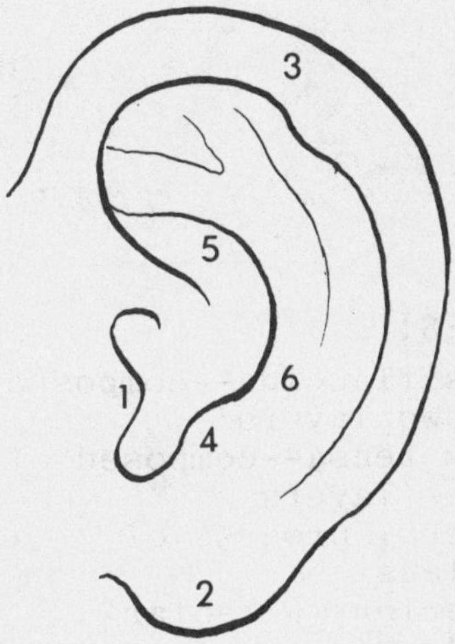

EXTERNAL AUDITORY MEATUS

The meatus--a canal--follows a reverse S shape approximately 2.5 cm. long in the adult; the canal is not in one horizontal plane and is straightened by drawing the pinna upwards, outwards and backwards. It is narrowed at the eardrum and again 1/3 of the way from the outer aspect where the cartilaginous and bony portions of the canal join. Ceruminous--wax--glands exist only in the cartilaginous portion of the tube. The length of the canal is longest in the anterior, inferior wall since the eardrum or tympanic membrane faces anteriorly and inferiorly. VASCULARIZATION is from the superficial temporal, posterior auricular and deep auricular branch of maxillary a. SENSORY INNERVATION of the meatus is from the auriculotemporal nerve--V3--and a branch of the Vagus nerve.

TYMPANIC MEMBRANE

Divided into quadrants along the handle of the malleus and perpendicular at the umbo. The handle is directed from anterior-superior to posterior-inferior.

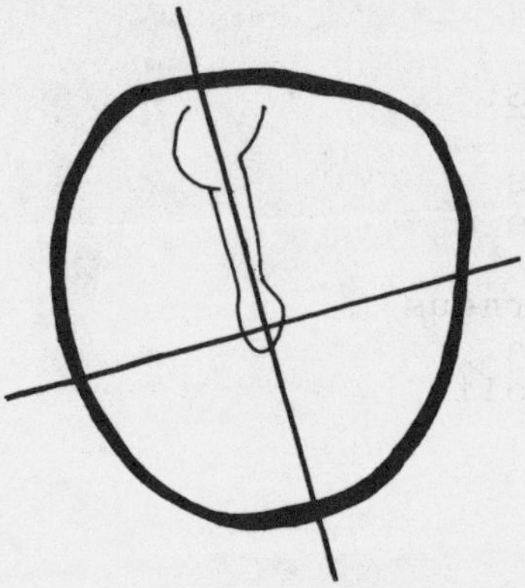

LEFT TYMPANIC MEMBRANE

STRUCTURES:

1) Pars flaccida--composed of two layers
2) Pars tensa--composed of three layers
3) Lateral process of malleus
4) Anterior malleolar fold
5) Posterior malleolar fold
6) Handle of malleus
7) Umbo
8) Cone of light

Observe how the membrane is divided into quadrants--the cone of light is in the anterior inferior quadrant.

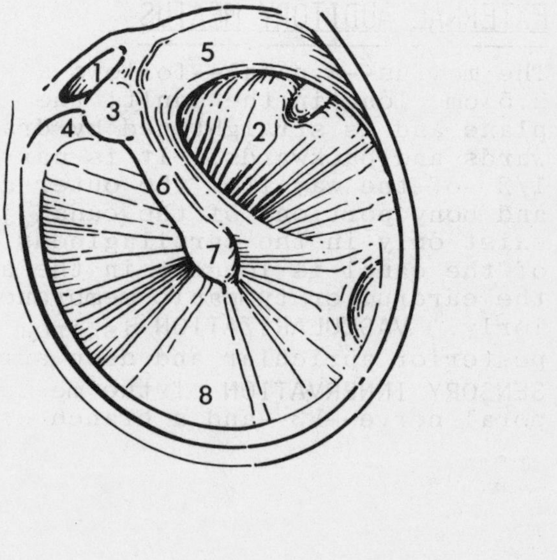

III. MIDDLE EAR

STRUCTURES IN THE TYMPANIC CAVITY

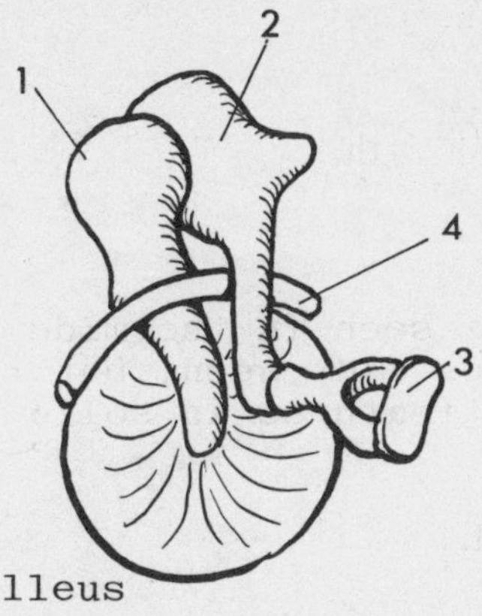

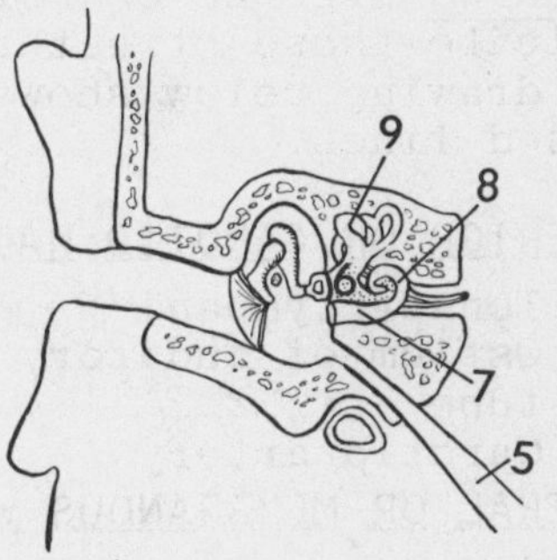

1) Malleus
2) Incus
3) Stapes
4) Chorda tympani--
   passes between malleus
   and incus.
5) Auditory tube

Tensor tympani--not pictured
innervated by V3, contracts
to prevent damage and to pick
up higher frequency sounds, inserts onto malleus
Stapedius--not pictured, innervated by VII, inserts onto
stapes, protects from loud sounds.

ALSO SEE:

6) Oval or vestibular window
7) Round or cochlear window--site of
   secondary tympanic membrane
8) Cochlea
9) Semicircular canals

THE EAR OSSICLES transduce sound vibrations to movement of fluid in the internal ear for the perception of sound. An inward movement of the malleus results in an inward movement of the baseplate of the stapes on the oval window.

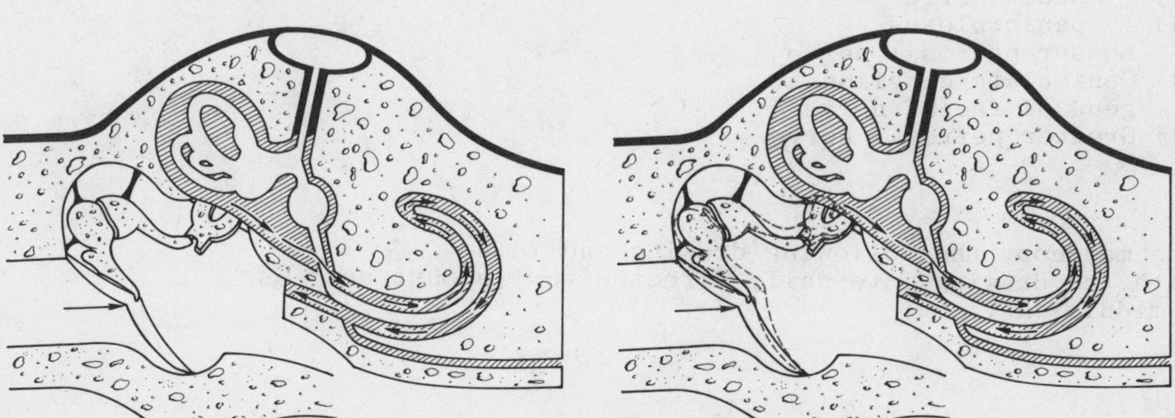

ACTION OF OSSICLES

FEATURES AND RELATIONS OF THE MIDDLE EAR

Features are the objects seen in the middle ear, while relations indicate those structures not seen, but related to the cavity. The drawing below shows features in solid lines, and relations in dotted lines.

ANTERIOR OR CAROTID WALL:

1) Tensor tympani
2) Ostium of auditory tube
3) Carotid artery

LATERAL OR MEMBRANOUS WALL:

4) Tympanic membrane
5) Malleus
6) Chorda tympani

POSTERIOR OR MASTOID WALL:

7) Mastoid antrum
8) Pyramid of stapedium
9) Facial nerve

MEDIAL OR LABYRINTHINE WALL:

10) Prominence of lateral semicircular canal
11) Prominence of facial canal
12) Oval window
13) Round window
14) Promontory of basal turn of cochlea
15) Tympanic nerve
16) Tympanic plexus
17) Lesser petrosal nerve
18) Geniculate ganglion at genu of facial nerve
19) Greater petrosal nerve

FLOOR OR JUGULAR WALL:

20) Bulb of internal jugular vein

ROOF OR TEGMEN TYMPANI:

Shows nothing of interest

One may copy the following drawing and fold with the drawing internally directed to 'reconstruct' a middle ear.

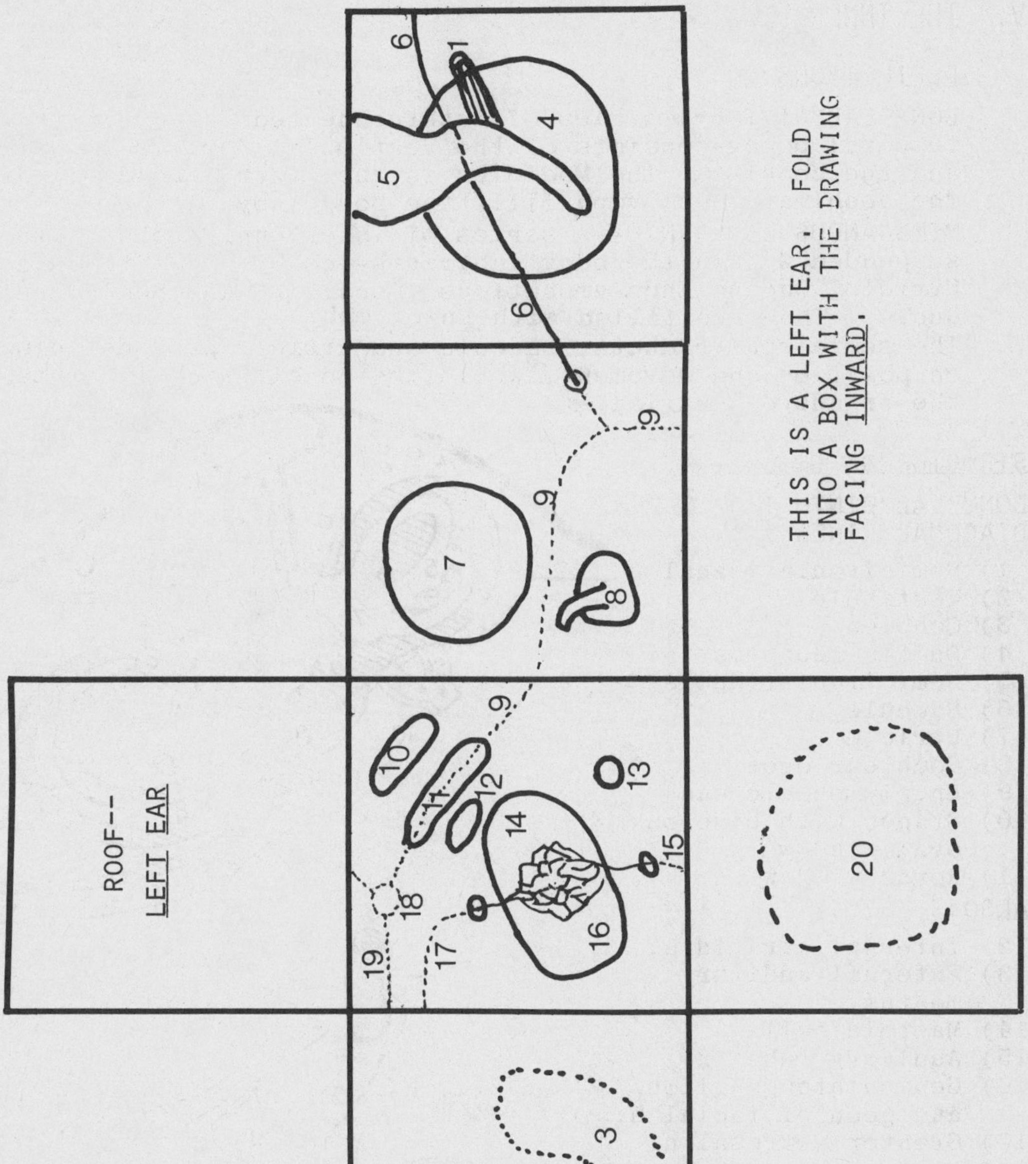

THIS IS A LEFT EAR. FOLD INTO A BOX WITH THE DRAWING FACING INWARD.

IV. THE INNER EAR

DEFINITIONS:

BONY LABYRINTH--a series of interconnected spaces in the petrous temporal bone--consists of the Vestibule, Semicircular canals and the cochlea. The Modiolus is the central bony pillar of the cochlea. Perilymph fills the bony labyrinth.

MEMBRANOUS LABYRINTH--a series of interconnected structures suspended within the bony labyrinth--consists of the Saccule, Utricle, Saccus Endolymphaticus, Cochlear and Semicircular ducts. They are filled with Endolymph.

The semicircular ducts, saccule and utricle provide information on position and movement, while the cochlear ducts contain the organs of hearing.

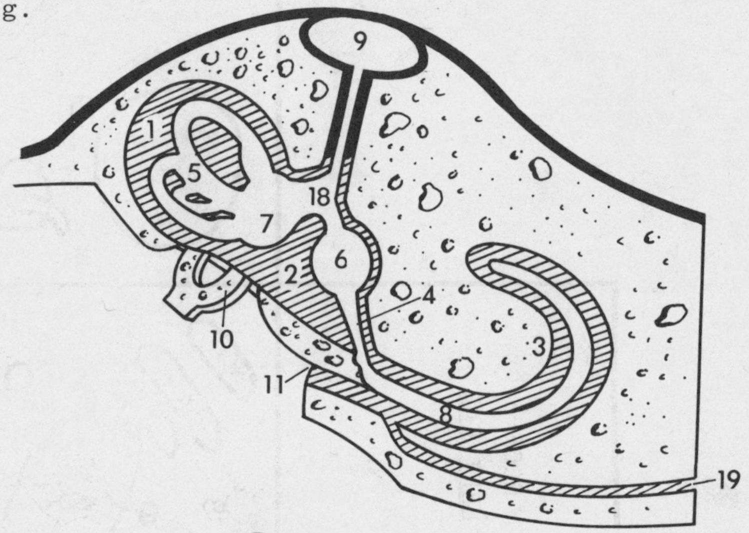

STRUCTURES:

BONY LABYRINTH IN DIAGONAL LINES

1) Semicircular canal
2) Vestibule
3) Cochlea
4) Ductus reuniens
5) Semicircular ducts
6) Saccule
7) Utricle
8) Cochlear duct
9) Endolymphatic sac
10) Stapes with base on oval window
11) Round window

ALSO:

12) Internal carotid a.
13) External auditory meatus
14) Mastoid cells
15) Auditory tube
16) Geniculate ganglion and genu of facial n.
17) Greater petrosal n.
18) Utriculosaccular duct
19) Cochlear aqueduct

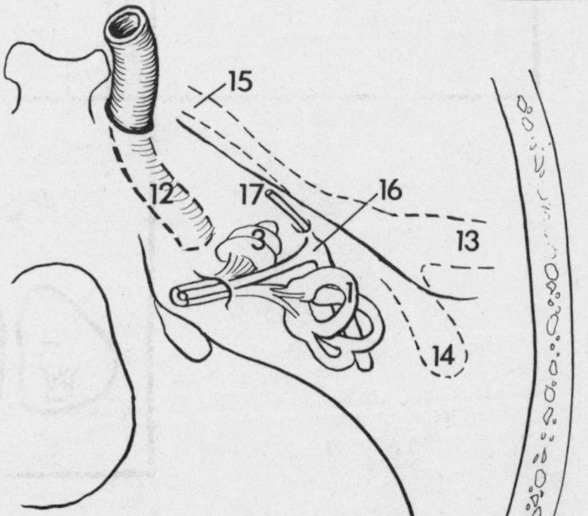

THE COCHLEAR DUCT

The cochlear duct contains transducers--the Organ of Corti--for converting fluid movement into impulses the brain perceives as sound. The sound vibrations were first transduced into fluid movement by action of the middle ear ossicles. The cochlear duct does not completely fill the bony cochlea. However, the manner in which the duct is contained in the cochlea partitions the cochlea into two compartments. Think of the cochlea as a tube--the cochlear duct forms a roughly triangular shaped entity with an apex on one side of the tube and its base opposite. The duct starts at the opening of the tube at one end, but doesn't reach the other end. This results in a sealed cochlear duct inside the tube, but the tube now has two longitudnal divisions continuous at the end or helicotrema. In essence the length of the original cochlea has doubled. The stapes will displace fluid in the portion of the cochlear duct--scala vestibuli--related to the oval window. The fluid movement will continue around the turn at the helicotrema and result in opposite movements at the round window, as transmitted through fluid in the scala tympani. In actuality the cochlea is not straight, but makes a spiral. It makes 2 3/4 turns, with the basal turn resulting in a promontory in the tympanic cavity. The walls of the triangular cochlear duct are the basilar membrane, the vestibular or Reissner's membrane and the spiral ligament. The basilar and vestibular membranes converge upon the modiolus at a bony projection called the limbus. Nerve fibers with cells bodies in the spiral ganglion are seen at the modiolus. The Organ of Corti is a specialization in the cochlear duct and consists of a large number of supporting cells and hair cells that serve as hearing receptors.

STRUCTURES:

1) Stapes
2) Helicotrema
3) Cochlear duct
4) Round window
5) Scala tympani
6) Scala vestibuli
7) Vestibular membrane
8) Basilar membrane
9) Spiral ligament
10) Organ of Corti
11) Limbus
12) Spiral ganglion
13) Cochlear nerve
14) Vestibular nerve
15) Vestibulo-cochlear --acoustic--nerve

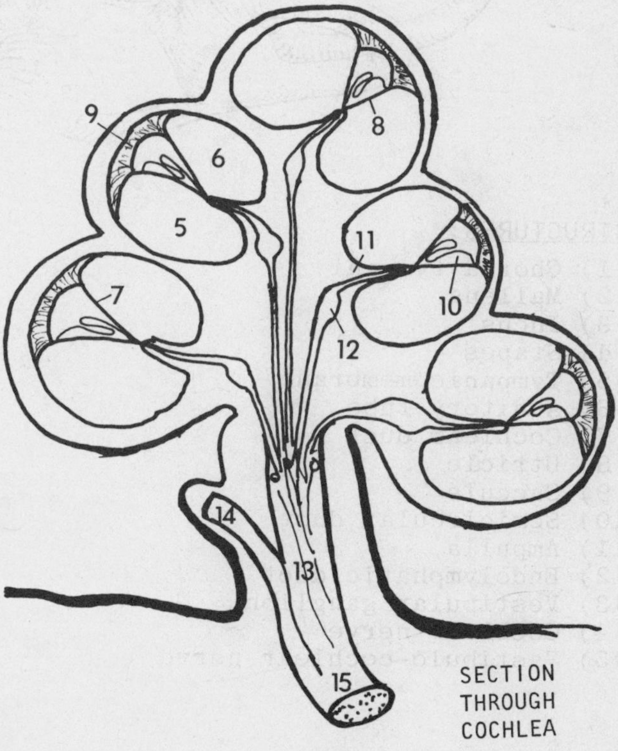

SECTION THROUGH COCHLEA

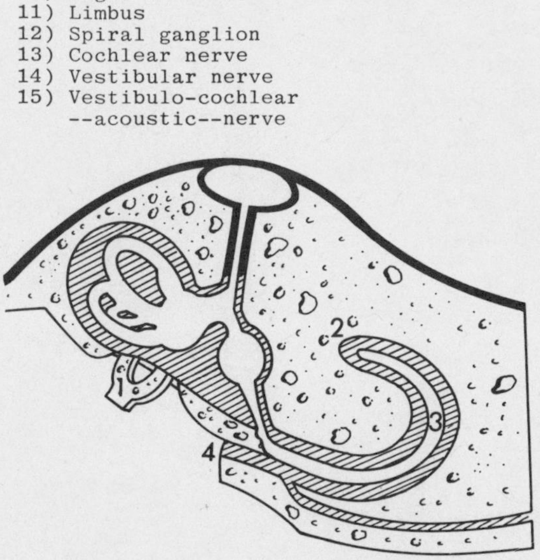

SCHEMATIC VIEW

PHANTOM VIEW MEMBRANOUS LABYRINTH PROJECTED ONTO TYMPANIC CAVITY

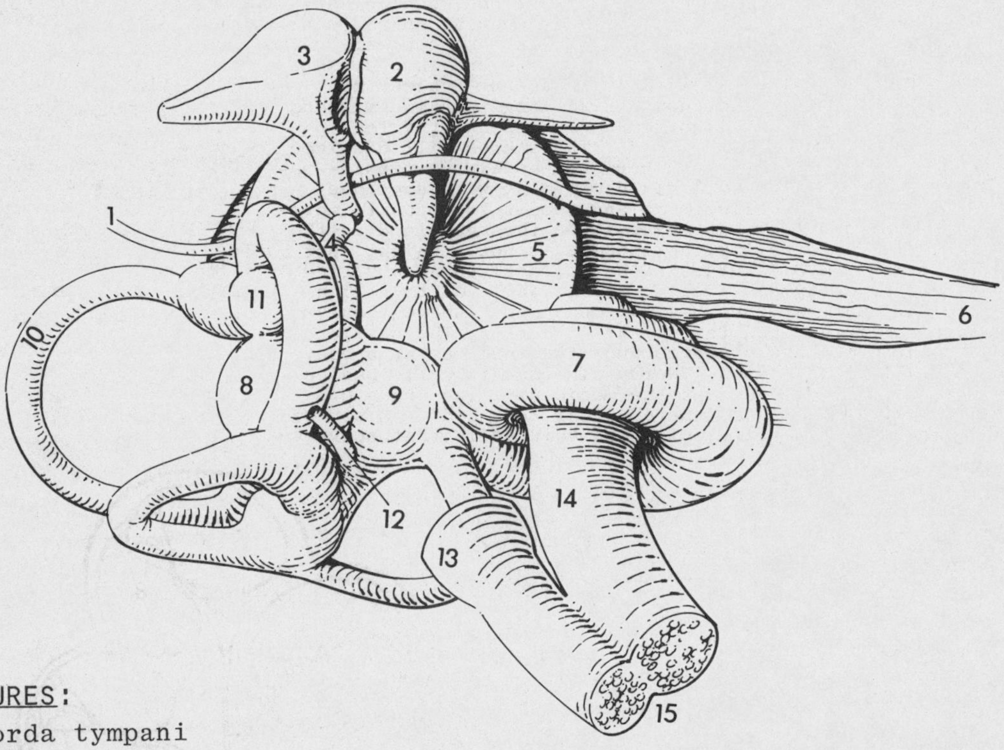

STRUCTURES:

1) Chorda tympani
2) Malleus
3) Incus
4) Stapes
5) Tympanic membrane
6) Auditory tube
7) Cochlear duct
8) Utricle
9) Saccule
10) Semicircular duct
11) Ampulla
12) Endolymphatic duct
13) Vestibular ganglion
14) Cochlear nerve
15) Vestibulo-cochlear nerve

COMMENTS

1. Since the external meatus is directed anteriorly, downward and medially, the pinna must be pulled superiorly, laterally and posteriorly to examine the tympanic membrane.
2. The eustachian tube in children is more horizontal than in adults and accounts for the great number of middle ear infections in children. An infection of the middle ear--otitis media--may perforate the tympanum or pass above it and dissect between the epithelium and wall of the canal itself; if necessary the eardrum may be lanced in the posterior inferior quadrant--this spares the chorda tympani, malleus and malleolar processes. Ruptured eardrums may result in infection and deafness.
3. Dura on the base of the cranial cavity is more adherent to the bone than it is to the calvaria and fractures of the base of the cranium routinely tear the dura and arachnoid, allowing CSF to leak. Thus, fractures of the petrous--rocklike--temporal bone can allow CSF to leak to the middle ear--otorrhea--and thence to the pharynx or externally.
4. Vagal contribution to the innervation of the ear canal combined with the phenomenon of referred pain explains why infections and manipulations in the canal may result in reflexes and sensations associated with the vagus--sneezing, coughing and nausea.
5. Mastoiditis--infection of mastoid air cells--was not rare before the advent of antibiotics. Commonly the mastoid process was crudely chiseled off or bored out. Such a procedure caused damage to the facial nerve and often resulted in erosion, necrosis and spread of the infection--particularly to the jugular fossa where the internal jugular vein was eroded and severe, sometimes fatal, hemorrhage resulted.
6. Meniere's disease--actually a syndrome--results from an increase in endolymphatic pressure which may rupture the membranous labyrinth. Signs are dizziness--vertigo--ringing in the ear--tinnitis--and deafness. Often, several recurrent episodes, each with more severe sequelae, take place. At present, a delicate operation to place a tube in the membranous labyrinth is used to alleviate the disease.
7. Repetitive loud sounds can result in deafness to high frequency sounds; other trauma induced by high pressure, such as 'boxing the ear,' industrial noises and explosions, results in compression waves and also damaged hearing.
8. Cerumen--ear wax--is found mainly in the cartilaginous portions of the canal and prevents softening and sloughing of the tissues by water and may repel attacks by insects.

NECK

I. SUPERFICIAL STRUCTURES

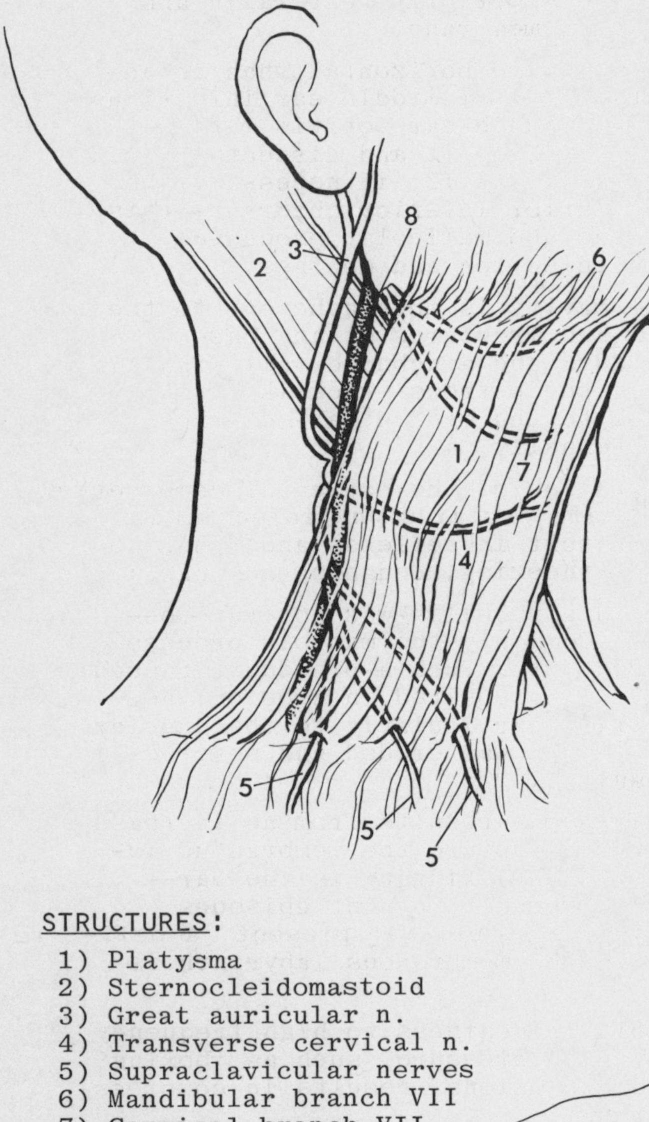

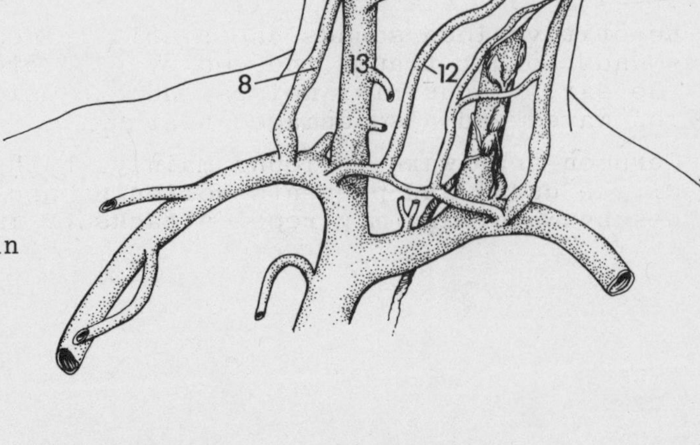

STRUCTURES:

1) Platysma
2) Sternocleidomastoid
3) Great auricular n.
4) Transverse cervical n.
5) Supraclavicular nerves
6) Mandibular branch VII
7) Cervical branch VII
8) External jugular vein
9) Facial vein
10) Retromandibular vein
11) Posterior auricular vein
12) Anterior jugular vein
13) Internal jugular vein

SURFACE ANATOMY BY VERTEBRAL LEVELS

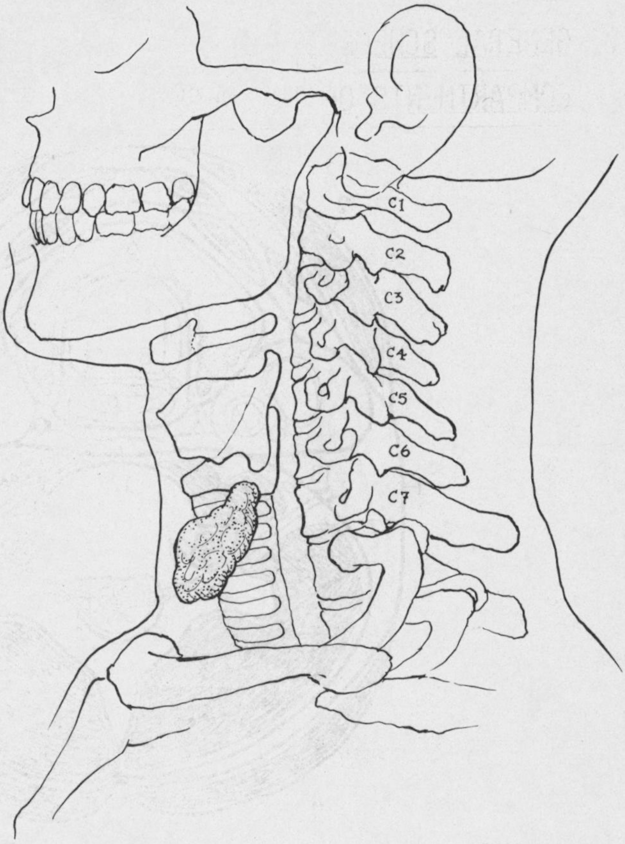

C3--Hyoid bone
C4,5--Bifurcation of common carotid artery; location of thyroid cartilage
C6--Cricoid cartilage; end of pharynx; beginning of esophagus; end of larynx, beginning of trachea; recurrent laryngeal n. reaches larynx; inferior thyroid artery turns posterior to neurovascular sheath containing carotid artery; middle cervical ganglion
C7--Isthmus of thyroid; most superior limit of thoracic duct

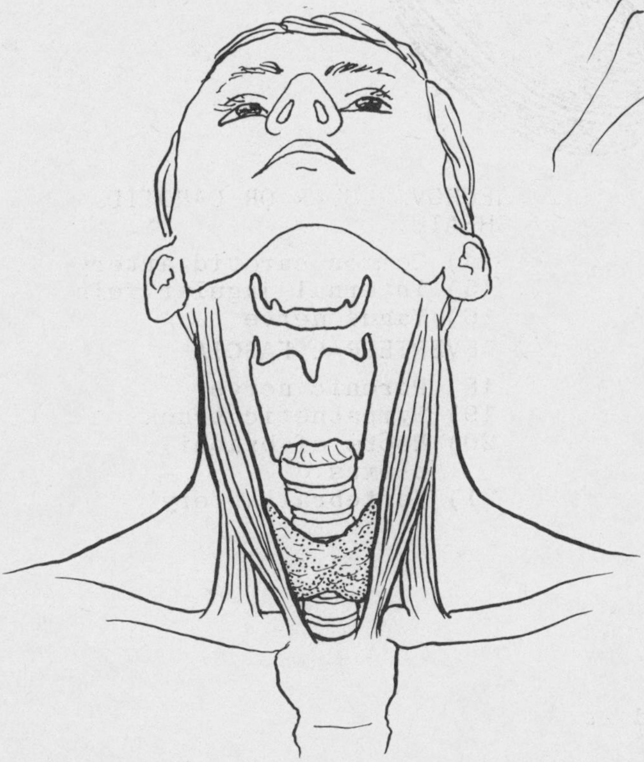

II. GENERAL SCHEME OF COMPARTMENTS OF THE NECK

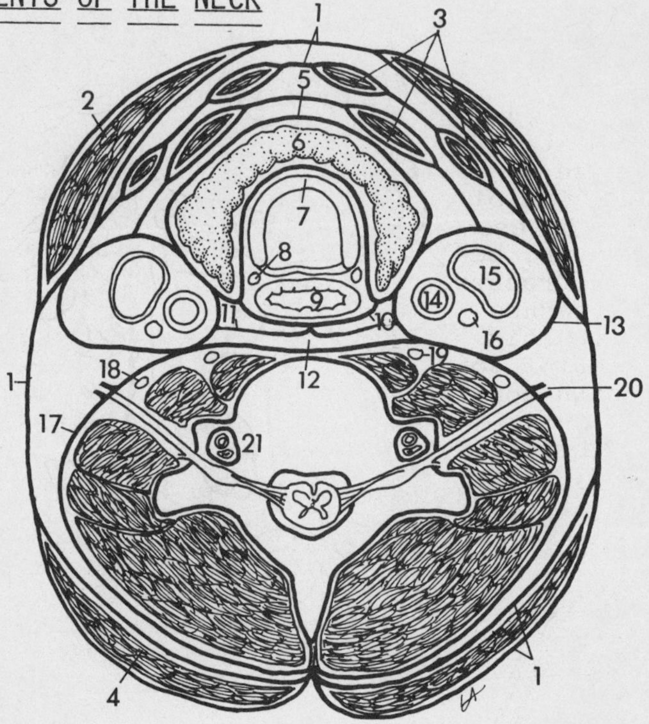

1) INVESTING FASCIA:

2) Sternocleidomastoid
3) Infrahyoid muscles
4) Trapezius

Observe potential pre-tracheal spaces.

5) VISCERAL OR PRETRACHEAL FASCIA:

6) Thyroid gland
7) Trachea
8) Recurrent laryngeal n.
9) Esophagus
10) Buccopharyngeal fascia
11) Alar fascia--present only in upper portions of pharynx, fuses with (10) T2-3
12) Retropharyngeal or retro-esophageal space, doubled at level of alar fascia

13) NEUROVASCULAR OR CAROTID SHEATH:

14) Common carotid artery
15) Internal jugular vein
16) Vagus nerve

17) PREVERTEBRAL FASCIA

18) Phrenic nerve
19) Sympathetic trunk
20) Roots of brachial plexus
21) Vertebral artery

III. TRIANGLES OF THE NECK

The muscles of the neck have traditionally been used to divide the neck into triangles; this is useful for locating deeper structures in relation to the surface.

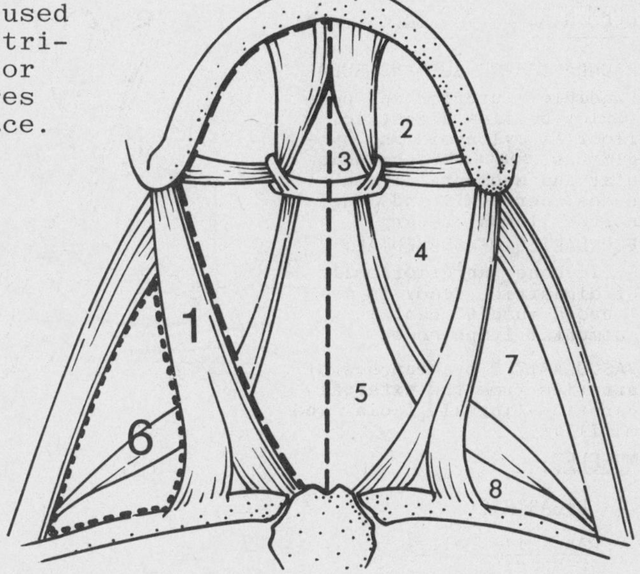

TRIANGLES:

1) ANTERIOR
   2) Submandibular or digastric
   3) Submental
   4) Carotid
   5) Muscular

6) POSTERIOR
   7) Occipital
   8) Subclavian

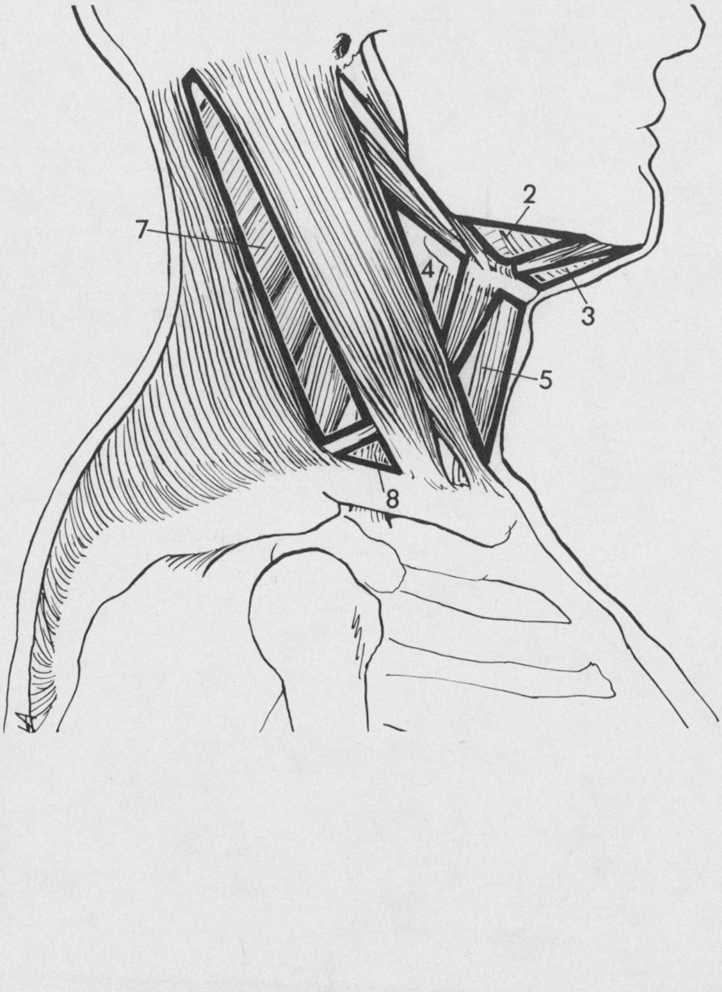

ANTERIOR TRIANGLE

SUBMANDIBULAR AND SUBMENTAL TRIANGLES

BOUNDARIES OF SUBMANDIBULAR:
Mandible, anterior and posterior bellies digastric, floor is mylohyoid and hyoglossus, contains submandibular and submental lymph nodes, nerve XII and lingual nerve, lingual artery.

BOUNDARIES OF SUBMENTAL:
Hyoid bone, anterior belly of digastric, floor is mylohyoid muscle, contains submental lymph nodes.

VASCULARIZED by branches of arteries from the external carotid--lingual, facial and maxillary.

MUSCLES:

1) DIGASTRIC

Anterior belly
O-Mandible
I-intermediate tendon
A-Depress mandible, elevate hyoid
N-V3, nerve to mylohyoid

Posterior belly:
O-Mastoid notch
I-Intermediate tendon
A-Same as above
N-VII

2) STYLOHYOID

O-Styloid process
I-Hyoid bone
A-Elevates and retracts hyoid bone
N-VII

3) MYLOHYOID

O-Mandible
I-Hyoid bone and midline raphe
A-Elevates floor of mouth, can elevate hyoid and depress mandible
N-V3, nerve to mylohyoid

4) HYOGLOSSUS

O-Hyoid bone
I-Tongue
A-Retracts tongue
N-XII

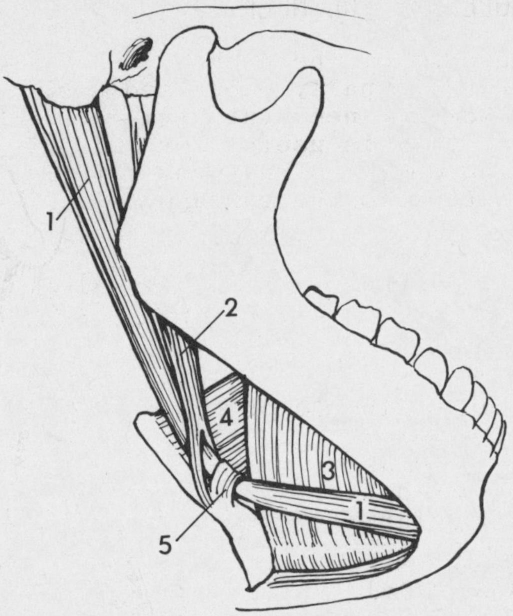

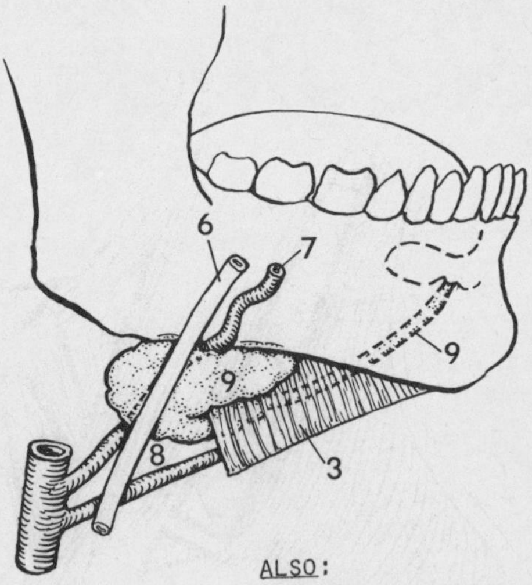

ALSO:
5) Fascial sling around intermediate tendon
6) Facial vein
7) Facial artery
8) Lingual artery
9) Submandibular gland and duct

MUSCULAR TRIANGLE

1) STERNOCLEIDOMASTOID

O-Sternum and clavicle
I-Mastoid process p36 Basmajian
A-Turn head to opposite side, tilt head
N-XI

INFRAHYOID OR STRAP MUSCLES:

2) STERNOHYOID

O-Sternum
I-Hyoid
A-Depress hyoid

3) THYROHYOID

O-Thyroid cartilage
I-Hyoid
A-Depress hyoid or raise larynx

4) STERNOTHYROID

O-Sternum
I-Thyroid cartilage
A-Depress larynx

5) OMOHYOID

O-Scapula
I-Hyoid
A-Depress hyoid

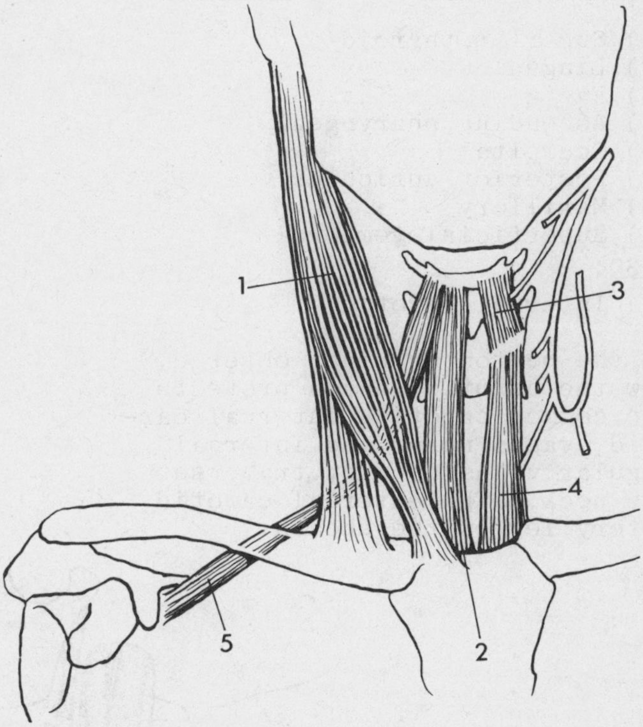

ALL of the strap muscles are INNERVATED by the ansa cervicalis, a derivative of the cervical plexus. Although fibers to the thyrohyoid may travel for a distance 'piggy back' with XII, they are of cervical origin

VASCULATURE TO THE AREA is derived from branches of the external carotid, mainly superior thyroid, and from the thyrocervical trunk from the subclavian artery.

CAROTID TRIANGLE AND CAROTID SHEATH

BRANCHES of the common carotid arteries and the first 5 of the external carotid branches listed below are found in the triangle.

BRANCHES OF THE EXTERNAL CAROTID:

1) Superior thyroid
2) Lingual
3) Facial
4) Ascending pharyngeal
5) Occipital
6) Posterior auricular
7) Maxillary
8) Superficial temporal

ALSO:

9) Internal carotid

In the botton drawing, observe how the carotid sheath protects the common carotid, internal carotid, vagus nerve and internal jugular veins as they traverse the neck. The external carotid artery leaves the sheath.

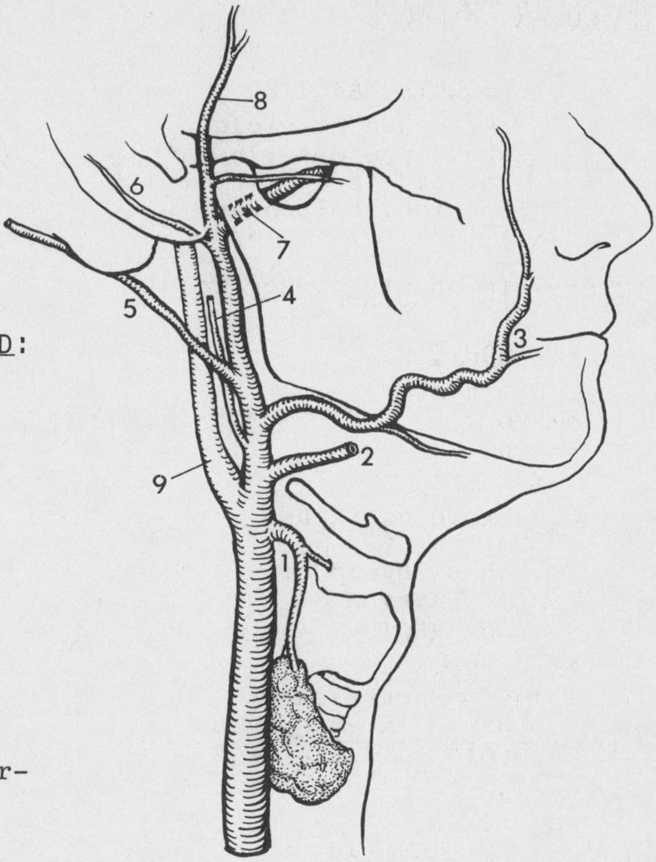

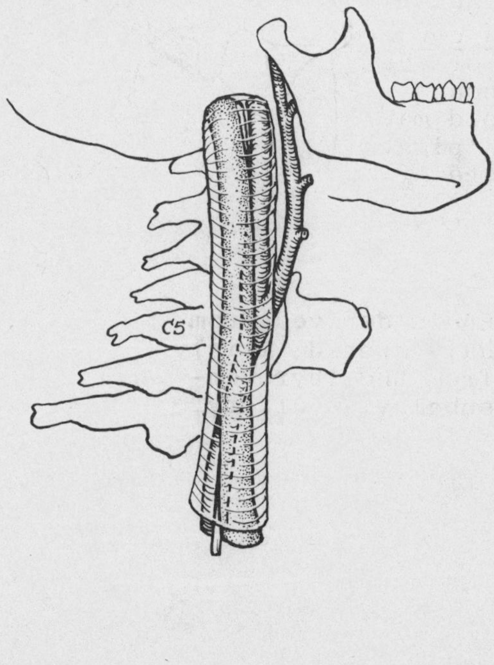

POSTERIOR TRIANGLE

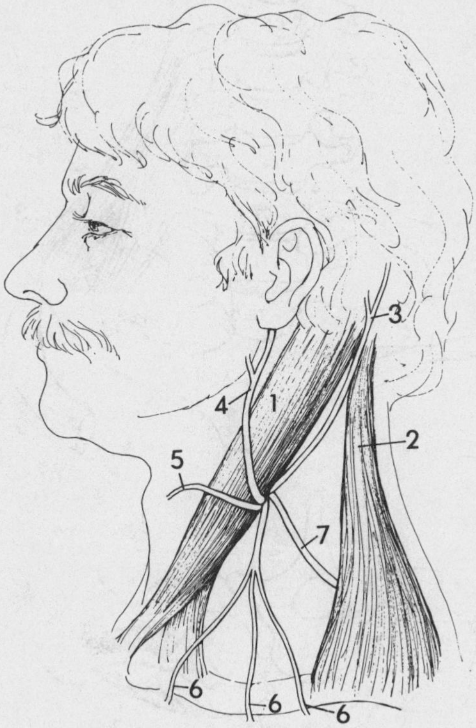

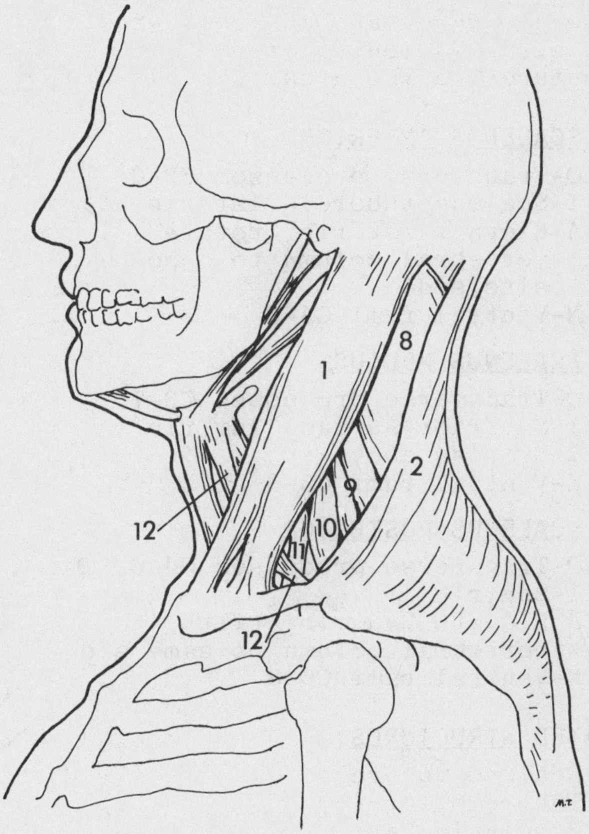

STRUCTURES:

1) Sternocleidomastoid
2) Trapezius
3) Lesser occipital n.
4) Great auricular n.
5) Transverse cervical n.
6) Supraclavicular n.
7) Spinal accessory n.
8) Splenius capitis
9) Levator scapulae
10) Middle scalene
11) Anterior scalene
12) Omohyoid

The accessory nerve divides the posterior triangle into a "carefree" region superiorly and a "careful" region inferiorly.

SCALENE MUSCLES

These are important muscles due to their relation to other structures in the neck.

1) SCALENUS ANTERIOR
   O-Transverse processes C3-6
   I-Scalene tubercle 1st rib
   A-Elevate 1st rib, rotate vertebral column to opposite side
   N-Ventral rami C4-6

2) SCALENUS MEDIUS
   O-Transverse processes C3-7
   I-Superior surface 1st rib
   A-Same as above
   N-Ventral rami C3-8

3) SCALENUS POSTERIOR
   O-Transverse processes C4-6
   I-2nd rib
   A-Elevate 2nd rib, tilt vertebral column to same side
   N-Ventral rami C6-8

RELATED STRUCTURES:

4) Scalene tubercle
5) Tubercle for (2)
6) Subclavian artery
7) Subclavian vein
8) Cupola of pleura
9) Brachial plexus

Observe the brachial plexus and subclavian artery between the middle and anterior scalene.

VASCULATURE for the scalenes is from the thyrocervical and costocervical trunks.

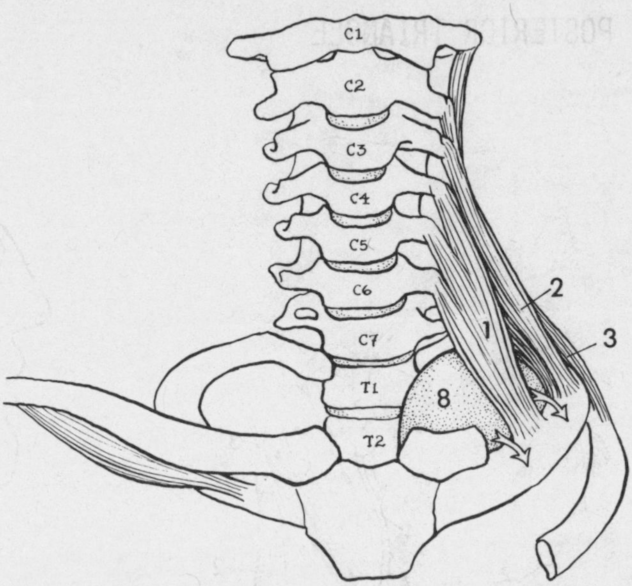

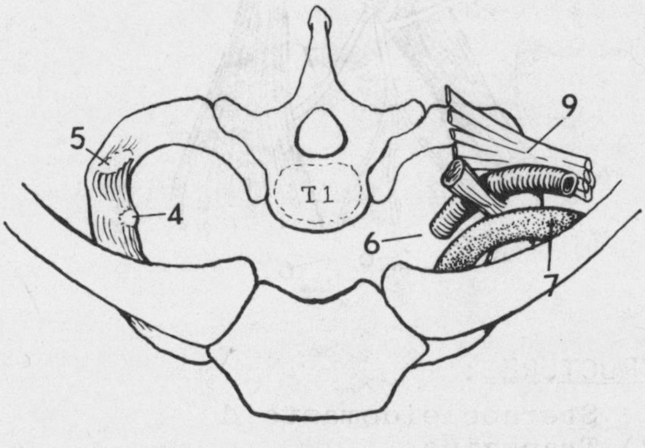

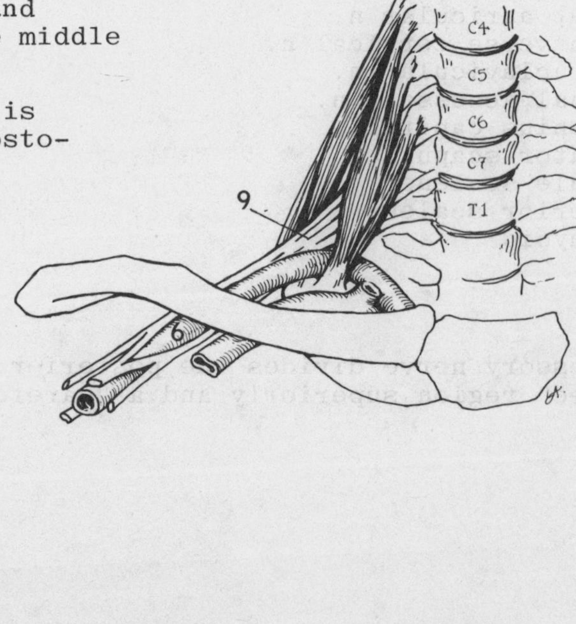

CERVICAL PLEXUS AND PHRENIC NERVE

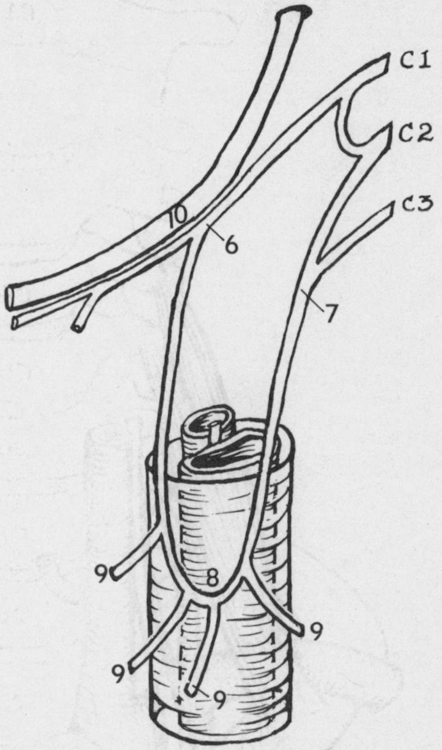

The cervical plexus is formed from ventral rami of C1-4--if the phrenic nerve is included in the plexus, as some sources place it, the plexus must also include C5.

COMPOSITION:

1) Lesser occipital-sensory
2) Great auricular-sensory
3) Transverse cervical-sensory
4) Supraclavicular-sensory
5) Phrenic-motor

ANSA CERVICALIS-Motor to strap muscles

6) Superior or hypoglossal ramus--travels with XII, but is not a part of it
7) Inferior or descending ramus
8) Communication between the rami to form the ansa--may be absent
9) Muscular branches
10) XII

The ansa usually forms superficial to the carotid sheath--see upper illustration--but may be deep to it. The ansa is highly variable.

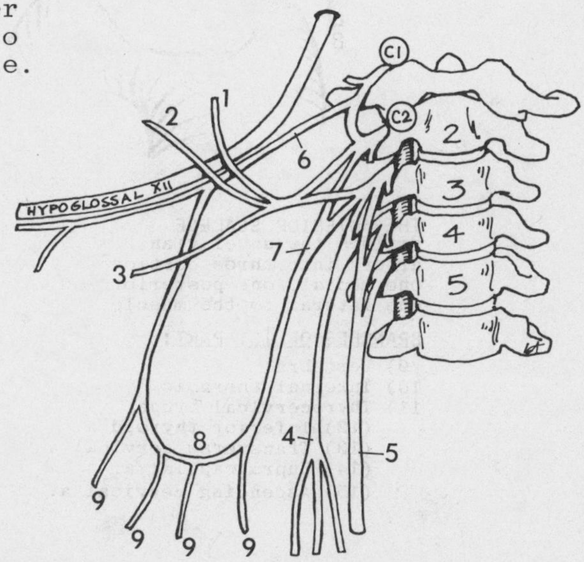

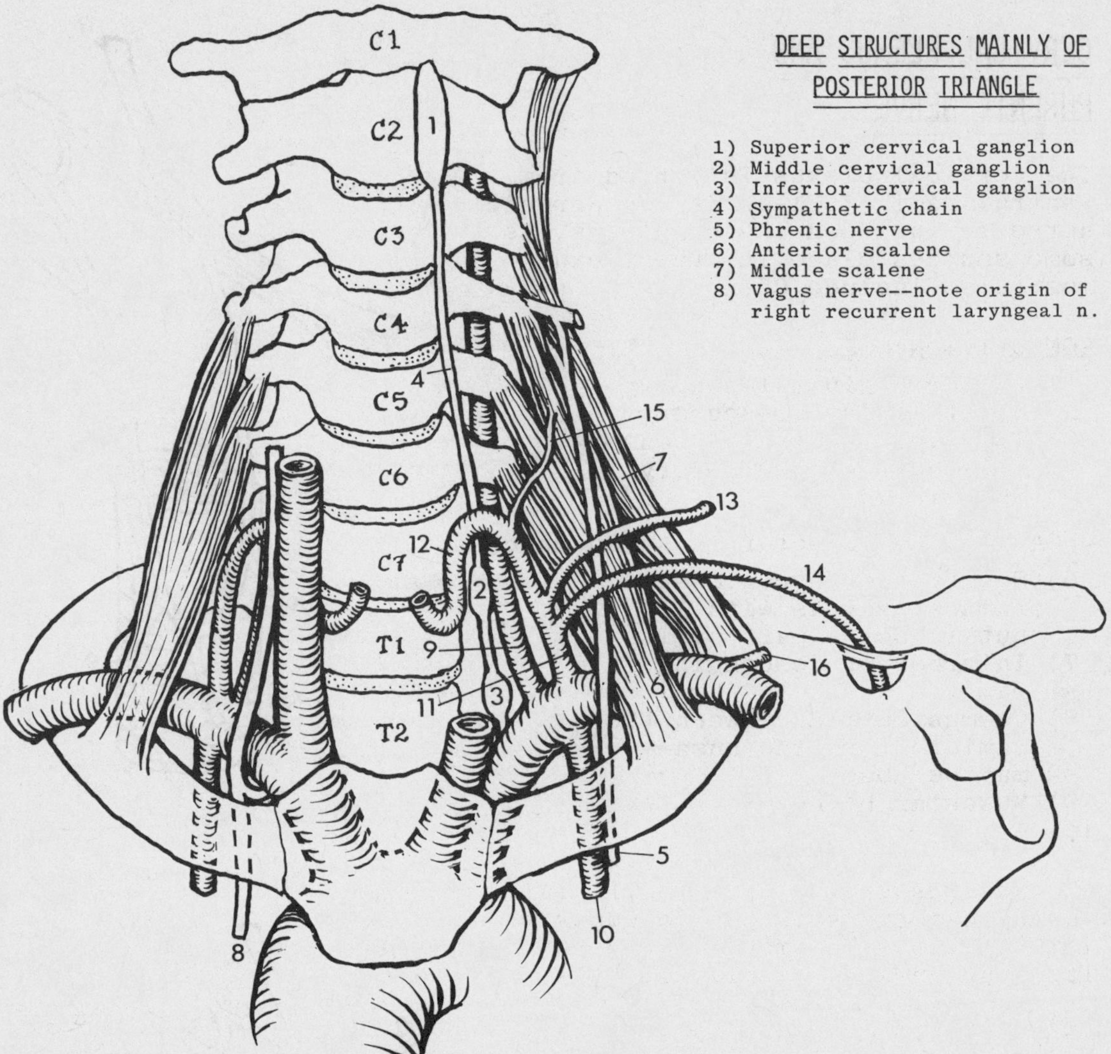

DEEP STRUCTURES MAINLY OF POSTERIOR TRIANGLE

1) Superior cervical ganglion
2) Middle cervical ganglion
3) Inferior cervical ganglion
4) Sympathetic chain
5) Phrenic nerve
6) Anterior scalene
7) Middle scalene
8) Vagus nerve--note origin of right recurrent laryngeal n.

THE ANTERIOR SCALENE divides the subclavian artery into three portions--one medial, one posterior and one lateral to the muscle.

BRANCHES OF 1ST PART:

9) Vertebral
10) Internal thoracic
11) Thyrocervical trunk
    (12) Inferior thyroid a.
    (13) Transverse cervical a.
    (14) Suprascapular a.
    (15) Ascending cervical a.

BRANCHES OF 2ND PART:

Costocervical trunk--not shown--gives superior intercostal and deep cervical

BRANCHES OF 3RD PART:

16) Dorsal scapular--variable, this artery may arise as a branch of the transverse cervical artery

REMEMBER: The subclavian artery becomes the axillary artery at the lateral border of the 1st rib.

IV. DEEP NECK STRUCTURES

STRUCTURES MAINLY OF THE ANTERIOR TRIANGLE

1) External laryngeal n.
2) Superior thyroid a.
3) Cricothyroid m.
4) External carotid a.
5) Recurrent laryngeal n.
6) Esophagus
7) Inferior thyroid a.
8) Inferior thyroid v.
9) Middle thyroid v.
10) Superior thyroid v.
11) Thyrocervical trunk
12) Trachea
13) Left brachiocephalic v.
14) Brachiocephalic a.
15) Internal carotid
16) Subclavian a.
17) Common carotid a.
18) Thyroid gland
19) Thyroid cartilage
20) Parathyroid glands
21) Vagus nerve
22) Internal jugular v.

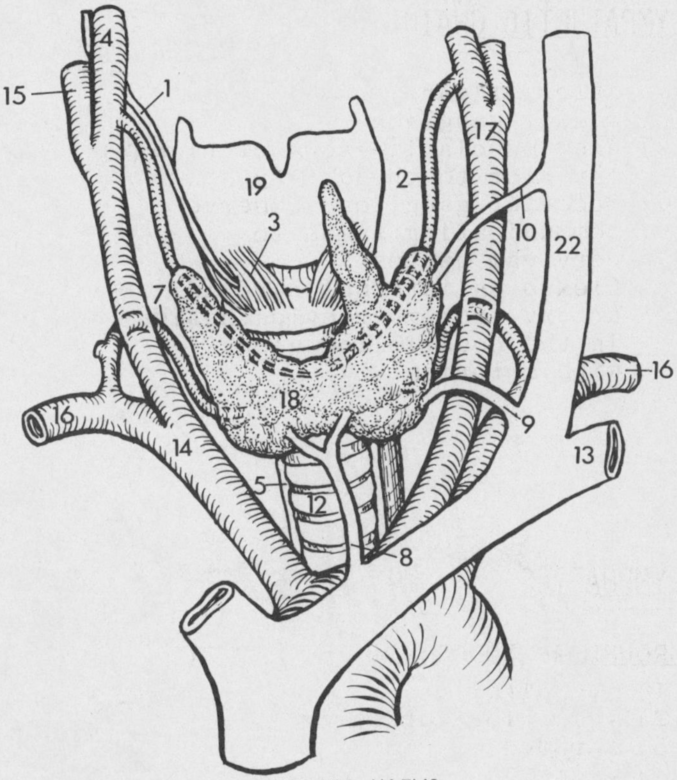

ANTERIOR VIEWS

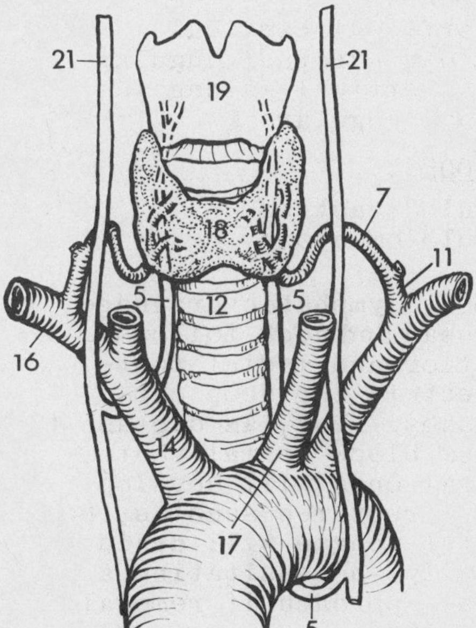

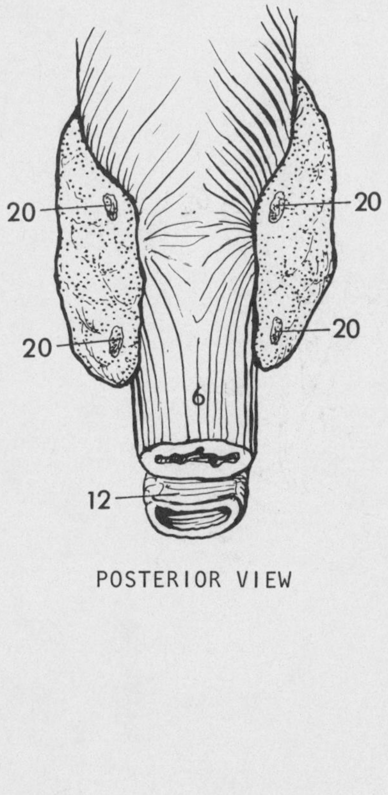

POSTERIOR VIEW

SYMPATHETIC CHAIN

1) Inferior ganglion
2) Middle ganglion
3) Ansa subclavia--connect (1 & 2) anterior to the artery
4) Superior ganglion--observe communicating ramus to cervical nerves
5) Plexus on internal carotid artery--remember sympathetics in the head follow along with arteries

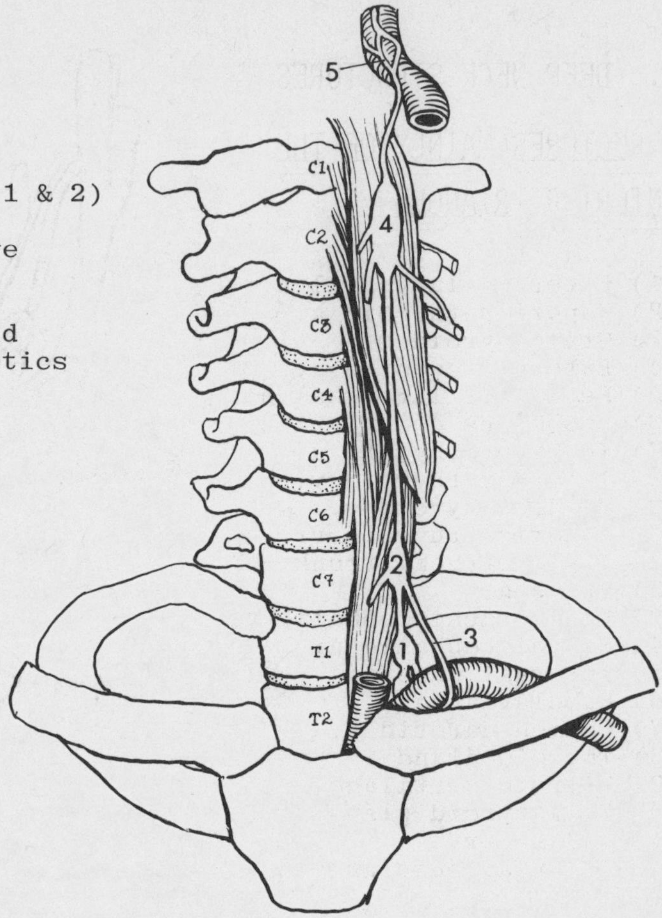

LYMPHATICS

GROUPS OF NODES:

1) Occipital
2) Posterior auricular
3) Parotid
4) Buccal
5) Submandibular
6) Submental
7) Superficial cervical --along external jugular vein
8) Deep cervical--along internal jugular

NAMED NODES:

9) Jugulodigastric
10) Jugulo-omohyoid

AS A RULE, lymphatic drainage of the head and neck follows an inferior and medial route to collect in the deep cervical nodes. But, as disease processes block normal routes, collateral ones will develop. Midline structures such as the tip of the tongue will drain bilaterally--necessitating a bilateral approach if removal is warranted.

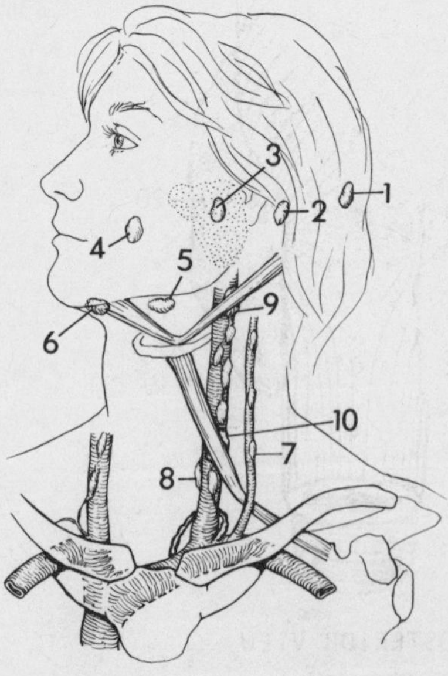

COMMENTS

1. The neck is a zone of transit for vitally important neural, vascular, digestive and respiratory structures. Quite a few of these structures do not have their actual origin in the neck but it is their location and relationship to nearby structures and surface projections that are important. Often the structures are traumatized or must be accessed during surgery. A basic knowledge of the fascial planes of the neck and the location of all structures is a necessity.

2. The posterior triangle is the site of neural structures which include the spinal accessory nerve, cervical plexus, brachial plexus and phrenic nerve. Although their superficiality at this point makes them readily accessible, it also makes them vulnerable to accidental or iatrogenic trauma; a moment's thought will bring the realization of injury to some or all of these structures.

3. Although the platysma is a muscle of the superficial tissues, it is important in maintaining tone and appearance of skin of the neck. It and its nerve supply is carefully treated during all surgeries of the neck for cosmetic purposes.

4. Branches of the external carotid artery may be reached in the carotid triangle and the triangle is one point pressure may be applied to control bleeding. One carotid artery may be ligated temporarily during surgery to minimize bleeding; in fact, both externals may be ligated for a short time if needed due to enormous collateral circulations. The internal carotid has no branches in the neck, but may be opened in order to 'strip the artery' of atherosclerotic plaques--endarterectomy.

5. Horner's Syndrome is the loss of sympathetic innervation to the head and neck--signs are partial ptosis, pupillary constriction, vasodilation and dry skin. Endophthalmos--sunken eye--may also be present but is poorly understood.

6. The thyroid gland develops in the pharynx and must descend to its normal level. Remnants of the gland may persist as accessory glands along the course of its descent or thyroglossal duct cysts may develop along the same path. Occasionally a cyst may open to the skin forming a sinus. A pyramidal lobe, if present, is a superior extension of tissue into the lower part of the thyroglossal duct. A thyroidea ima artery--lowest thyroid --may occur in the midline and interfere with surgical procedures inferior to the thyroid.

7. An awareness of the proximity of the thyroid, parathyroids, inferior thyroid artery and recurrent laryngeal nerve is important since that area is often the site of surgical intervention. The parathyroid glands may range from the thyroid cartilage to the superior mediastinum, though they are usually found on the posterior aspect of the thyroid. Their accidental removal or compromise can result in death by tetany; the recurrent laryngeal nerve supplies all but 1 laryngeal muscle; bilateral destruction can result in suffocation, unilateral can change the quality of the voice.

NASAL AND ORAL CAVITIES, PHARYNX AND LARYNX

I. DEFINITIONS

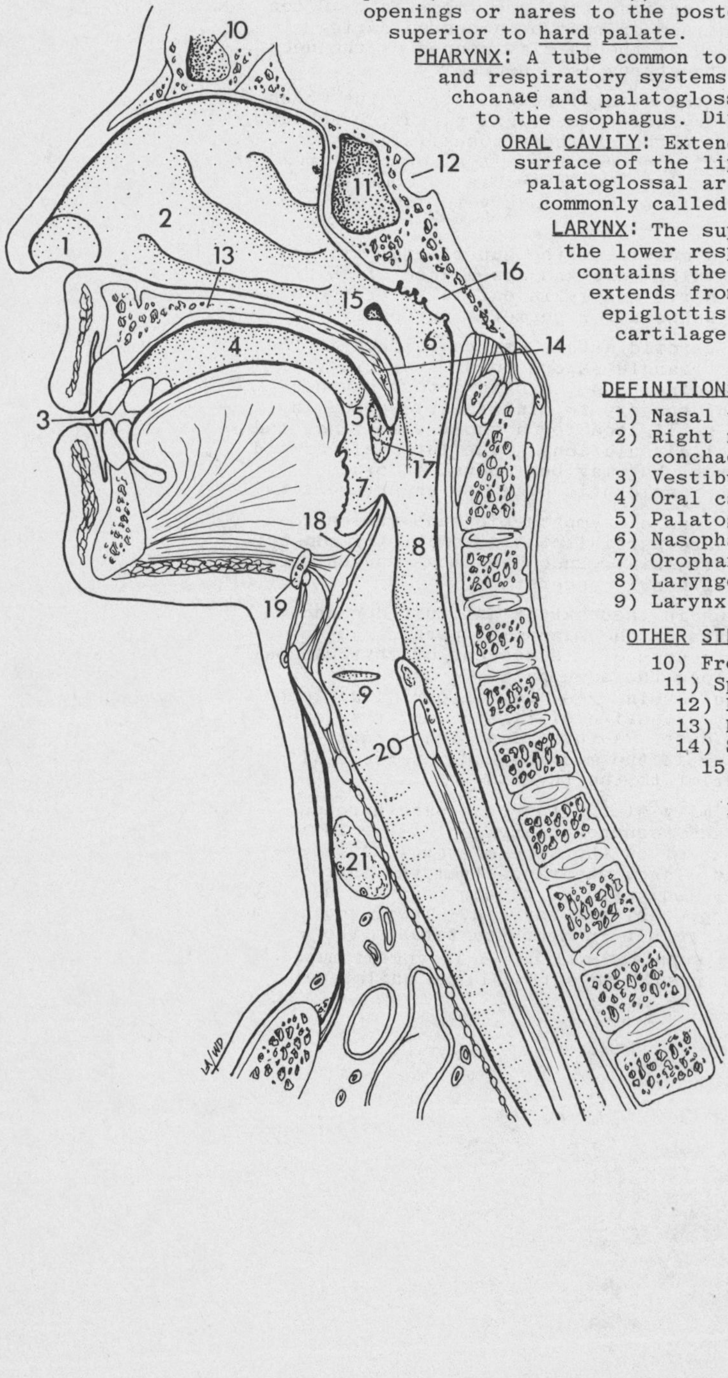

NASAL CAVITY: Divided into left and right cavities by a septum (not shown below), extends from the anterior openings or nares to the posterior nares or choanae, superior to hard palate.

PHARYNX: A tube common to both the digestive and respiratory systems, extends from the choanae and palatoglossal arch inferiorly to the esophagus. Divided into 3 regions.

ORAL CAVITY: Extends from the inner surface of the lips and gums to the palatoglossal arch, has a vestibule commonly called the buccal cavity.

LARYNX: The superior portion of the lower respiratory system, contains the vocal apparatus, extends from the margin of the epiglottis to the cricoid cartilage.

DEFINITIONS OF REGIONS:

1) Nasal vestibule
2) Right nasal cavity with conchae or turbinates
3) Vestibule of oral cavity
4) Oral cavity proper
5) Palatoglossal arch
6) Nasopharynx
7) Oropharynx
8) Laryngopharynx
9) Larynx

OTHER STRUCTURES

10) Frontal sinus
11) Sphenoid sinus
12) Pituitary fossa
13) Hard palate
14) Soft palate
15) Ostium of auditory tube
16) Adenoid
17) Tonsil
18) Epiglottis
19) Hyoid bone
20) Cricoid cartilage
21) Thyroid gland

II. NOSE AND NASAL CAVITY

The nose can be divided into the external nose and nasal cavity. The external nose has a bony and cartilaginous skeleton; it is INNERVATED by V1 and V2--sensory--and VII--muscles. It is VASCULARIZED by branches of the maxillary, ophthalmic and facial arteries; veins drain to the facial vein or ophthalmic.

The nasal cavity is basically triangular in frontal section. VASCULARIZATION of the cavity is from the anterior and posterior ethmoidal arteries--ophthalmic--the sphenopalatine --from maxillary--the greater palatine, the facial, and from arteries supplying the sinuses. INNERVATION sensory mainly V2, some V1, olfaction from I, along with postganglionic parasympathetic fibers from the pterygopalatine ganglion.

<u>STRUCTURES</u>:

SEPTAL OR MEDIAL WALL-

1) Septal cartilage
2) Vomer
3) Perpendicular plate of ethmoid
4) Maxilla
5) Palatine
6) Olfactory filia through cribiform plate
7) Sphenoid
8) Nasal
9) Frontal

LATERAL WALL-

Made of maxillary, palatine, lacrimal, ethmoid and nasal bones

10) Superior concha
11) Middle concha
12) Inferior concha--a separate bone
13) Vestibule
14) Sphenoethmoidal recess

MEATAL STRUCTURES-

15) Superior meatus--ostium for posterior ethmoid cells
16) Middle meatus--the bulge or ethmoid bulla has ostium for middle ethmoid cells; the opening or hiatus semilunaris leads to a recess--ethmoidal infundibulum--containing the ostia of the frontal, anterior ethmoidal and maxillary sinuses from superior to inferior.
17) Inferior meatus--opening of nasolacrimal duct
18) Sphenoid sinus opens into sphenoethmoidal recess

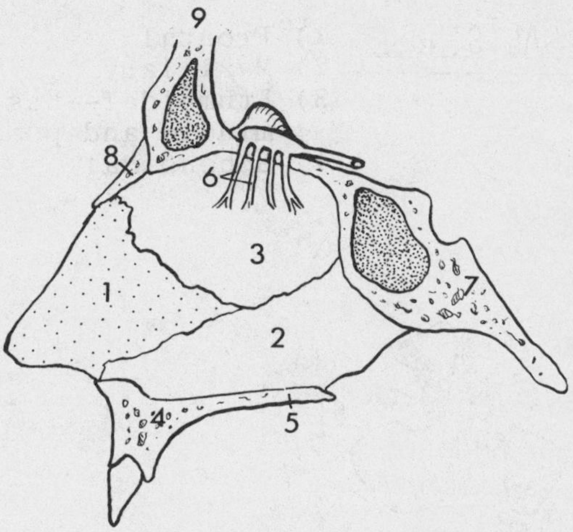

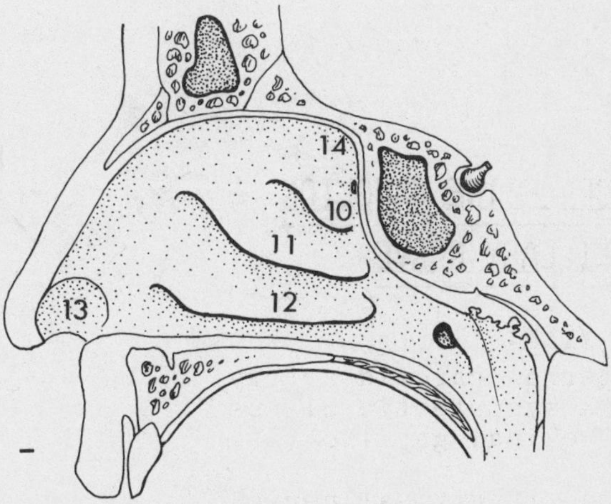

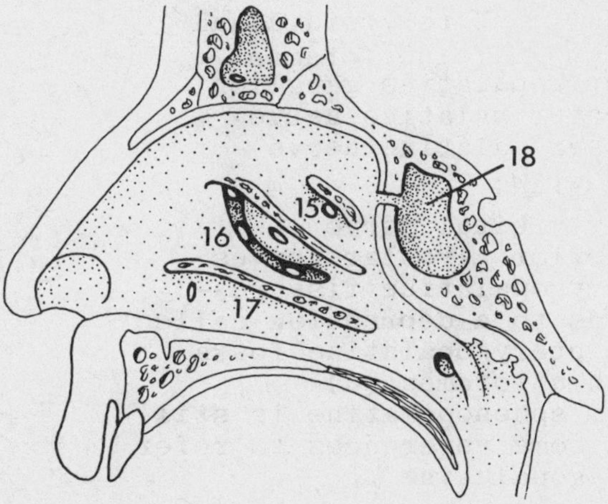

PARANASAL SINUSES

1) Frontal
2) Maxillary
3) Ethmoidal--has anterior middle and posterior cells
4) Sphenoidal

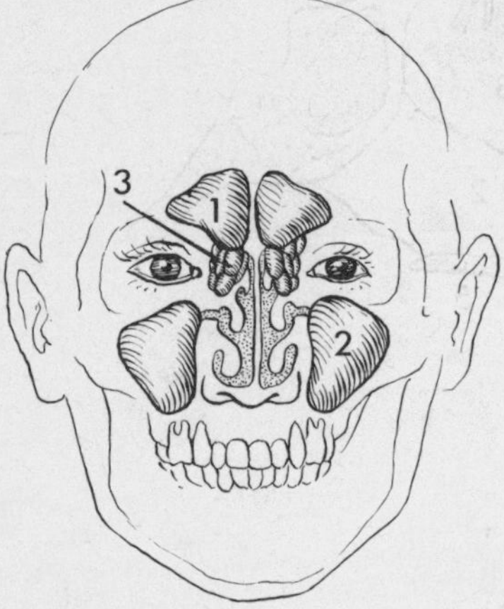

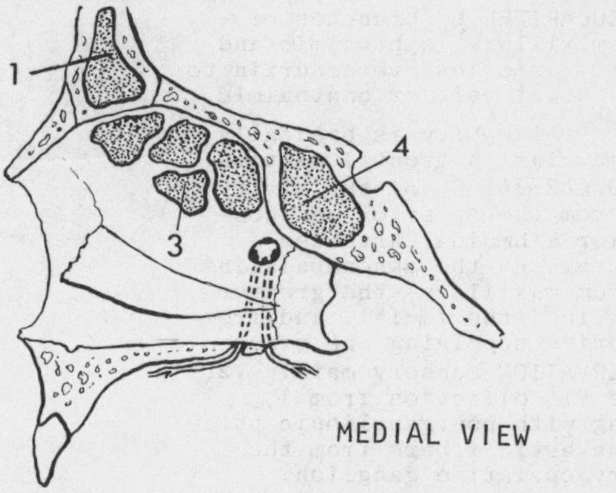

MEDIAL VIEW

PTERYGOPALATINE GANGLION AND MAXILLARY NERVE

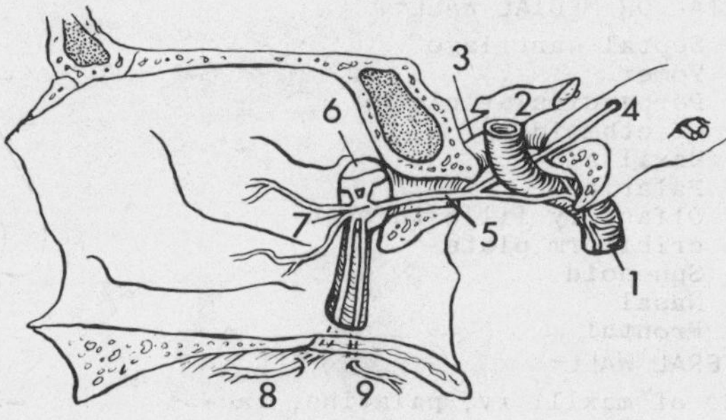

MEDIAL VIEW

MEDIAL VIEW:

1) Internal carotid artery with sympathetic plexus
2) Trigeminal ganglion
3) V2
4) Greater petrosal nerve
5) Nerve of the pterygoid canal--in pterygoid canal
6) Sphenopalatine foramen
7) Pterygopalatine ganglion
8) Greater palatine nerve
9) Lesser palatine nerve

LATERAL VIEW:

10) Infraorbital nerve
11) Superior alveolar nerves
12) Pterygopalatine fissure--leads to a depression called the pterygopalatine fossa with a foramen (6)

The term sphenopalatine is still used in some references to refer to pterygopalatine.

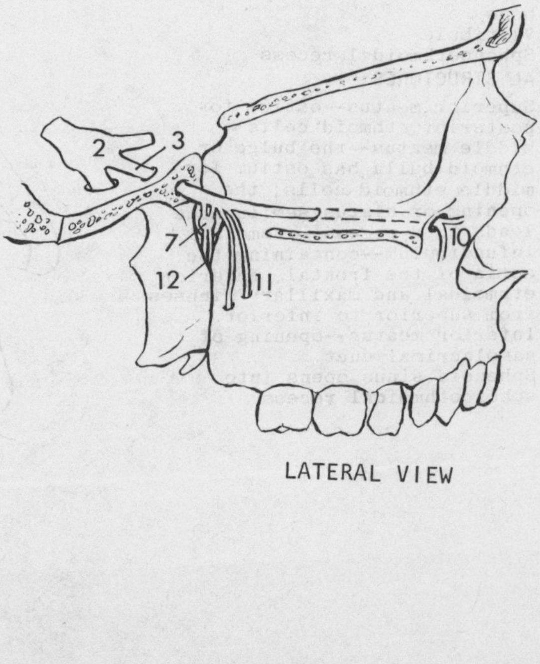

LATERAL VIEW

III. ORAL CAVITY

This cavity is divided into a vestibule and oral cavity proper. It is bounded by the palate, oropharynx, cheeks, lips and the floor of the mouth. The genio- and mylohyoid muscles form a deep muscular support or subfloor for the cavity. It contains the teeth, tongue and openings of salivary glands.

<u>STRUCTURES</u>:

1) Vestibule
2) Frenulum
3) Sublingual fold--with openings of sublingual gland
4) Opening of submandibular duct
5) Lingual nerve
6) Submandibular duct
7) Sublingual gland
8) Submandibular gland
9) Hyoid bone
10) Mylohyoid muscle
11) Geniohyoid muscle
12) Palatoglossal arch--posterior limit of oral cavity
13) Uvula
14) Palatine tonsil

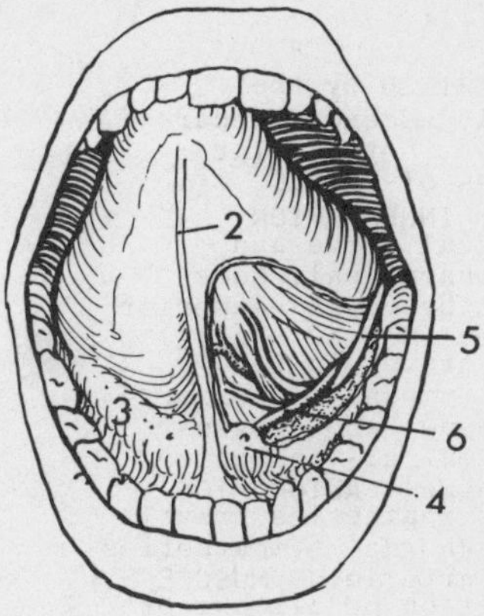

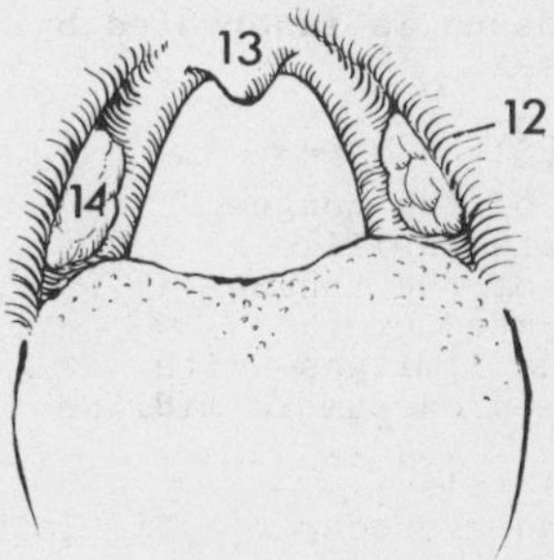

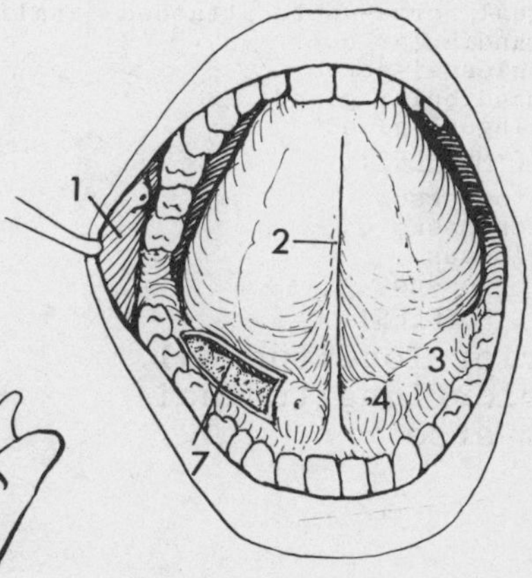

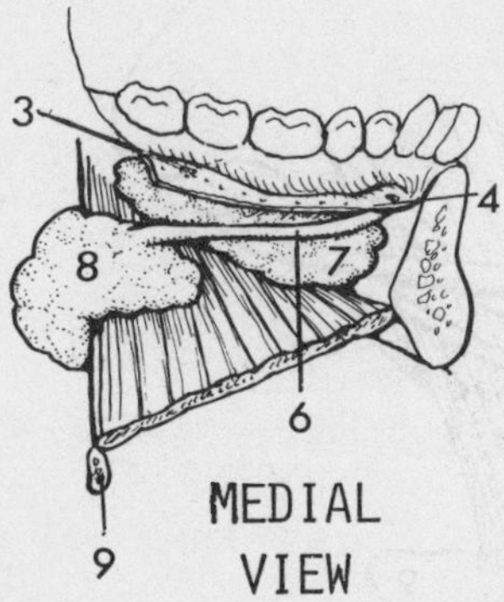

MEDIAL VIEW

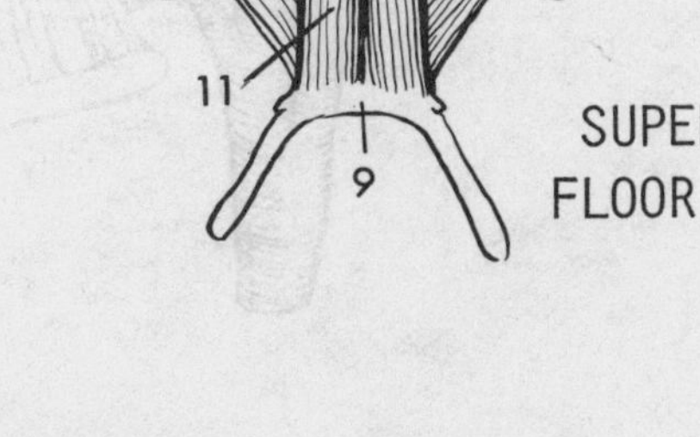

SUPERIOR VIEW FLOOR OF MOUTH

TONGUE

VASCULARIZED by the lingual, ascending pharyngeal, and tonsillar branches of facial. SENSORY INNERVATION by lingual nerve and glossopharyngeal, some at root by vagus--superior laryngeal nerve. TASTE fibers travel in the chorda tympani--anterior 2/3 tongue--and glossopharyngeal n.--posterior 1/3 tongue. AUTONOMICS --parasympathetics travel in the chorda, sympathetics follow arteries. MUSCLE innervation of <u>all</u> <u>except</u> the palatoglossus is by the Hypoglossal Nerve--XII; the palatoglossus is innervated by the Vagus-X.

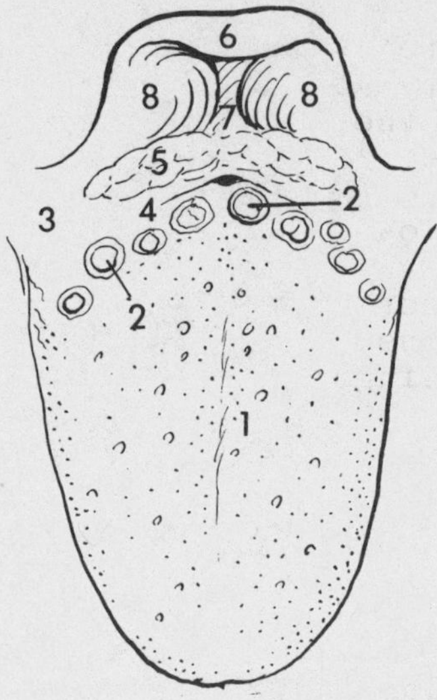

SUPERIOR VIEW

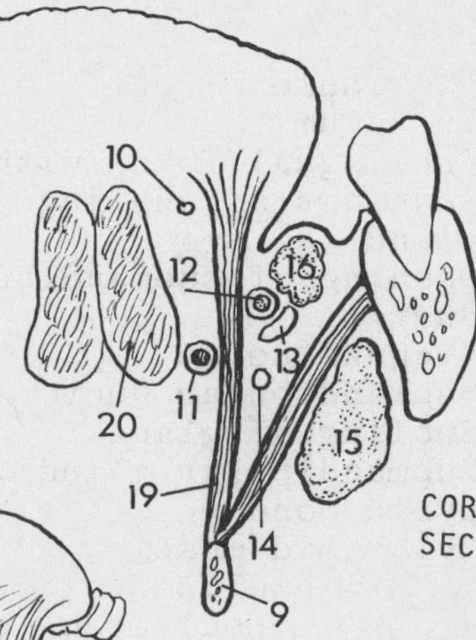

CORONAL SECTION

<u>STRUCTURES</u>:

1) Body of the tongue
2) Vallate papillae
3) Root of the tongue--separated by (2)
4) Sulcus limitans--with foramen caecum in midline
5) Lingual tonsil
6) Epiglottis
7) Median glossoepiglottic fold
8) Vallecula--a recess
9) Hyoid bone
10) Glossopharyngeal n.
11) Lingual artery
12) Lingual nerve--note attached ganglion
13) Submandibular duct
14) Hypoglossal nerve
15) Submandibular gland
16) Sublingual gland

<u>EXTRINSIC MUSCLES</u>:

17) Palatoglossus
18) Styloglossus
19) Hyoglossus
20) Genioglossus

<u>INTRINSIC MUSCLES</u>:

Superior longitudinal
Inferior longitudinal
Transverse

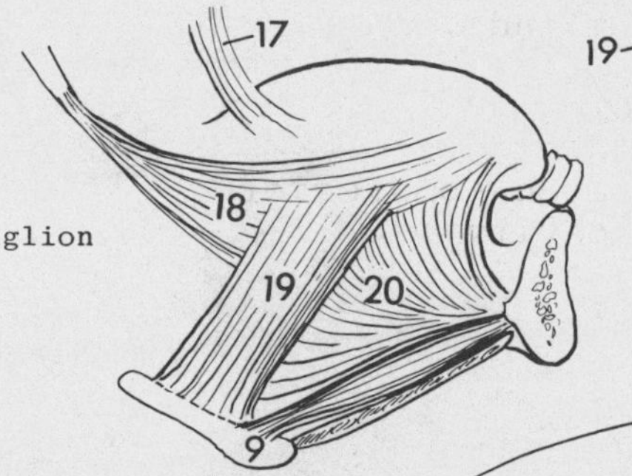

LATERAL VIEWS

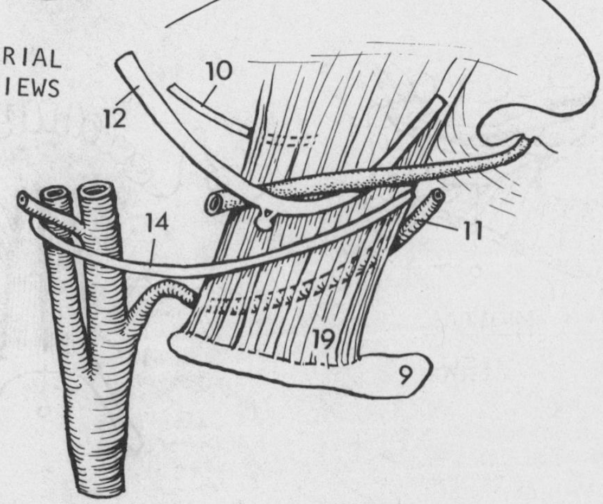

IV. SOFT PALATE

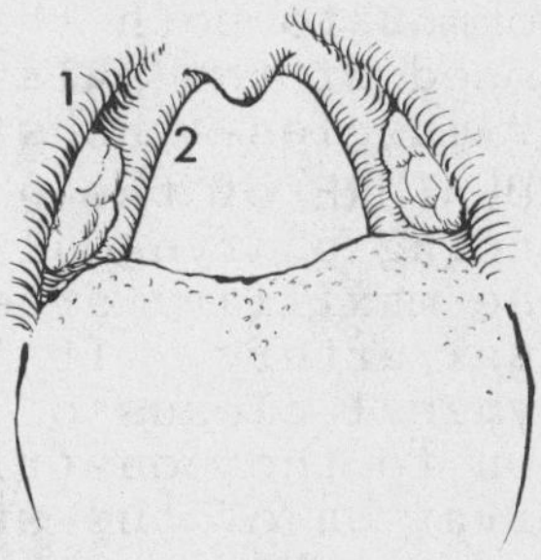

The soft palate, although discussed separately here, is actually a portion of the pharynx. It is VASCULARIZED by the palatine branch of the ascending pharyngeal, ascending palatine branch of facial and greater palatine branch of the maxillary artery. It is INNERVATED by sensory branches of V2 and glossopharyngeal nerves. ALL MUSCLES EXCEPT the tensor are innervated by the vagus nerve through a pharyngeal plexus. The tensor is innervated by V3.

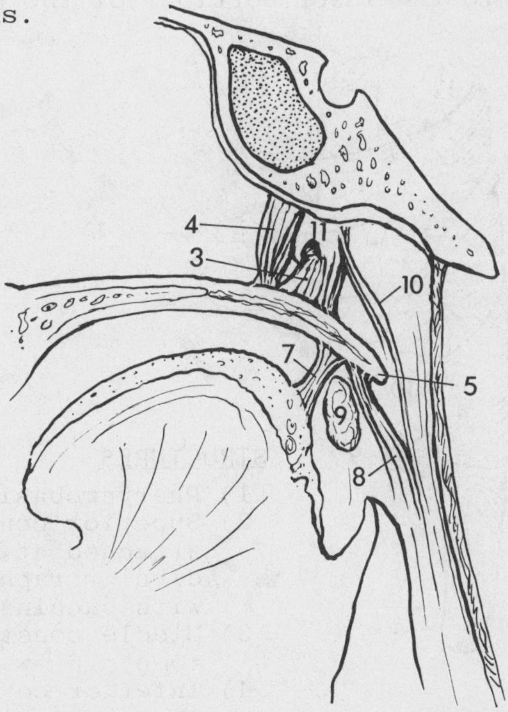

STRUCTURES:

1) Palatoglossal arch
2) Palatopharyngeal arch --both have underlying muscles of the same name
3) Levator veli palatini
4) Tensor veli palatini
5) Uvular muscle
6) Hook of hamulus
7) Palatoglossal muscle
8) Palatopharyngeal muscle
9) Palatine tonsil
10) Salpingopharyngeus--discussed with the pharynx
11) Cartilage of auditory tube, produces an elevation called the torus tubarius
12) Choana

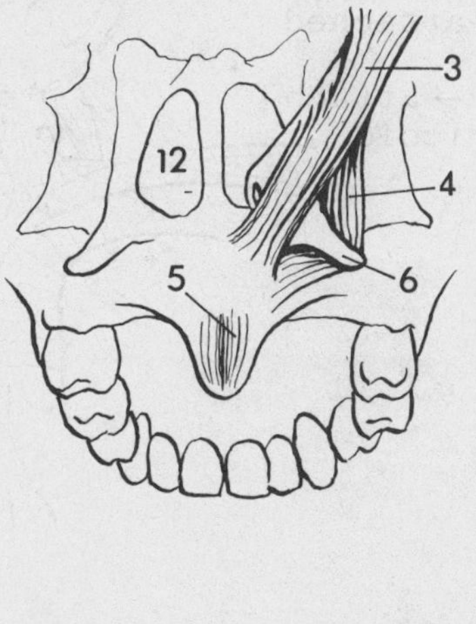

V. PHARYNX

The pharynx is basically a muscular tube, open anteriorly, that is common to both the respiratory and digestive systems. It is composed of the palatal muscles, three constrictors and three other muscles--the stylo, salpingo and palatopharyngeus muscles. VASCULATURE of the pharynx arises from the ascending palatine, ascending pharyngeal and tonsillar branch of facial, branches of the maxillary artery and dorsal lingual branches of the lingual artery. It is INNERVATED through a loosely knit pharyngeal plexus of the IX and X cranial nerves. Motor innervation to the constrictors can be traced to the Vagus, while innervation of the stylopharyngeus is from IX. Sensory innervation of the nasopharynx is from V2, IX is sensory to the upper portions of the rest of the pharynx, while X is sensory to the lower portions of the pharynx.

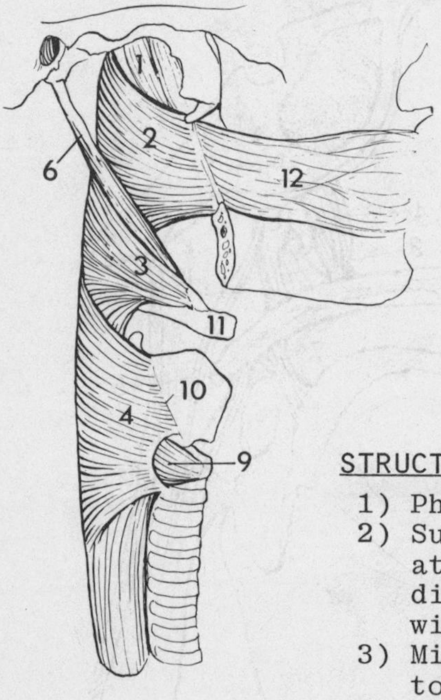

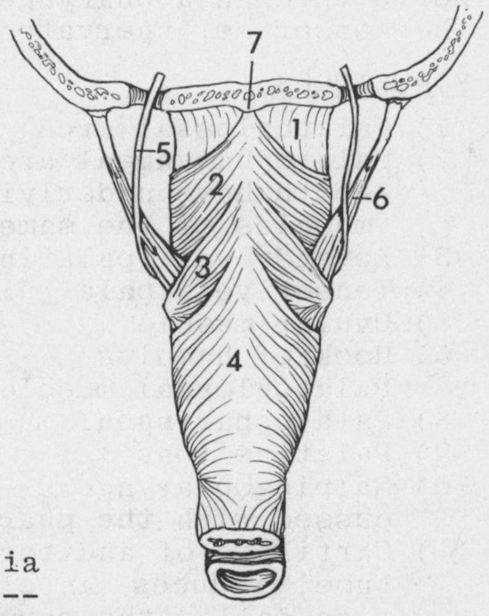

STRUCTURES:

1) Pharyngobasilar fascia
2) Superior constrictor--attached at pterygomandibular raphe, in common with buccinator
3) Middle constrictor--attached to hyoid
4) Inferior constrictor--attached to thyroid and cricoid cartilages
5) Nerve IX
6) Stylopharyngeus
7) Pharyngeal tubercle
8) Palatopharyngeus
9) Cricothyroid
10) Thyroid cartilage
11) Hyoid bone
12) Buccinator

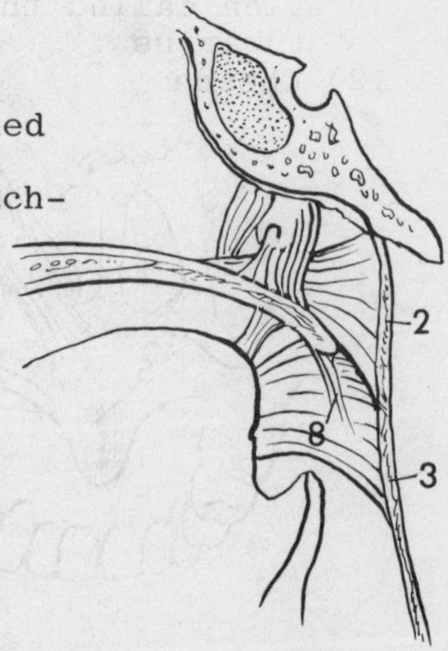

INTERIOR OF THE PHARYNX

STRUCTURES:

1) Nasopharynx--with choanae, extends to palatopharyngeal arch
2) Oropharynx--contains root of tongue, extends between the arches of the tonsil, some of posterior wall of pharynx
3) Laryngopharynx--posterior to the larynx, along upper border of the epiglottis, also called hypopharynx
4) Piriform recess
5) Aryepiglottic fold--note bulges of cuneiform and corniculate cartilages of larynx
6) Internal laryngeal nerve
7) Cornu of hyoid bone
8) Thyroid cartilage
9) Uvula

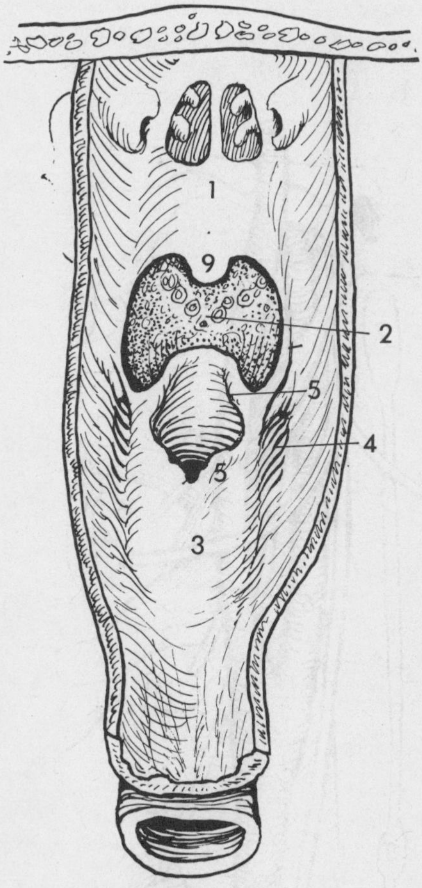

SUPERFICIAL

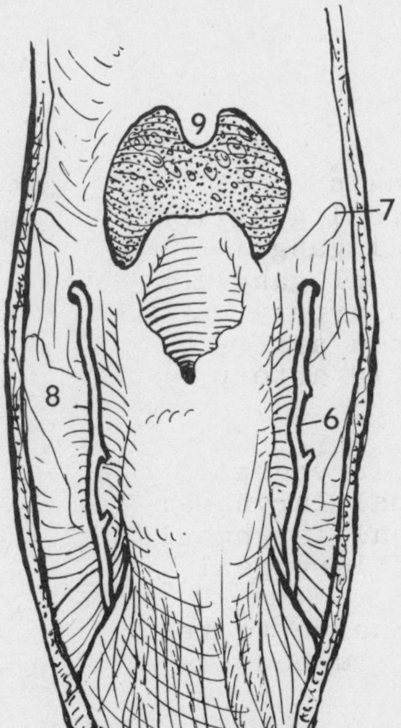

MUCOSA REMOVED

RELATIONS OF THE PHARYNX

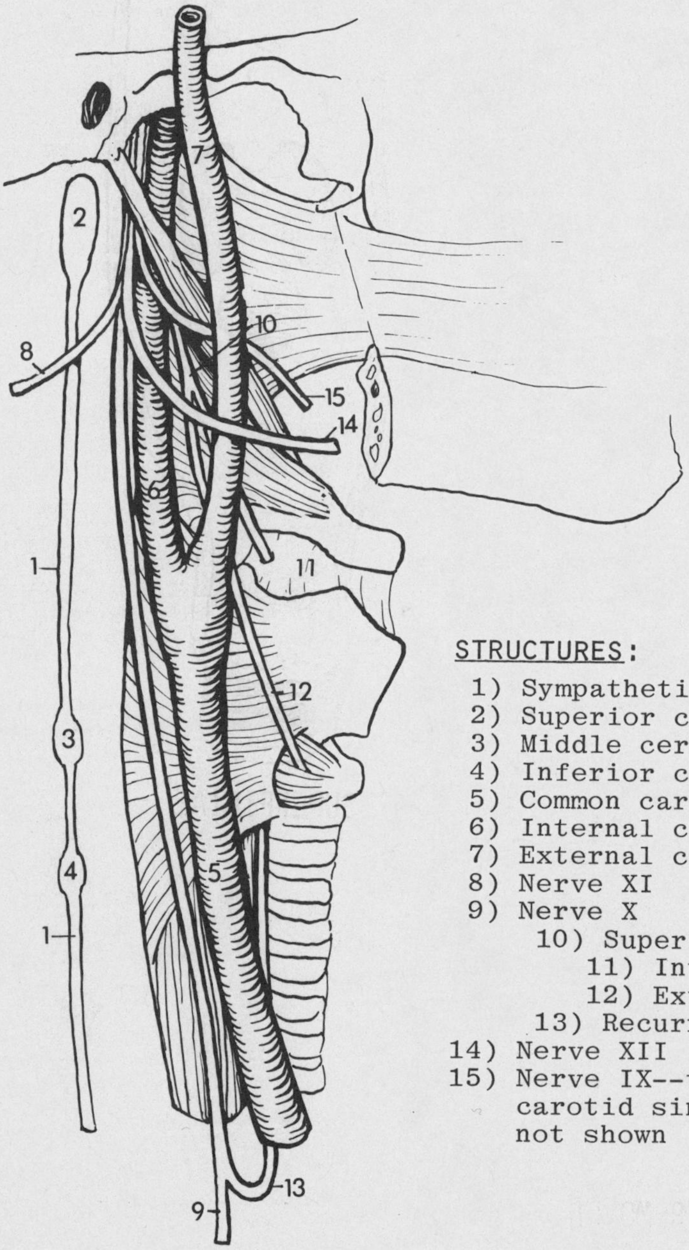

STRUCTURES:

1) Sympathetic chain
2) Superior cervical ganglion
3) Middle cervical ganglion
4) Inferior cervical ganglion
5) Common carotid artery
6) Internal carotid artery
7) External carotid artery
8) Nerve XI
9) Nerve X
    10) Superior laryngeal
        11) Internal laryngeal
        12) External laryngeal
    13) Recurrent laryngeal
14) Nerve XII
15) Nerve IX--the nerve to the carotid sinus and body is not shown

VI. LARYNX

The larynx is commonly called the voice box.

STRUCTURES:

1) Epiglottis
2) Medial glosso-epiglottic fold
3) Lateral glosso-epiglottic fold
4) Vallecula
5) Root of tongue
6) Vestibular fold--false cords
7) Vocal folds--true cords
8) Aryepiglottic fold
9) Cuneiform tubercle
10) Corniculate tubercle
11) Rima glottidis
12) Vestibule
13) Ventricle or laryngeal sinus
14) Subglottic space
15) Vocal ligament and vocalis muscle in vocal fold

The glottis refers to the combination of (7 & 11)

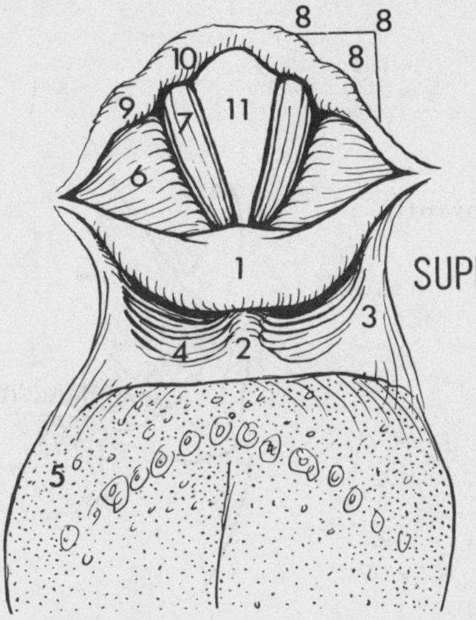

SUPERIOR VIEW

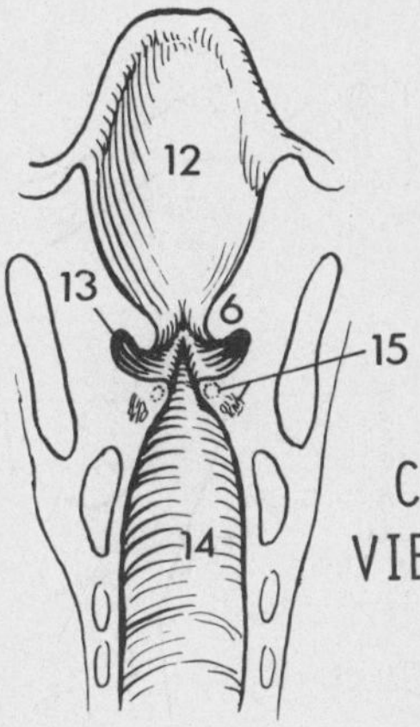

CORONAL SECTION VIEWED POSTERIORLY

SKELETON OF THE LARYNX

1) Hyoid bone
2) Epiglottal cartilage
3) Thyroid cartilage
4) Cricoid cartilage
5) Arytenoid cartilage
6) Corniculate cartilage
7) Tracheal cartilages

The cuneiform cartilages, not shown, are within the aryepiglottic fold.

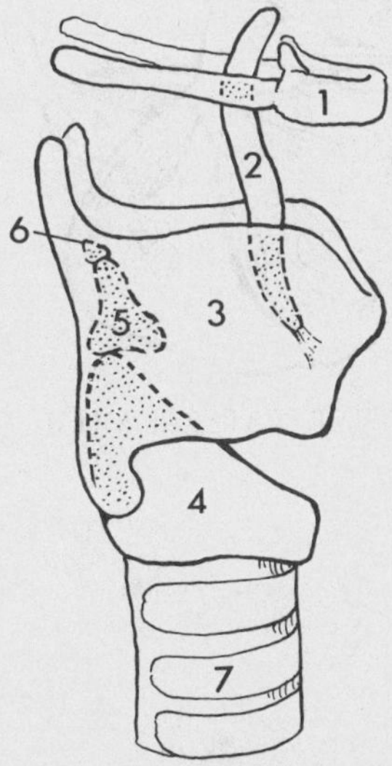

LATERAL VIEW

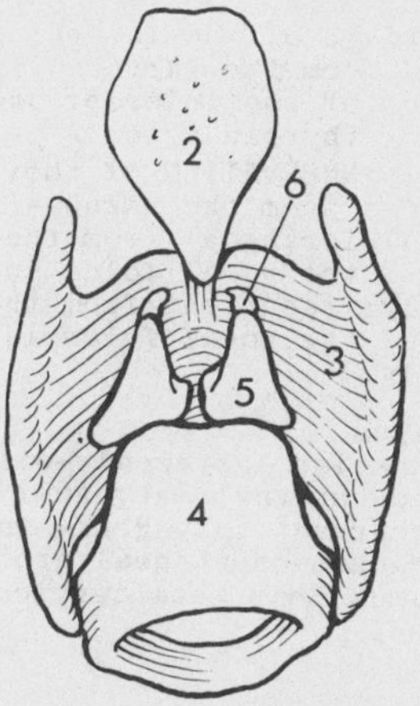

POSTERIOR VIEW

MEMBRANES

1) Thyrohyoid
2) Quadrangular
3) Vestibular fold
4) Median cricothyroid ligament
5) Vocal ligament
6) Cricothyroid ligament

(5 & 6) Constitute the Conus elasticus

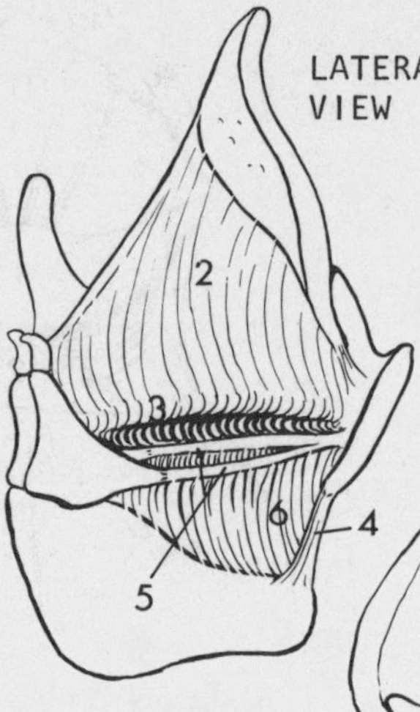

LATERAL VIEW

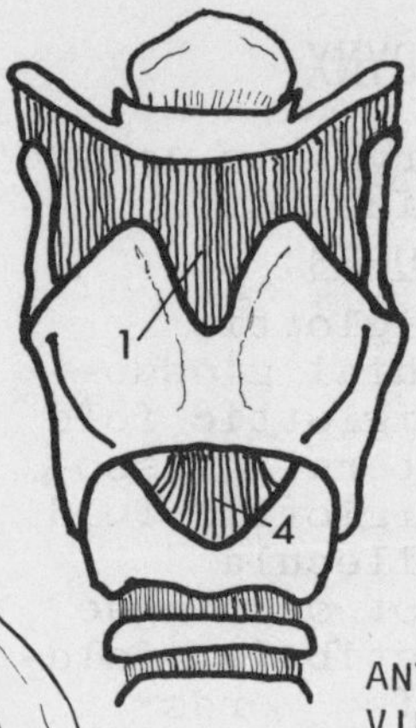

ANTERIOR VIEW

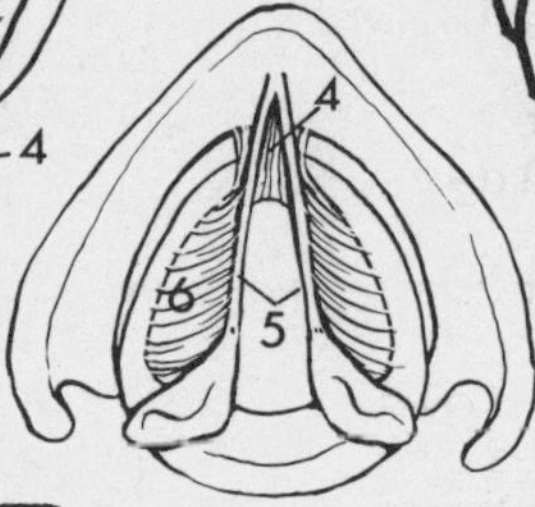

SUPERIOR VIEW

MUSCULATURE AND INNERVATION

1) Thyrohyoid
2) Cricothyroid
3) Aryepiglotticus
4) Thyroarytenoid
5) Arytenoideus, oblique and transverse
6) Posterior cricoarytenoid
7) Lateral cricoarytenoid
8) Inferior constrictor

The thyroepiglotticus is not shown.
The Thyrohyoid muscle is innervated by the ansa cervicalis. All other muscles are innervated by the Vagus. However, The cricothyroid muscle is innervated by the external laryngeal while the rest are innervated by the recurrent laryngeal. This is important.

VASCULATURE of the larynx is derived from the laryngeal branches of the inferior and superior thyroid arteries.
SENSORY INNERVATION of the larynx is from the vagus--internal laryngeal from the level of the vocal folds superiorly; recurrent from the level of the vocal folds inferiorly.

9) Vagus
10) Superior laryngeal--with internal and external laryngeal nerves
11) Recurrent laryngeal--found in tracheo-esophageal groove
12) Ramus from ansa cervicalis

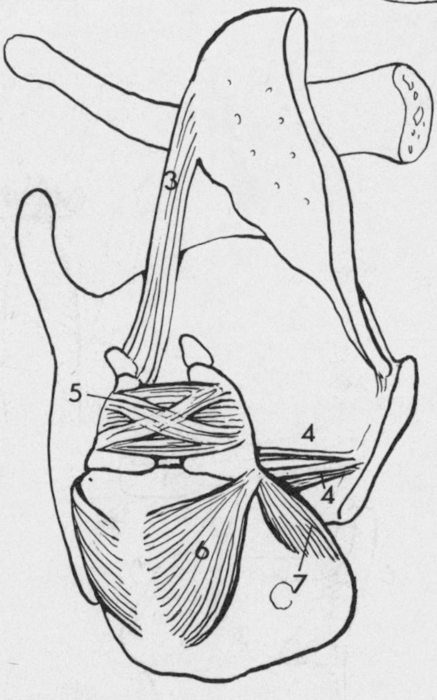

POSTERIOR-LATERAL VIEW

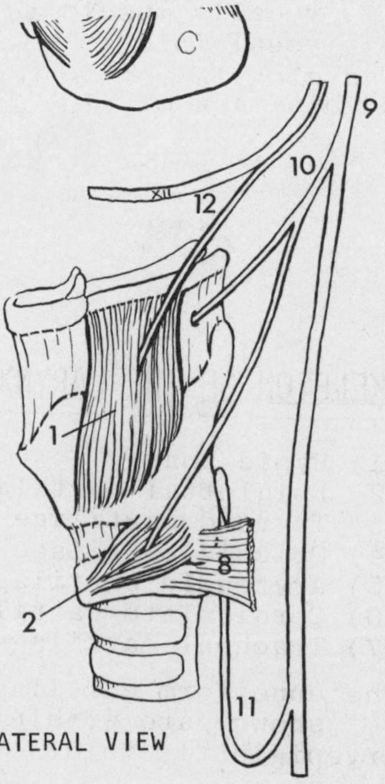

LATERAL VIEW

ACTIONS OF LARYNGEAL MUSCLES

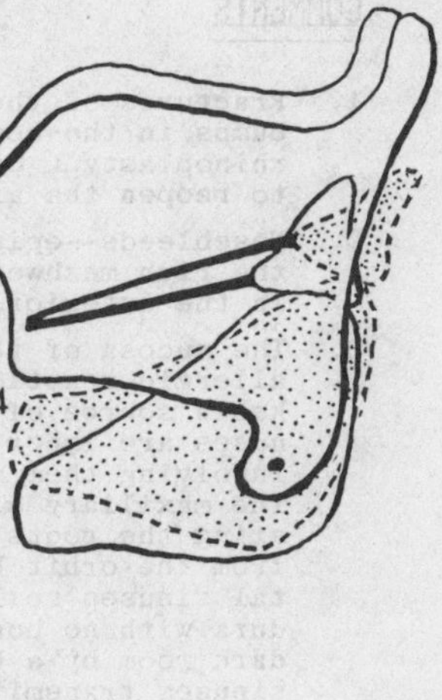

1) Lateral cricoarytenoids adduct or approximate the folds
2) Posterior cricoarytenoids abduct or open the folds
3) Vocalis muscle shortens or varies tension of folds
4) Cricothyroid tenses the vocal ligaments and folds, adducts also, by rocking the thyroid cartilage forward and the cricoid superiorly.

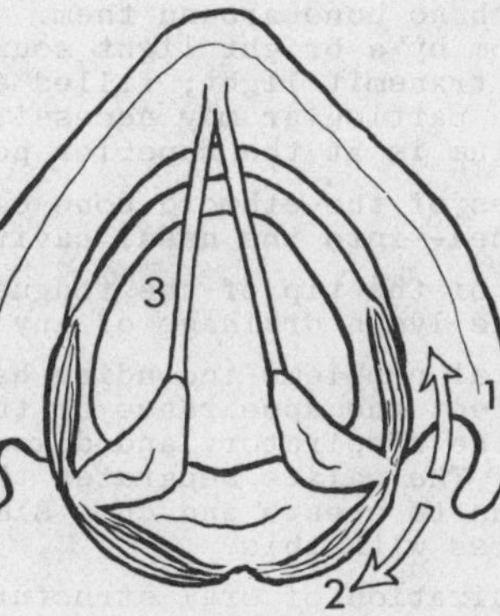

The thyroarytenoid, not shown, shortens and relieves tension on the vocal folds. The vocalis muscle is considered to be a deep derivative of the thyroarytenoid.

The recurrent laryngeal nerve innervates all intrinsic laryngeal musculature except the cricothyroid. If bilaterally injured, abduction of the folds is lost. The cricothyroids then tense and may slightly adduct the folds; however the folds may become rather flaccid due to their paralysis; on inspiration, the rushing of air past the folds may approximate them and close the glottis. Suffocation may result. In this case the vocal folds are acting as flap valves--positive pressure beneath will open them, but pressure from above will close them. Unilateral injury may result in hoarseness and change in quality of voice, sometimes permanent.

COMMENTS

1. Fractures of the nose, deviated septum and unsightly humps and bumps in the nose make it a popular site of plastic surgery--rhinoplasty. Often the rhinoplasty is for practical purposes, to reopen the airway, as well as cosmetic reasons.

2. Nosebleeds--epistaxis--commonly result from traumatization of the rich meshwork of arteries known as Keisselbach's Plexus in the anterior portions of the cavity.

3. The mucosa of the nasal cavity may engorge with blood during allergic reactions, colds or sinus infections and is a well known source of trouble to virtually everyone. Paranasal sinuses are important because infections, abscesses and fractures involving them may affect the maxillary teeth, orbit and brain. The maxillary sinuses may enlarge to the point of incorporating the roots of teeth; the ethmoidal sinus may be separated from the orbit by paper thin bone; the authors have seen frontal sinuses so large they interposed between the calvaria and dura with no bone around them. Sinuses may be visualized in a dark room by a bright light source in the closed mouth. Open sinuses transmit light, filled ones do not. The maxillary sinus in particular may necessitate artificial drainage since its ostium is at the superior portion of the sinus.

4. Fractures of the ethmoid bone can result in leakage of CSF --otorrhea--into the nasal cavity.

5. Cancers of the tip of the tongue can metastasize bilaterally; as a rule lymph drainage of any midline structures is bilateral.

6. Congenital problems including hare lip and cleft palate not only affect the appearance of the face but impair the functioning of the respiratory and digestive systems, as well as phonation. The palate separates the nose and oral cavity and allows us to breath and chew simultaneously--a cleft palate interferes with this.

7. Anesthetization of oral structures by blockage of branches of the mandibular and maxillary divisions of the trigeminal nerve is a common procedure. Anesthesia that affects the genioglossus muscles must be counteracted by intubation; otherwise the tongue may recede and compromise a clear airway.

8. The function of the hypoglossal nerve may be examined by asking the patient to protrude the tongue--impairment of the hypoglossal nerve will be detected by deviation of the tongue to the ipsilateral--same--side as the lesion.

9. The number of important structures related to the pharynx should serve to emphasize the importance of injuries and the delicacies of surgeries on this structure. Infections and abscesses may spread in the retropharyngeal space from the base of the skull to the diaphragm.

10. Removal of pharyngeal tonsils--adenoids--and palatine tonsils, commonly called 'the tonsils,' is not as carefree as is thought due to the possibility of injury to nearby structures. Occasionaly lingual tonsils must also be scraped.

11. Much has been made in the past of the pharyngeal lymphatic 'Ring of Waldemeyer' consisting of the adenoids, tonsils, lingual tonsils and tubal tonsils, as sentinels in the interception of foreign bodies and production of antibodies. The exact importance of this 'ring' is in question.

12. The auditory tube offers an 'expressway' to the middle ear for the spread of infections. It is also important in the equalization of pressure between the middle ear and outside. With each swallow, the levator and tensor veli palatini, possibly assisted by a dilator tubae muscle, rocks the tube open. Yawning in planes and elevators and increasing intra-pharyngeal pressure by 'blowing' underwater also accomplish the same purpose.

13. Foreign objects may lodge in the piriform recesses and sometimes in the valleculae. The internal laryngeal nerve--sensory to the larynx--may be injured by the objects or in retrieval of the objects from the piriform recess.

14. If food becomes lodged in the larynx, trachea or laryngopharynx--cafe coronary--the Heimlich Maneuver may clear the passageway. The fists are clenched one around the other from behind the choking individual. A rapid thrust posteriorly and superiorly directed between the umbilicus and xiphoid can build up intra-thoracic pressure to expel the object. It may be self induced by leaning on a chair, bannister or other nearby structure.

15. In the complex development of the pharynx, the branchial gill clefts may persist in the adult as patent tubes--branchial fistulas--or form cysts along the partially occluded course of the branchial clefts.

STUDY QUESTIONS: HEAD AND NECK

1. The calvaria is made up of the _____, _____, and _____ bones.	1. frontal, occipital, parietal
2. The emissary veins connect the veins of the scalp and the _____.	2. dural sinuses
3. Branches of the ophthalmic artery supply parts of the scalp. T/F	3. True
4. The lesser occipital nerve is made up of nerve fibers from the posterior primary ramus of C_2. T/F	4. False
5. The dura in the cranium is composed of two components. What are they?	5. endosteal; meningeal
6. The epidural space is a (true/potential) space in the cranium.	6. potential - since it only becomes apparent when it fills with blood
7. Bleeding can occur into the _____ space when the veins entering the sinuses are torn.	7. subdural
8. CSF fills the _____ space.	8. subarachnoid
9. Falx cerebri separates right and left cerebral hemispheres. T/F	9. True
10. The tentorium intervenes between the parietal lobes and occipital lobes. T/F	10. False - it separates cerebrum and cerebellum.
11. The diaphragm sella roofs the area of the pituitary gland. T/F	11. True
12. The main sensory innervation to the meninges is from the _____ cranial nerve.	12. trigeminal - 5th
13. Which of the following does NOT pass through the cavernous sinus: A. internal carotid artery B. VII C. VI D. V_1 E. V_2	13. B
14. The internal carotid and vertebral arteries anastomose at the ________.	14. Circle of Willis
15. The basilar artery lies on the anterior aspect of the brainstem. T/F	15. True
16. The middle and posterior fossae are separated by the petrous ridge. T/F	16. True
17. The temporal poles of the cerebrum fill the posterior fossa. T/F	17. False - The cerebellum fills the posterior fossa.

18-25 Matching

Fossa:

A. Anterior
B. Middle
C. Posterior

18. sella turcica ____ — 18. B
19. olfactory bulbs ____ — 19. A
20. internal acoustic meatus ____ — 20. B
21. foramen spinosum ____ — 21. B
22. foramen lacerum ____ — 22. B
23. VII ____ — 23. C
24. crista galli ____ — 24. A
25. clivus ____ — 25. C

26. The _____ cranial nerve is the only one which leaves the brain-stem on its posterior aspect. — 26. IV

27. Nerve I rests on the _____ on the floor of the anterior fossa. — 27. cribiform plates

28. The _____, _____, & _____ cranial nerves exit via the jugular foramen. — 28. IX, X, and XI

29. The _____, _____, and _____ nerves exit the cranium via the superior orbital fissure. — 29. III, IV, VI

30. The _____ and _____ nerves exit the cranium via the internal auditory meatus. — 30. VII, VIII

31. V_2 exits via _____. — 31. foramen rotundum

32. V_3 exits via _____. — 32. foramen ovale

33. The purely motor cranial nerves are: — 33. III, IV, VI, XI, XII

34. The mixed cranial nerves are: — 34. V, VII, IX, X

35-39 Matching

A. II
B. III
C. IV
D. V_1
E. V_2
F. V_3
G. VI

35. Muscles of mastication ____ — 35. F
36. Sensory to upper lip ____ — 36. E
37. Exits cranium via optic canal ____ — 37. A
38. Motor and autonomic function ____ — 38. B
39. Sensory around chin ____ — 39. F

40-49 Matching

A. VII
B. VIII
C. IX
D. X
E. XI
F. XII

40. Geniculate ganglion ____ — 40. A
41. Balance ____ — 41. B
42. Motor posterior digastric ____ — 42. A
43. Taste anterior tongue ____ — 43. A
44. Motor to pharyngeal constrictors ____ — 44. D
45. Motor sternocleidomastoid muscle ____ — 45. E
46. Motor tongue ____ — 46. F
47. Sensory to pharynx ____ — 47. C
48. "Wanderer" ____ — 48. D
49. Divides within parotid gland ____ — 49. A

50. Which scalp layers form the "movable" unit?
50. Skin, connective tissue, and galea aponeurosis

51. Why are superficial scalp lacerations very bloody?
51. Because vessels are embedded in dense connective tissue and don't readily collapse.

52. What is the "danger area" of the scalp?
52. loose connective tissue layer

53. What is a fontanelle?
53. membranous spaces between incompletely formed skull bones in newborns

54. An epidural bleed may arise from hemorrhage of the _______ artery.
54. middle meningeal

55. A subdural bleed is usually from _______.
55. ruptured cerebral veins as they enter the dural sinuses

56. What is a CSF cistern?
56. enlarged subarachnoid space

57. What is the trigeminal cave (Meckel's cave)?
57. depression in the sphenoid bone in which the trigeminal ganglion (Gasserian) lies

58. The angle between the lateral walls of the orbit is ___ degrees while the medial walls are ______.
58. 90°, parallel

59. Arterial supply of the orbit is mostly via the _____ artery, while the venous drainage is into the _____ or _____.
59. ophthalmic, cavernous sinus, facial veins

60. The muscular annulus is the origin for the _____ muscles.
60. rectus

61. What is the sympathetic innervation of the pupil?
61. from superior cervical ganglion (travels along internal carotid)

62. What is Horner's syndrome?
62. lesion of the cervical sympathetic nerves resulting in ptosis, miosis, anhydrosis.

63. What is the ora serrata?
63. anterior border of the visual(light sensitive) retinal epithelium

64. What is the collateral arterial circulation to the retina?
64. NONE

65. What is the area "raised" during papilledema -- choked disc?
65. optic disc

66. What is the fluid filling the anterior chamber?

67. What nerve exits via the infraorbital foramen?

68. What is the muscle (and innervation) involved in opening the eyelid?

69. Trace the route of aqueous humor circulation.

70. An increase in thickness and curvature of the lens is involved in _____.

71. Name the region of great visual acuity.

72. Name the two locations of the cell bodies for the sympathetic nerves involved in tearing.

73. What is the sensory innervation to the external auditory canal?

74. The pars flaccida of the eardrum is located on the _____ surface.

75. The cone of light is located in the _____ quadrant.

76. The _____ bone attaches to the eardrum.

77. The chorda tympani passes between the _____ and _____.

78. The tensor tympani inserts on the _____, is supplied by a branch of the _____ nerve, and tenses during _____ noises.

79. The _____ inserts on the stapes, is innervated by a branch of _____, and protects the ear from _____.

80. Stapes attaches to the _____ window.

81. _____ is the fluid of the bony labyrinth, while _____ is the fluid in the membranous labyrinth.

82. The basilar turn of the cochlea results in the _____ in the _____ wall of the tympanic cavity.

83. The walls of the triangular cochlear ducts are the _____, _____, and _____.

84. The specialized structure in the cochlear duct involved with hearing is the _____.

85. What are the nerve fibers which comprise the nerve of the pterygoid canal? Which nerves converge to form the nerve of the pterygoid canal?

86. The chorda tympani contains two types of fibers which are _____ and _____.

66. aqueous humor

67. infraorbital nerve (V_2)

68. levator palpebrae superioris muscle, III nerve

69. epithelium of ciliary processes→posterior chamber→pupil→anterior chamber→trabecular meshwork→canal of Schlemm

70. accommodation

71. macula (esp. the fovea centralis)

72. superior salivatory nucleus and pterygopalatine ganglion

73. primarily, the auriculotemporal nerve (V) and secondarily, a branch of X. Pressure on the external canal may lead to vomiting.

74. superior

75. anterior-inferior

76. malleus (causes a defect in the eardrum called the umbo)

77. malleus, eardrum

78. malleus, V_3, loud

79. stapedius, VII, loud noises

80. oval

81. perilymph, endolymph

82. promontory, medial

83. basilar membrance, spiral ligament, vestibular or Reissner's membrane

84. organ of Corti

85. preganglion parasympathetic and postganglion sympathetic; greater petrosal and deep petrosal

86. taste anterior 2/3 of tongue, preganglionic parasympathetic fibers for the submandibular and sublingual salivary gland

87. What is the epitympanic recess or attic?

87. the part of the tympanic cavity which extends superior to the tympanic membrane

88. What is the principal anterior relation of the tympanic cavity?

88. internal carotid artery

89. What is the sensory innervation to the occiput?

89. greater occipital nerve (cervical dorsal rami C_2&C_3)

90. Motor innervation to the muscles of facial expression is from the ____ nerve.

90. VII

91. The occipitalis m. and frontalis m. insert into the ____.

91. galea aponeurosis

92. The ____ nerve, which runs parallel to the parotid (Stenson's) duct is a motor nerve branch from ____ nerve.

92. buccal, VII

93. The muscle of mastication most involved with protraction at the mandible is the ____ muscle.

93. lateral pterygoid

94. The sphenomandibular ligament is attached to the ____ and ____.

94. spine of sphenoid, lingula of mandible

95. The ____ nerve passes between the sphenomandibular ligament and the mandible and is sensory for the ____.

95. inferior alveolar (V_3), lower jaw and teeth

96. The area of the upper lip, lower nose, lower eyelids and "lines" which connect them is called the ____ area of the face because this area may spread an infection into the ____ sinus.

96. danger, cavernous

97. The nerve to mylohyoid is a branch of ____.

97. V_3

98. Innervation to the anterior belly of the digastric is from ____.

98. V_3

99. General sensation of the anterior 2/3 of tongue is via the ____ nerve.

99. lingual (branch of V_3)

100. Most of the lymph from the scalp and face drains into what nodes?

100. deep cervical nodes along the internal jugular vein

101. A hard, enlarged submental node should make you suspect pathology in what area?

101. the lower lip or tip of tongue

102. What is the facial modiolus?

102. a point lateral to each labial commissure where several muscles attach - the location of "dimples"

103. Where is the philtrum?

103. the small depression above the upper lip, inferior to the septum of the nose

104. List the borders of the anterior triangle.	104. midline, mandible, sternocleidomastoid muscle
105. List the major subdivisions of the anterior triangle.	105. submental, submandibular, muscular, carotid
106. The investing fascia of the neck loses its organization as it passes into the face, except where it splits to invest the _____.	106. parotid gland
107. The pretracheal fascia enwraps the brachial plexus as axillary fascia (axillary sheath) T/F	107. False - The axillary sheath is considered part of the prevertebral fascia.
108. The retropharyngeal space is a potential space between the pretracheal and prevertebral fascia through which infection may spread from the pharynx into the mediastinum. T/F	108. True - very important concept
109. The carotid sheath contains the phrenic nerve. T/F	109. False
110. The common carotid artery is enclosed in the carotid sheath. T/F	110. True
111. The external and internal jugular veins are separated by what structure?	111. sternocleidomastoid muscle
112. Name the branches of the internal carotid artery in the neck.	112. NONE
113. Is the laryngeal arterial supply <u>directly</u> from the external carotid artery?	113. No - the laryngeal artery is a branch of the superior thyroid artery.
114. Where is the superior cervical ganglion?	114. behind the ramus of the mandible anterior to the transverse process of C_2
115. The posterior triangle is subdivided by which muscle?	115. omohyoid (inferior belly)
116. Before it inserts onto the hyoid bone, the stylohyoid muscle is "split" by what structure?	116. tendon of the digastrics
117. The right recurrent laryngeal nerve loops around the _____, while the left laryngeal nerve loops around the _____.	117. subclavian artery, aorta between the left common carotid and left subclavian arteries
118. Which artery goes deep to the neck of the mandible and enters the infratemporal fossa?	118. maxillary artery
119. The anterior scalene muscle separates the _____ anteriorly from the _____ posteriorly.	119. subclavian vein, brachial plexus
120. The _____ muscle divides the subclavian artery into three segments.	120. anterior scalene
121. Which artery supplies the first intercostal space?	121. a branch of the costocervical trunk, the "highest" intercostal artery
122. The stellate ganglion is commonly made up of the _____ and _____.	122. inferior cervical ganglion, first thoracic ganglion
123. Does the accessory nerve cross the posterior triangle?	123. yes
124. How is the platysma related to the investing fascia?	124. The platysma lies superficial to the investing fascia.

125. How is the phrenic nerve related to the prevertebral fascia in the neck?	125. It is deep to the prevertebral fascia in the neck
126. Which sinus has the most "inefficient drainage"?	126. Maxillary sinus: its canal opens at its superior-medial aspect which may lead to difficulty in drainage (drains into the semilunar hiatus).
127. Where does the frontal sinus drain into the nasal cavity?	127. middle meatus
128. What nerve supplies the upper molars and maxillary sinus?	128. alveolar branches of V_2
129. All extrinsic muscles of the tongue are innervated by the XII nerve. T/F	129. False - the palatoglossus m. is innervated by X.
130. The genioglossus muscle protrudes the tongue. T/F	130. True
131. The lingual artery lies deep to the hyoglossus muscle. T/F	131. True
132. The lingual nerve lies deep to the hyoglossus muscle. T/F	132. False, it lies superficial to the muscle and crosses the submandibular (Wharton's) duct twice!!
133. The XII nerve crosses the internal and external carotid arteries, passing below the occipital artery. T/F	133. True
134. Taste to the posterior 1/3 of tongue is via the _____ nerve.	134. glossopharyngeal
135. The palatine tonsil lies between the _____ anteriorly and _____ posteriorly.	135. palatoglossal arch, palatopharyngeal arch
136. Most of the palatine muscles are supplied by the _____ nerve except the tensor veli palatini which is supplied by _____.	136. vagus, V_3
137. Which muscles act on the soft palate and open the auditory tube?	137. tensor and levator palatini muscles
138. Which muscle passes between the superior and middle constrictor ?	138. stylopharyngeus muscle
139. Which nerve is motor to MOST of the laryngeal muscle?	139. recurrent laryngeal
140. The false vocal cord is the free edge of _____.	140. the vestibular fold
141. The posterior cricoarytenoid muscle (abducts/adducts) the vocal cords.	141. abducts
142. What is Kiesselbach's plexus?	142. extensive anastomosis of the blood vessels in the anterior inferior portion of the nasal septum; a common area for nosebleeds
143. What is the laryngeal vestibule?	143. the part of the laryngeal cavity between the superior aperture and false cords
144. What structure passes through the gap below the inferior constrictor?	144. recurrent laryngeal nerve

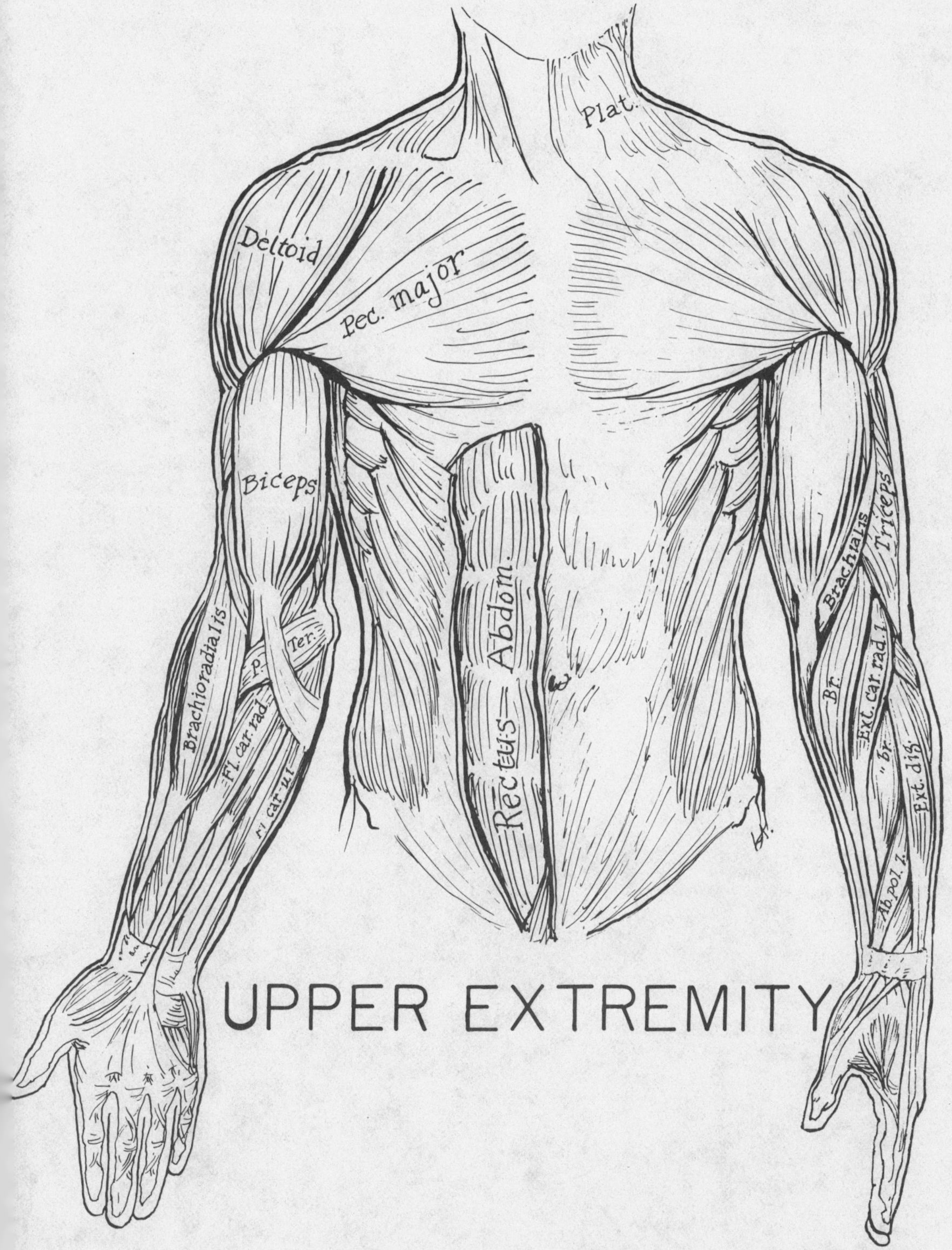

UPPER EXTREMITY

UPPER EXTREMITY

I. PECTORAL GIRDLE

OSTEOLOGY

SCAPULA

1) Lateral border
2) Medial border
3) Superior border
4) Inferior angle
5) Superior angle
6) Glenoid fossa
7) Spine
8) Acromion
9) Coracoid process
10) Supraspinous fossa
11) Infraspinous fossa
12) Scapular notch
13) Subscapular fossa

CLAVICLE

14) Medial end
15) Lateral end
16) Anterior border
17) Posterior border
18) Facet for sternum
19) Facet for acromion

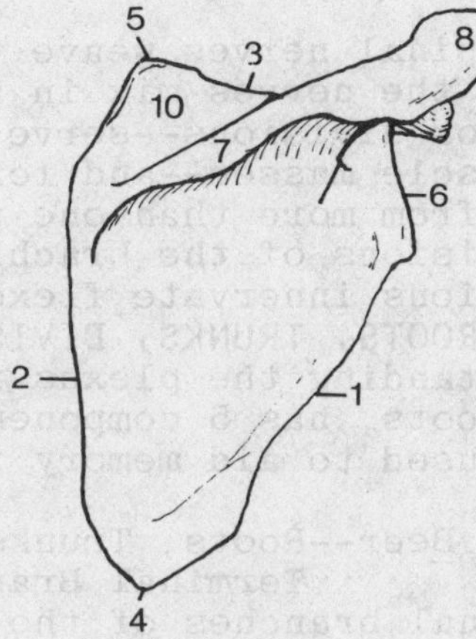

POSTERIOR

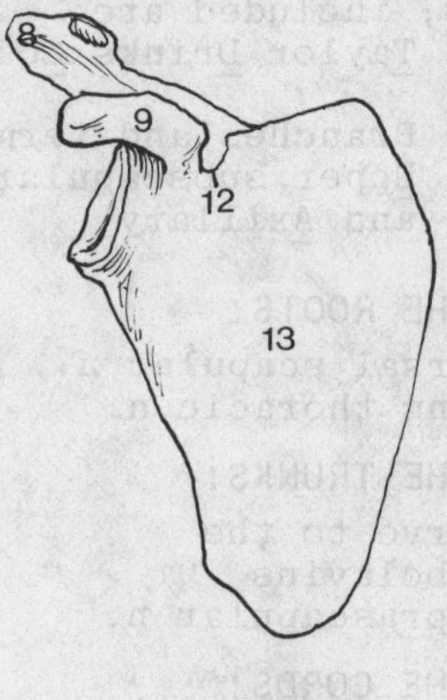

ANTERIOR

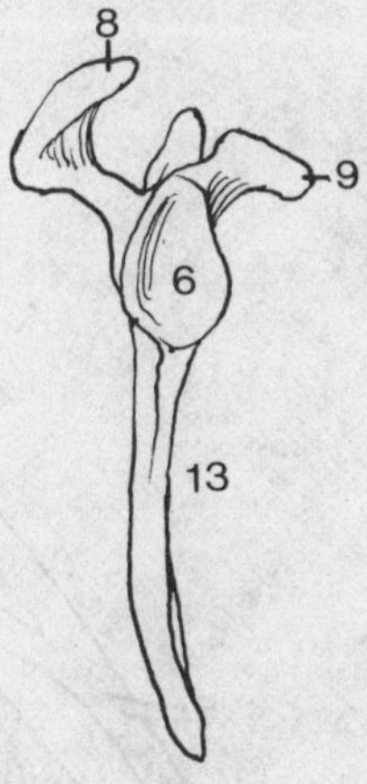

LATERAL

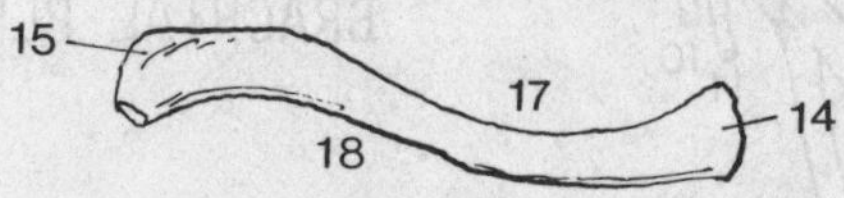

SUPERIOR

INFERIOR

BRACHIAL PLEXUS

VENTRAL RAMI of certain spinal nerves weave together to form a spinal nerve plexus. Generally, the nerves mix in a definite pattern, sort into anterior and posterior divisions--serve muscles derived from anterior and posterior muscle masses--and form a few, often large nerves with nerve fibers from more than one ventral ramus. With few exceptions, posterior divisions of the brachial plexus innervate extensors; anterior divisions innervate flexors. The brachial plexus is made of 5 components: ROOTS, TRUNKS, DIVISIONS, CORDS AND TERMINAL BRANCHES. A key to understanding the plexus is the number 5. The plexus is formed from 5 roots, has 5 components--roots, trunks, etc. There are many MNEMONICS used to aid memory in learning the brachial plexus; included are:

Robert Taylor Drinks Cold Beer--Roots, Trunks, Divisions, Cords, Terminal Branches

ULTRA: Branches and terminal branches of the posterior cord--Upper subscapular, Lower subscapular, Thoracodorsal, Radial and Axillary

FROM THE ROOTS:

1) Dorsal scapular n.
2) Long thoracic n.

FROM THE TRUNKS:

3) Nerve to the subclavius
4) Suprascapular n.

FROM THE CORDS:

--Named by relation to axillary artery

5) Lateral pectoral n.
6) Medial pectoral n.
7) Upper subscapular n.
8) Middle subscapular n. or Thoracodorsal n.
9) Lower subscapular n.
10) Medial brachial cutaneous n.
11) Medial antebrachial cutaneous n.
12) Medial head median n.
13) Lateral head median n.
14) Median nerve*
15) Ulnar nerve*
16) Musculocutaneous nerve*
17) Radial nerve*
18) Axillary nerve*

* Terminal branches

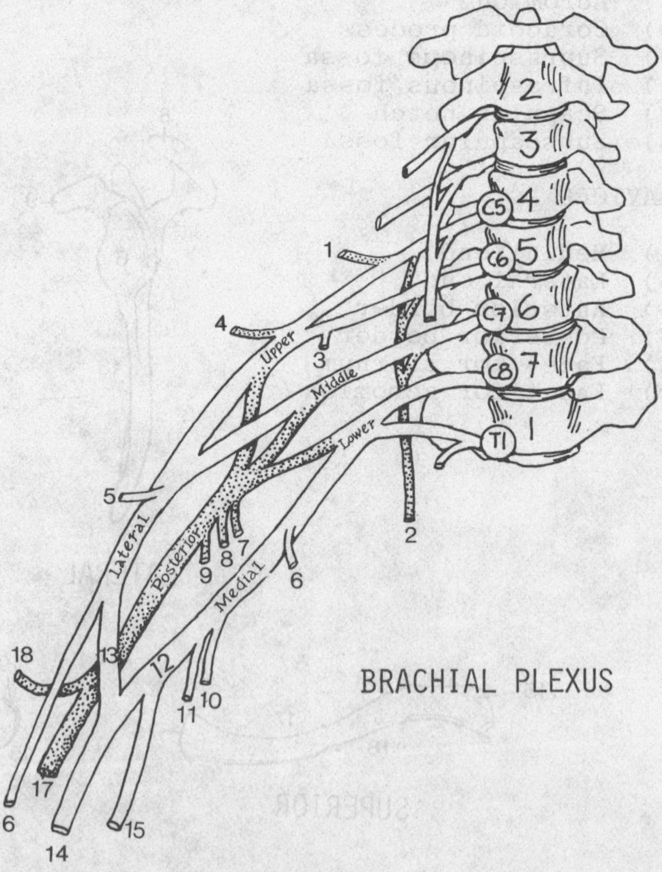

BRACHIAL PLEXUS

SCAPULAR REGION

SUPERFICIAL

1) TRAPEZIUS

O-Spines, supraspinous ligaments thoracic vertebrae, occipital bone, ligamentum nuchae
I-Clavicle, spine of scapula and acromion
A-In parts can elevate, retract or depress scapula
N-Spinal accessory and C3,4
B-Transverse cervical

2) LATISSIMUS DORSI

O-Spines T6-12, lumbar and sacral vertebrae via the thoracolumbar fascia
I-Humerus, intertubercular groove
A-Adduct and medially rotate arm "swimmers m."
N-Thoracodorsal n.
--middle subscapular
B-Thoracodorsal a.

OTHER STRUCTURES

3) Deltoid
4) Triceps
5) Teres major
6) Teres minor
7) Infraspinatus
8) Rhomboideus major
9) External oblique
10) Triangle of auscultation
11) Lumbar triangle
12) Iliac crest

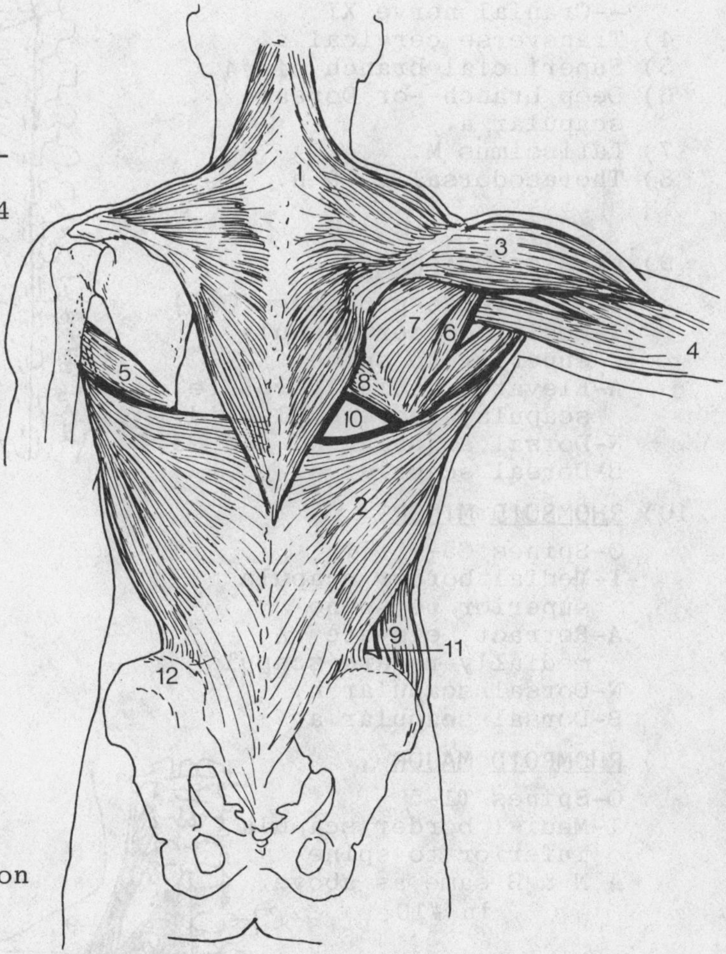

SUPERFICIAL BACK MUSCLES

DEEP SCAPULAR REGION

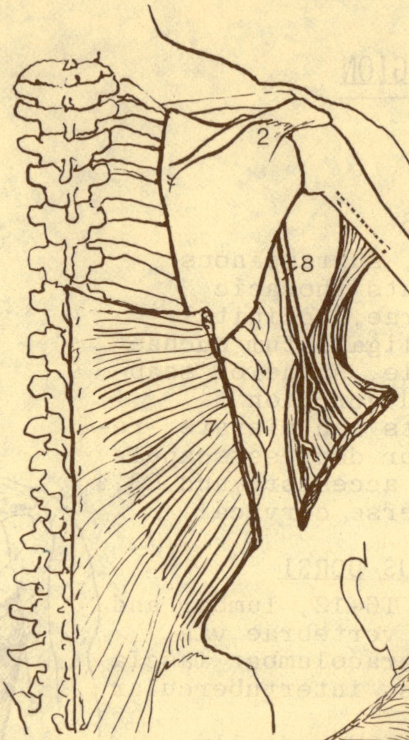

STRUCTURES

1) Splenius capitis
2) Spine of scapula
3) Spinal accessory n.
   --Cranial nerve XI
4) Transverse cervical a.
5) Superficial branch of #4
6) Deep branch--or Dorsal
   scapular a.
7) Latissimus M.
8) Thoracodorsal a. & n.

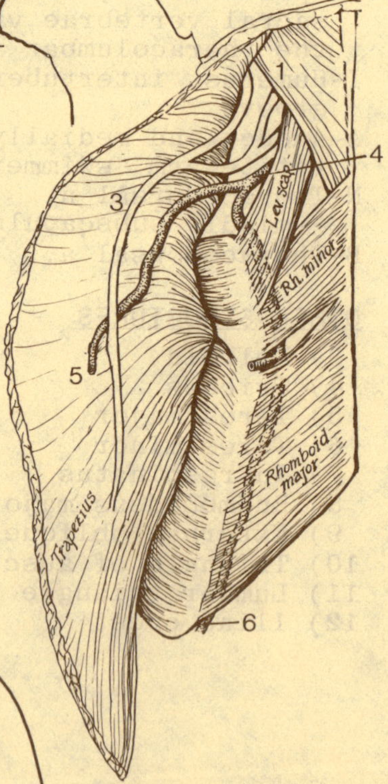

9) LEVATOR SCAPULAE

O-Transverse processes C1-4
I-Medial border scapula,
  superior to spine
A-Elevate, medially rotate
  scapula, depress glenoid
N-Dorsal scapular n. and C3,4
B-Dorsal scapular a.

10) RHOMBOID MINOR

O-Spines C5-7, T1
I-Medial border scapula
  superior to spine
A-Retract, elevate, &
  medially rotate scapula
N-Dorsal scapular n.
B-Dorsal scapular a.

11) RHOMBOID MAJOR

O-Spines T1-5
I-Medial border scapula
  inferior to spine
A,N & B same as above
  in #10

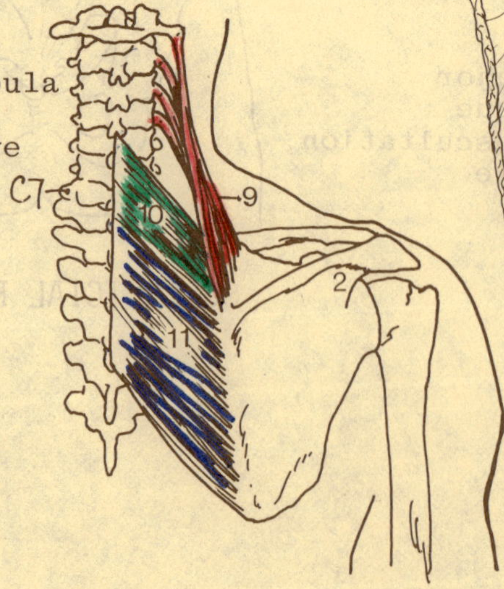

POSTERIOR PROJECTIONS
DEEP SCAPULAR
REGIONS

STRUCTURES:

1) Coracoid process
2) Acromion
3) Scapular ligament
4) Suprascapular a.
5) Suprascapular n.--
   remember by "Army
   over Navy"

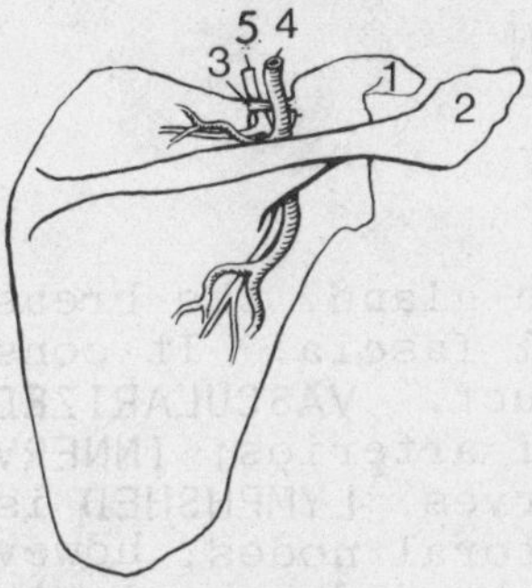

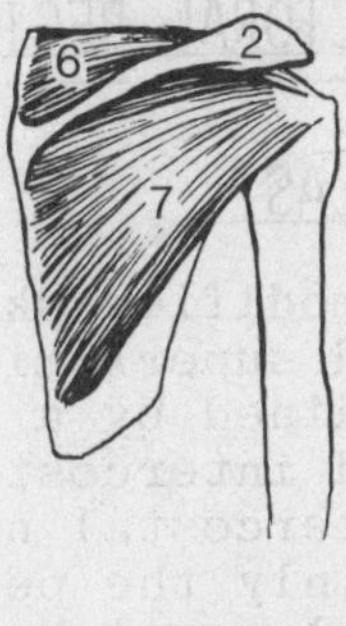

6) SUPRASPINATUS
   O-Supra~~scapular~~ spinous fossa
   I-Greater tubercle humerus
   A-Initiates abduction of arm
   N-Suprascapular n.
   B-Suprascapular a.

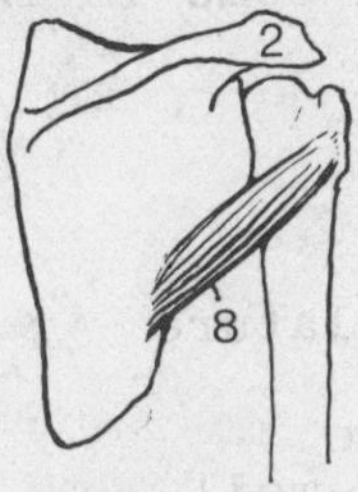

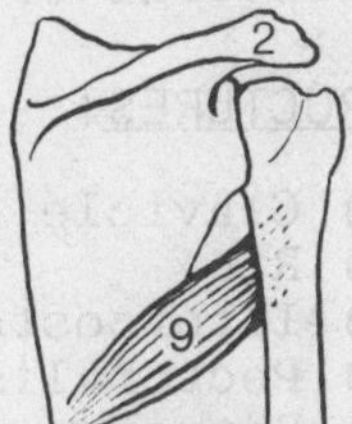

7) INFRASPINATUS
   O-Infra~~scapular~~ spinous fossa
   I-Greater tubercle humerus
   A-Lateral rotation, horizontal
     abduction of arm
   N-Suprascapular n.
   B-Suprascapular a.,
     circumflex scapular a.

8) TERES MINOR
   O-Lateral edge scapula
   I-Greater tubercle humerus
   A-Lateral rotation,
     horizontal abduction
   N-Axillary--N. to teres minor
   B-Circumflex scapular a.

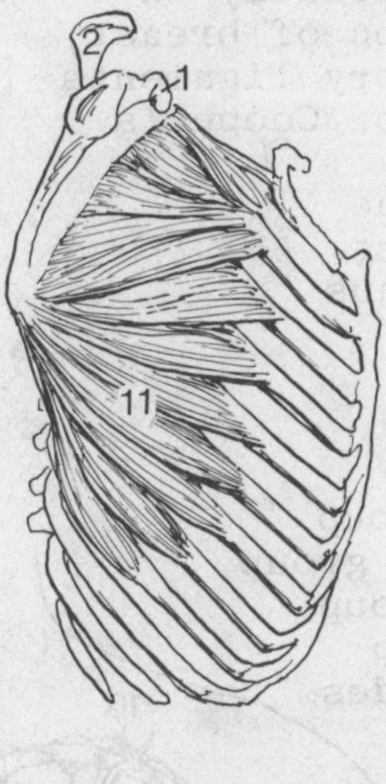

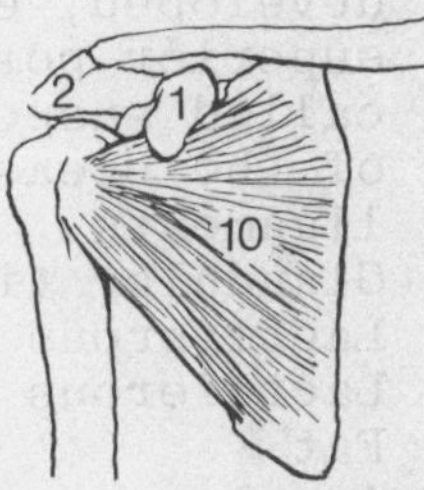

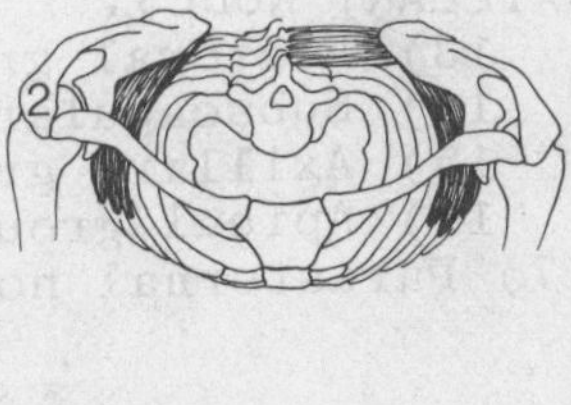

9) TERES MAJOR
   O-Inferolateral infra-
     spinous fossa scapula
   I-Humerus, bicipital groove
   A-Adduct, extend, medially
     rotate arm
   N-Lower subscapular n.
   B-Circumflex scapular a.

10) SUBSCAPULARIS
   O-Subscapular fossa
   I-Lesser tubercle humerus
   A-Medial rotation arm
   N-Upper and lower sub-
     scapular nn.
   B-Subscapular a.

11) SERRATUS ANTERIOR
   O-Ribs 1-8
   I-Medial border scapula
   A-Lateral rotation scapula,
     stabilization of scapula
   N-Long thoracic nerve--
     "C5,6,7 keep your wings
     from heaven"
   B-Lateral thoracic and
     thoracodorsal aa.

PECTORAL REGION

BREAST

A modified skin gland, the breast exists entirely within the skin and superficial fascia. It consists of 15-20 lobules, each lobule drained by a duct. VASCULARIZED by lateral thoracic, internal thoracic and intercostal arteries; INNERVATED by cutaneous branches T2-T6 intercostal nerves. LYMPHSHED is to the axillary group of nodes, mainly the pectoral nodes; however some is retrosternal to the contralateral breast and some is inferior to the umbilical region.

STRUCTURES:

1) Clavicle
2) Rib
3) Intercostal musculature
4) Pectoralis major
5) Retromammary space
6) Retinacula cutis--well developed, especially in superior portion of breast, called suspensory ligaments of the breast or Cooper's ligaments
7) Glandular tissue
8) Lactiferous duct
9) Lactiferous sinus
10) Fat
11) Areola
12) Nipple

AXILLARY NODES:
13) Pectoral group
14) Subscapular group
15) Axillary group
16) Apical group
17) Parasternal nodes

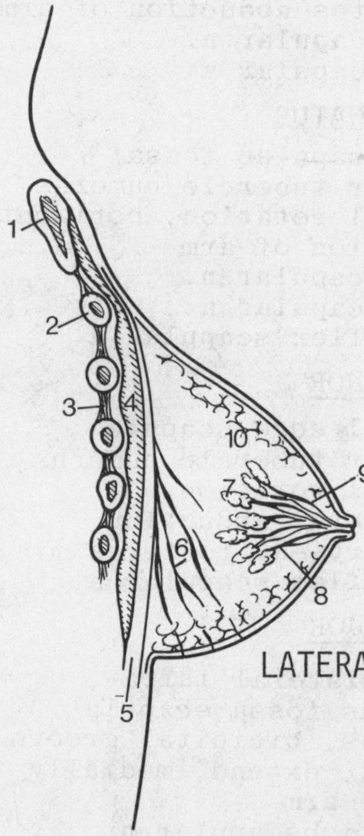

LATERAL

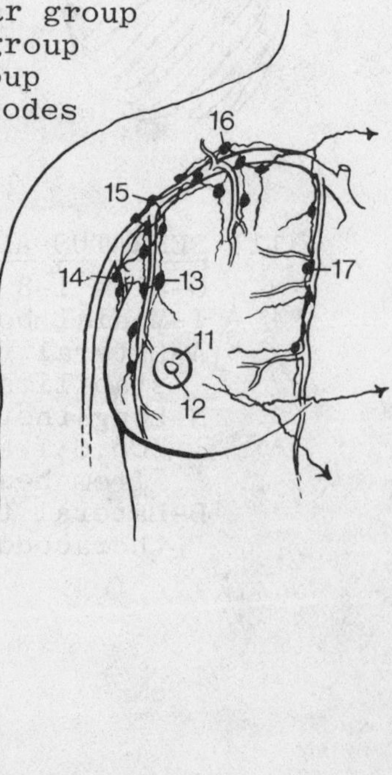

FRONTAL

SUPERFICIAL

1) Serratus anterior M.
2) External abdominal oblique M.
3) Biceps M.
4) Cephalic vein
5) Deltoid branch, thoracoacromial artery
6) Deltopectoral triangle or groove--bounded by deltoid, pectoralis major and clavicle, contains #4 & 5

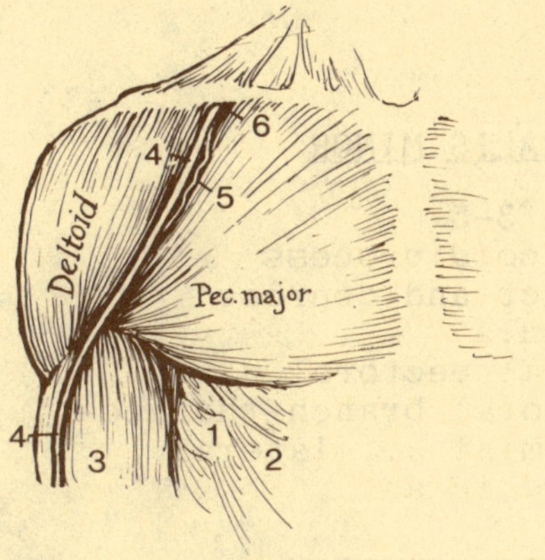

7) DELTOID
   O-Lateral 1/3 clavicle, spine and acromion of scapula
   I-Deltoid tuberosity of humerus
   A-Abduct arm from about 15° to 90°, portions can flex, medially rotate and horizontally adduct; extend, laterally rotate and horizontally abduct
   N-Axillary nerve (8)
   B-Posterior humeral circumflex artery (9)

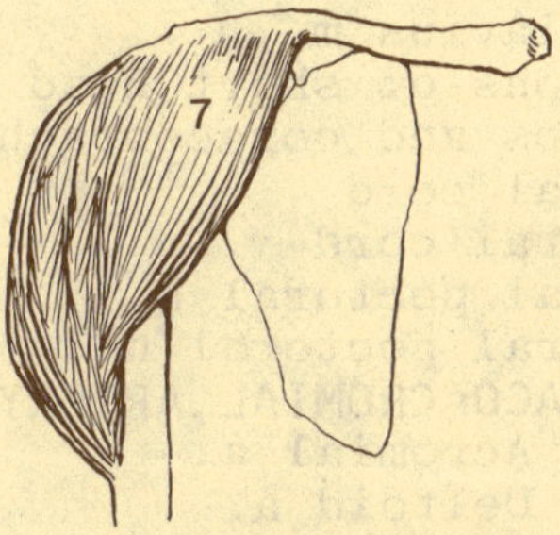

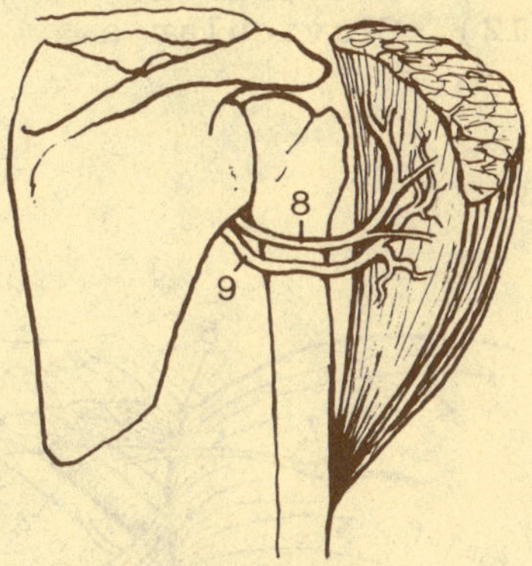

10) PECTORALIS MAJOR
   O-Medial 2/3 clavicle, sternum, cartilage ribs 1-6, abdominal aponeurosis
   I-Humerus--lateral lip bicipital groove
   A-Adduction medial rotation arm
   N-Medial and lateral pectoral n.
   B-Pectoral branch thoracoacromial a.

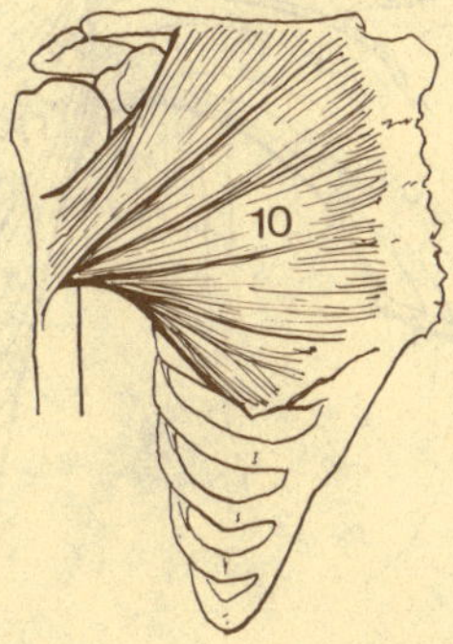

DEEP

1) PECTORALIS MINOR

O-Ribs 3-5
I-Coracoid process
A-Abduct and depress scapula
N-Medial pectoral n.
B-Pectoral branch thoracoacromial a., lateral thoracic a.

OTHER STRUCTURES:

2) Subclavius m.
3) Tendons of short head of biceps and coracobrachialis
4) Medial cord
5) Lateral cord
6) Medial pectoral n.
7) Lateral pectoral n.
8) THORACOACROMIAL ARTERY:
   9) Acromial a.
   10) Deltoid a.
   11) Pectoral a.
   12) Clavicular a.

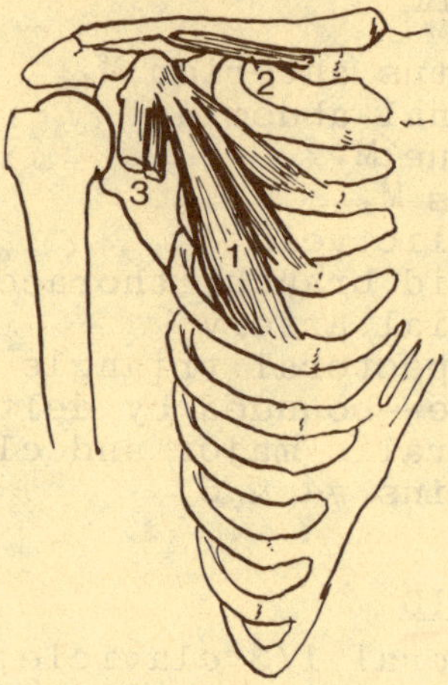

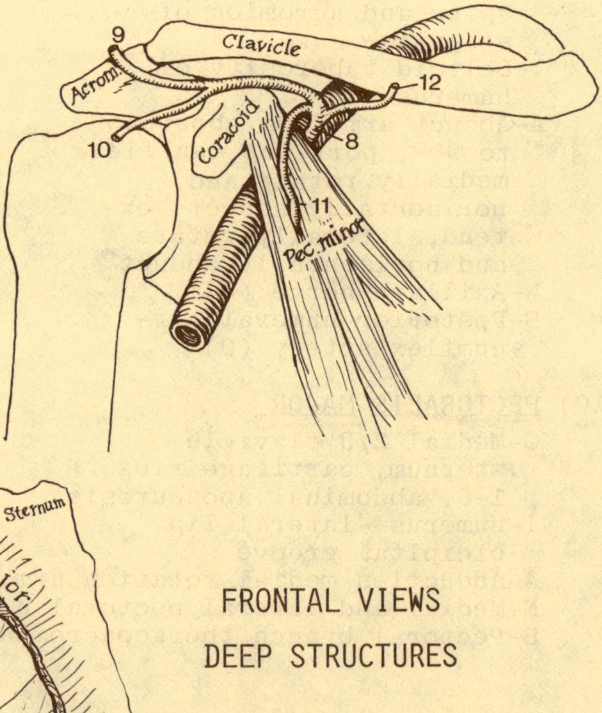

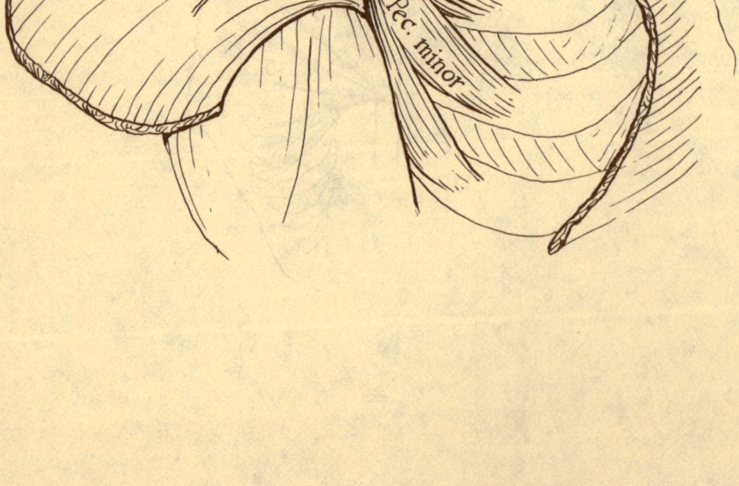

FRONTAL VIEWS
DEEP STRUCTURES

COMMENTS

1. The connective tissue septa are particularly well developed in the upper half of the breast and are called the Ligaments of Cooper.

2. A 'milk line' on either side of the body extends from the axilla medially to the midclavicular line and inferiorly to the medial surface of the proximal thigh. Supernumerary nipples--polythelia--or breasts--polymastia--may occur along this line.

3. An axillary extension of breast tissue--The Tail of Spence--is normal and should not be mistaken for a tumor.

4. Cancers of the breast are common and the lymphatic drainage should be understood. 75% of drainage is to the axillary nodes, a small percentage drains retrosternally or crosses the midline. Palpation of the breast--divided into quadrants for diagnostic purposes--and xerography--a soft tissue X-ray technique--are important for detection of cancer.

5. Orange peel skin--peau d'orange--is a dimpled appearance of skin of the breast caused by contracture of the suspensory ligaments when invaded in carcinoma. Mobility of the breast is also lost as the tissue becomes adherent to the pectoralis fascia and muscle.

6. Removal of the breast is known as mastectomy. It may be simple--merely removal of the gland--or radical--including removal of axillary tissue and nodes, pectoral fascia and muscles and sometimes adjacent muscles. The innervation of the serratus anterior muscle--the long thoracic nerve--should be spared.

7. Clavicular fractures are common and result in a drooping of the upper extremity on the affected side. Stability of the upper extremity is lost since its only articulation with the axial skeleton is the sternoclavicular joint. The clavicle is most commonly fractured at a point 2/3 of its length laterally.

8. At times a sternalis muscle, a persistent cranial continuation of the rectus abdominis, is found in the midline.

SHOULDER

POSTERIOR

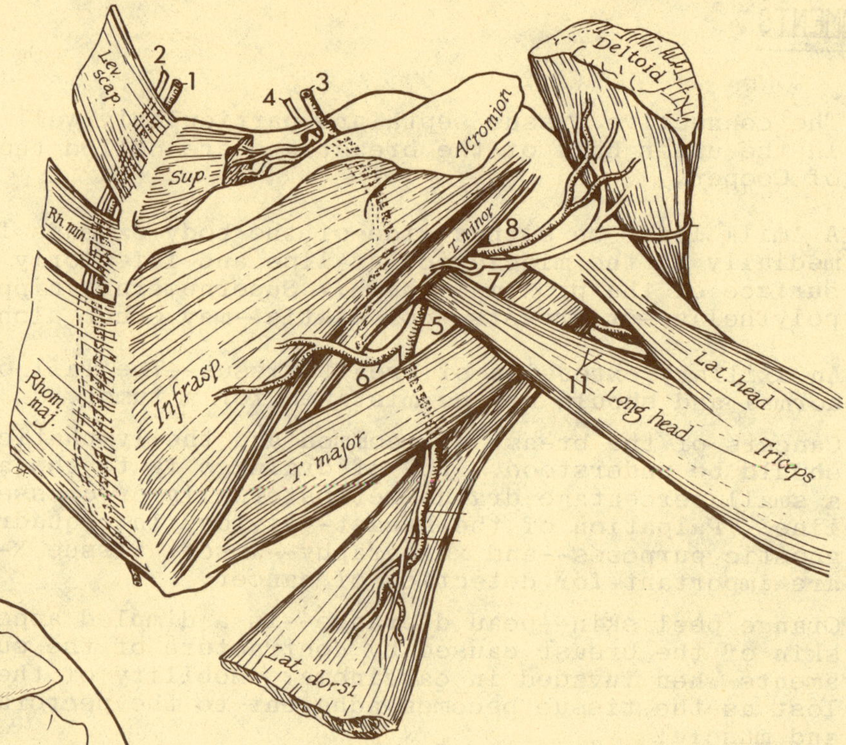

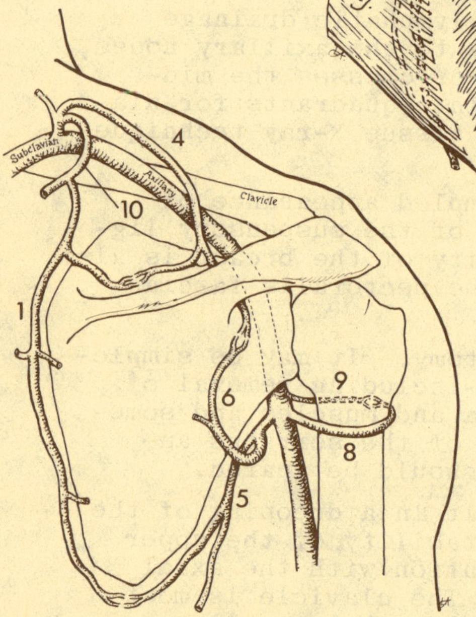

1) Dorsal scapular a.
2) Dorsal scapular n.
3) Suprascapular a.
4) Suprascapular n.
5) Subscapular a.
6) Circumflex scapular a.
7) Axillary n.
8) Posterior humeral circumflex a.
9) Anterior humeral circumflex a.
10) Transverse cervical a.
11) Radial n. and a.

TRIANGULAR SPACE:
Bounded by teres minor, teres major, long head triceps--contains circumflex scapular a.

QUADRANGULAR SPACE:
Bounded by teres minor, teres major, long head triceps, humerus--transmits axillary n. and posterior humeral circumflex a.

TRIANGULAR INTERVAL:
Bounded by long head triceps, lateral head triceps, teres major--contains radial nerve and deep (profound) brachial artery

ANTERIOR

1) Teres major
2) Latissimus dorsi
3) Long head biceps
4) Pectoralis major
5) Supraspinatus
6) Subscapularis

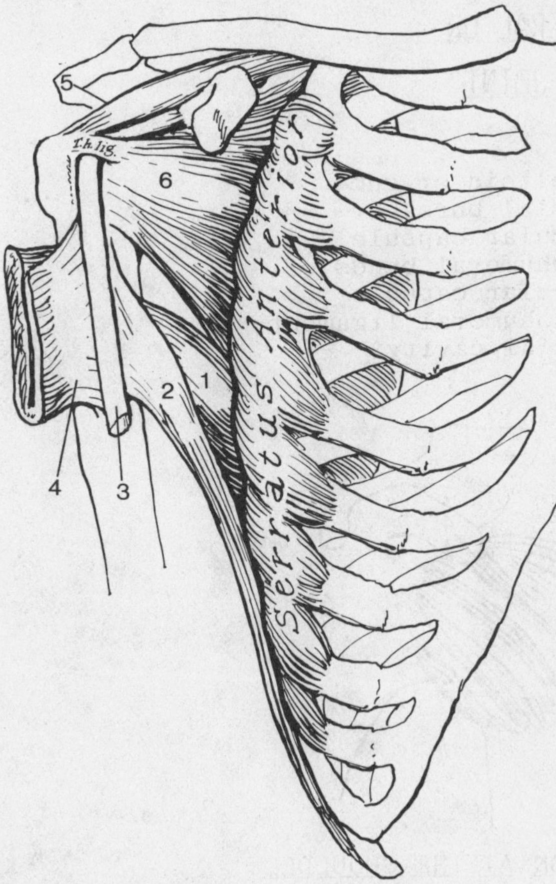

ARTHROLOGY

STERNOCLAVICULAR JOINT:

7) Articular disc
8) Interclavicular ligament
9) Sternoclavicular ligament
10) Costoclavicular ligament

ACROMIOCLAVICULAR JOINT:

11) Acromioclavicular ligament--contains an articular wedge, not a complete disc
12) Coracoacromial ligament
Coracoclavicular ligament
 13) Trapezoid ligament
 14) Conoid ligament

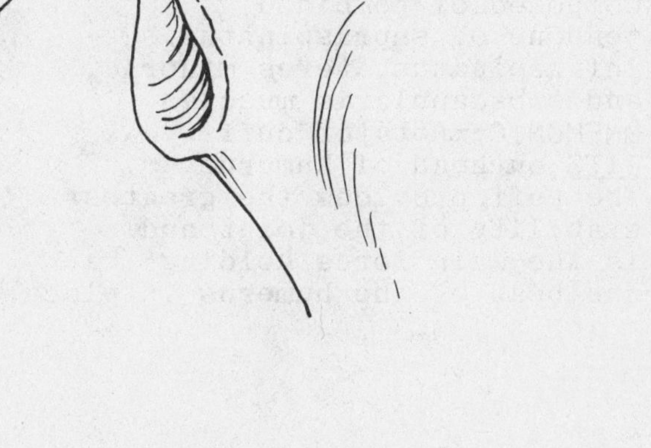

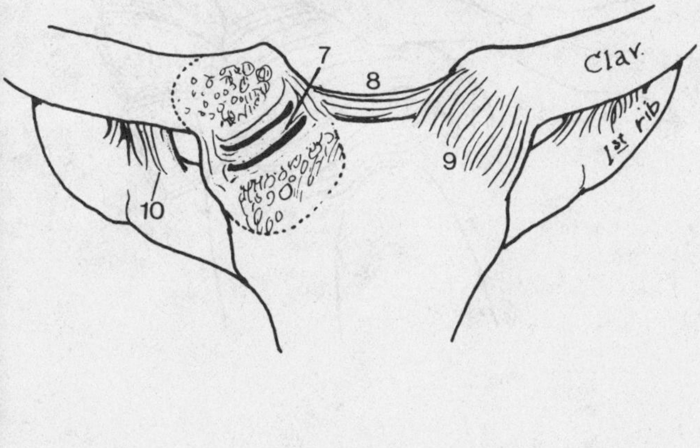

GLENOHUMERAL OR SHOULDER JOINT

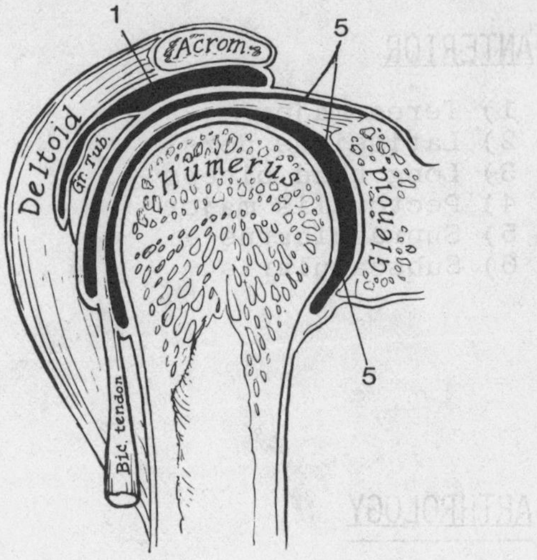

1) Subdeltoid or subacromial bursa
2) Articular capsule
3) Glenohumeral bands of articular capsule
4) Coracohumeral ligament
5) Synovial cavity

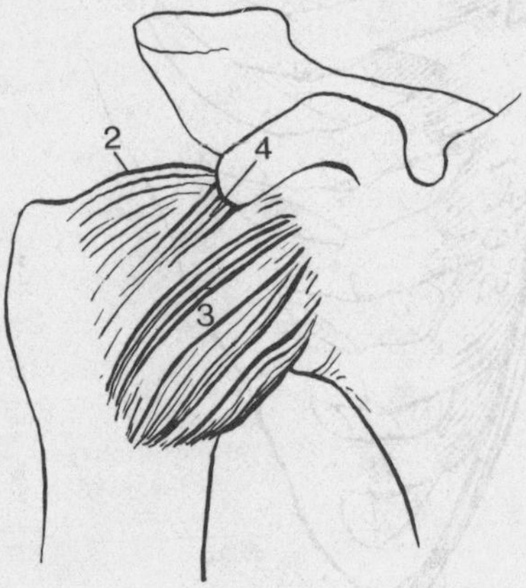

ABDUCTION AT THE SHOULDER:

0 - 15°	Supraspinatus
15 - 90°	Deltoid aided by supraspinatus
90 - 180°	Rotation of scapula by serratus anterior and trapezius--the acromion prevents further abduction solely by the deltoid

ROTATOR CUFF:

Composed of combined tendons of supraspinatus, infraspinatus, teres minor and subscapularis muscles. MNEMONIC--Rotator cuff SITS on head of humerus. The cuff provides the greatest stability of the joint and is the main force holding the head of the humerus in place.

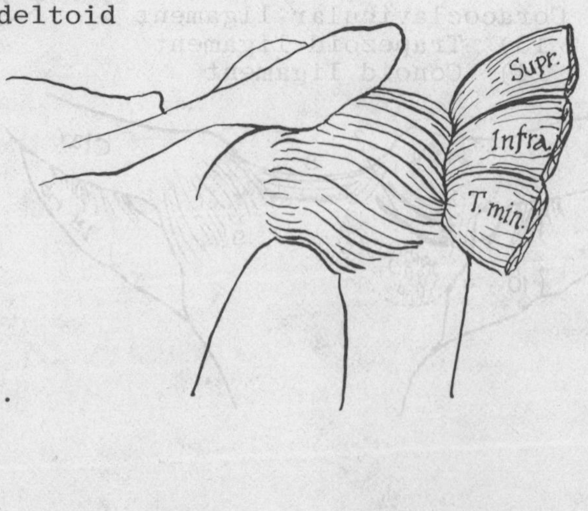

COMMENTS

1. The triangle of auscultation is an area bounded by the borders of the trapezius, latissimus and rhomboideus major/scapula. When the scapula is laterally rotated or protracted, and to a slight degree at rest, only skin and intercostal muscles intervene between a stethoscope and the pleural cavity; this facilitates auscultation.

2. Damage to the long thoracic nerve results in paralysis of the serratus anterior and is shown clinically by 'winging of the scapula'--protrusion of the inferior angle of the scapula--thus the phrase "C 5,6,7 keep your wings from heaven." The nerve may be injured by direct trauma or carrying heavy weights on the shoulder or inadvertently transected in mastectomy.

3. Only the supraspinatus, <u>not</u> the deltoid, can initiate abduction. In paralysis of the supraspinatus, the individual may lean to the afflicted side or knock the upper extremity laterally with the hip to initiate abduction. Even the deltoid, however, cannot abduct more than 90°--rotation of the scapula must assist.

4. Bursitis is an inflammation of a bursa. Around the shoulder are several bursae, including the subdeltoid, supraspinatus, infraspinatus and subscapularis bursae, which prevent friction when a muscle and/or its tendon crosses a joint or protuberance. The etiology of bursitis is not known; however the condition may become so severe the bursa ruptures.

5. A common 'law' states that a joint is innervated by the nerves which innervate muscles that act on that joint; another 'law' states that muscles which cross more than one joint will act at more than one joint.

6. The rotator cuff is the main source of stability of the glenohumeral joint and maintains the head of the humerus in place. Portions of the cuff may be ruptured or torn and this usually results in instability of the joint.

   A shoulder separation is a tearing of the acromioclavicular ligaments, while a dislocation--subluxation--is the displacement of the head of the humerus. Although the cuff is weakest inferiorly, most dislocations are anterior and inferior.

II. THE AXILLA

THE AXILLA is a pyramidal shaped region between the arm and body. It contains structures in transit between the body and upper extremity.

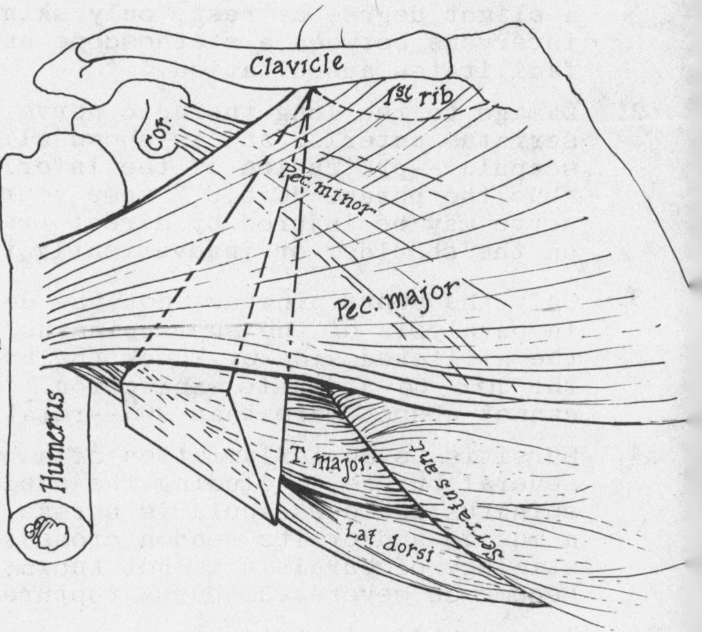

BOUNDARIES:

Base: Skin of the axilla or armpit
Anterior: Pectoral muscles
Lateral: Bicipital groove, long head biceps, humerus
Posterior: Latissimus dorsi, teres major, subscapularis
Medial: Serratus anterior, ribs and intercostal mm
Apex: Clavicle and 1st rib

THE CERVICOAXILLARY CANAL is the 'conduit' bounded by the clavicle, first rib and scapula.

HORIZONTAL SECTION:

1) Serratus anterior
2) Subscapularis
3) Pectoralis major
4) Pectoralis minor
5) Tendons of latissimus dorsi and teres major
6) Tendon long head biceps
7) Short head of biceps
8) Coracobrachialis
9) Axillary sheath
10) Axillary artery
11) Axillary vein
12) Medial cord
13) Posterior cord
14) Lateral cord

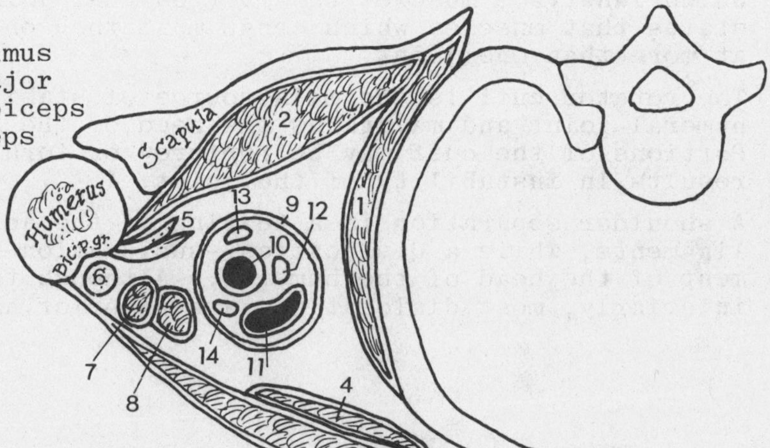

THE AXILLARY SHEATH is derived from the prevertebral layer of cervical fascia as the nerves leave the vertebral column and travel to the axilla

AXILLARY CONTENTS

THE AXILLA CONTAINS the axillary artery, vein, fat, cords of the brachial plexus, lymph nodes and fascia.

AXILLARY ARTERY

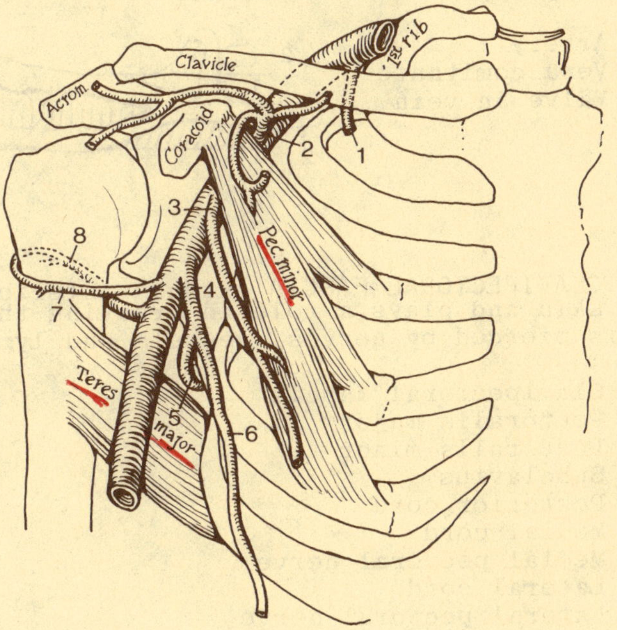

THE SUBCLAVIAN ARTERY ends at the lateral border of the 1st rib and becomes the AXILLARY ARTERY. The axillary artery ends at the inferior border of the teres major and is continuous with the brachial artery

PARTS OF THE AXILLARY ARTERY:

1st Part: Medial to the pectoralis minor
- 1) Supreme thoracic

2nd Part: Posterior to the pectoralis minor
- 2) Thoracoacromial a.
- 3) Lateral thoracic a.

3rd Part: Lateral to the pectoralis minor
- 4) Subscapular a.
  - 5) Circumflex scapular a.
  - 6) Thoracodorsal a.
- 7) Anterior humeral circumflex a.
- 8) Posterior humeral circumflex a.

OBSERVE there is one branch of the first part, 2 of the second and 3 from the third.

OTHER STRUCTURES:

9) Cords of the brachial plexus--note communication between medial and lateral pectoral nn.
10) Axillary vein
11) Scalenus medius
12) Scalenus anterior

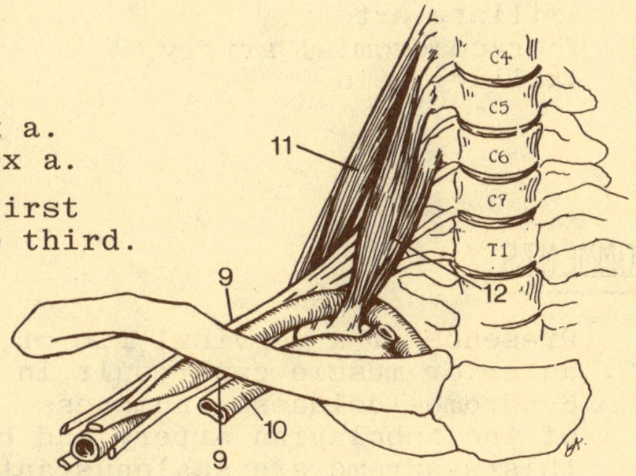

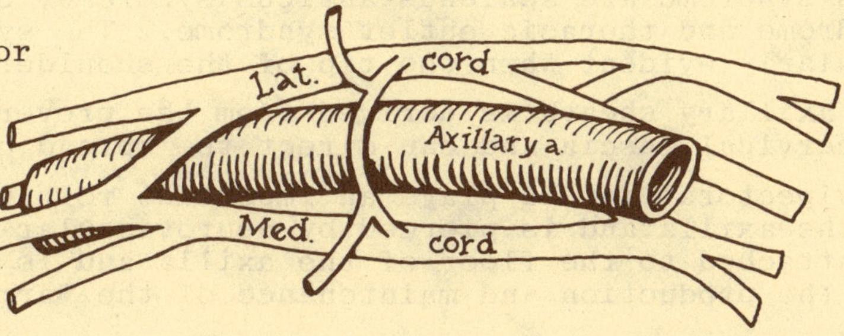

VENAE COMITANTES are often encountered in vasculature of the extremities. Instead of one major vein, i.e., axillary, the vein breaks up into a series of veins around the artery. This arrangement serves to conserve heat--an exchange between the warm artery and cooler vein--and to facilitate venous return--pumping action of artery propels blood in veins.

1) Artery
2) Vena comitante
3) Valve in vein

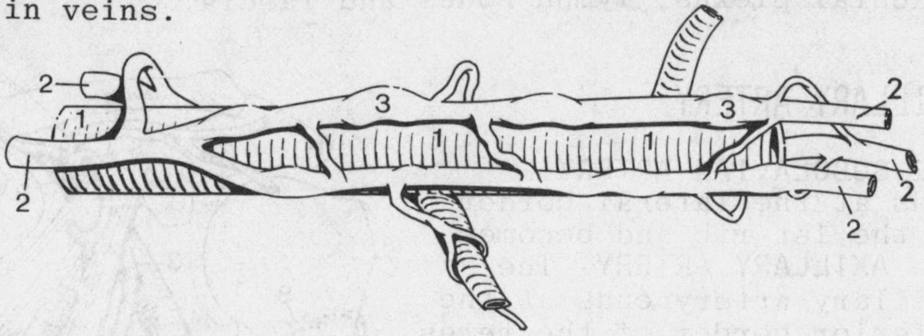

THE CLAVIPECTORAL FASCIA extends inferiorly from the clavicle to the skin and plays a role in creating the depression called the armpit. It is pierced by nerves, vessels and lymphatics.

1) Clavipectoral fascia
2) Pectoralis major
3) Pectoralis minor
4) Subclavius
5) Posterior cord
6) Medial cord
7) Medial pectoral nerve
8) Lateral cord
9) Lateral pectoral nerve
10) Axillary artery
11) Thoracoacromial artery
12) Axillary vein

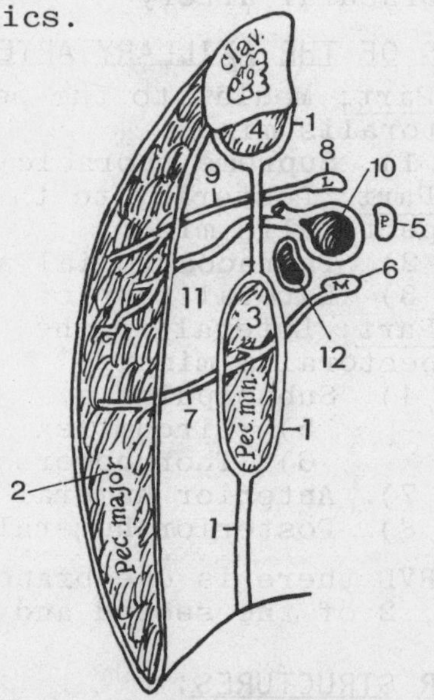

COMMENTS

1. Presence of a cervical rib or occasionally a large scalenus anterior muscle can result in a Neurovascular Compression Syndrome--coldness, numbness, weakness--due to compression of the subclavian artery and brachial plexus. Older terms for this syndrome are scalenus anticus syndrome, cervical rib syndrome and thoracic outlet syndrome. The syndrome is particularly evident when the tip of the shoulder is depressed.
2. The axillary sheath is derived from the prevertebral layer of cervical fascia and can direct the spread of infections.
3. Clavipectoral fascia plays an important role in the support of the axilla and is pierced by neurovascular elements. It is attached to the floor of the axilla and is responsible for the production and maintenance of the 'armpit'.

III. BRACHIUM (ARM)

THE ARM extends from the gleno-humeral joint to the elbow or olecranon; it contains one bone, 4 muscles, terminal branches of the brachial plexus and the brachial artery with its branches.

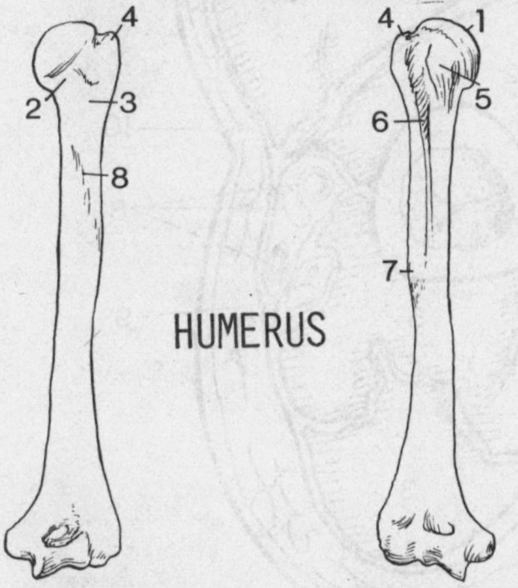

POSTERIOR ANTERIOR

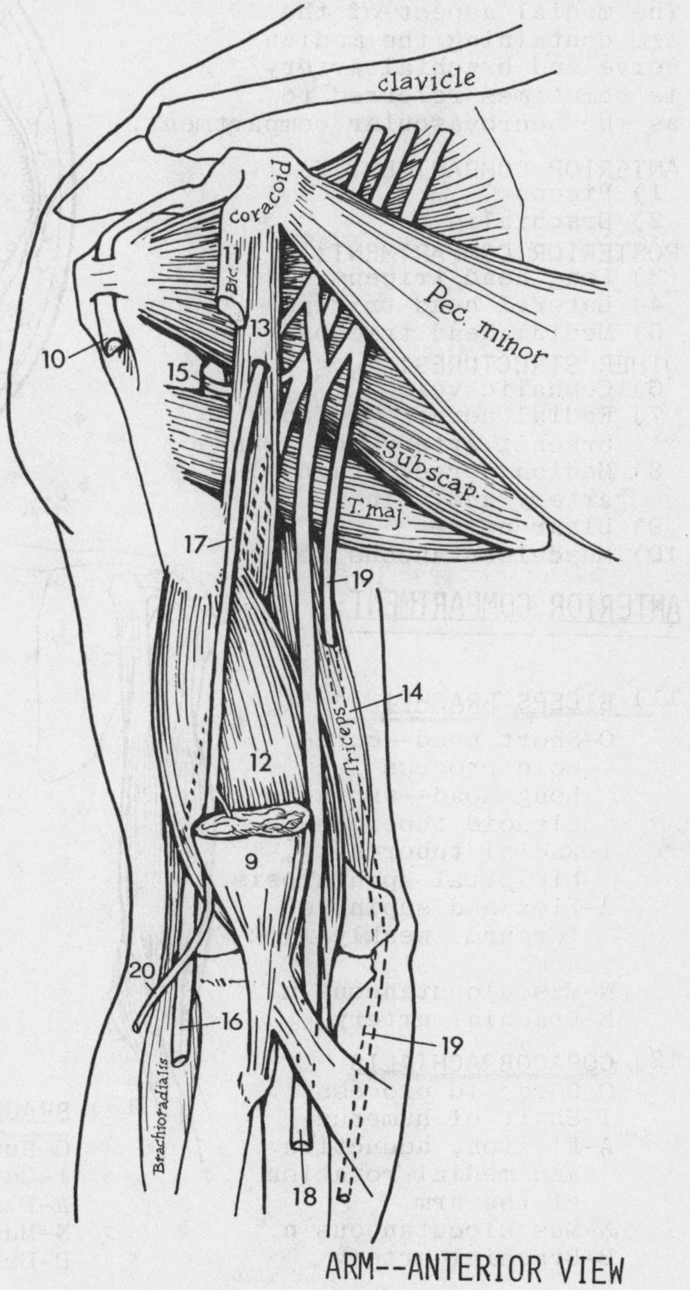

ARM--ANTERIOR VIEW

STRUCTURES:

1) Head of humerus
2) Anatomical neck
3) Surgical neck
4) Greater tubercle
5) Lesser tubercle
6) Bicipital or inter-tubercular groove
7) Deltoid tuberosity
8) Radial groove
9) Biceps brachii--belly
10) Tendon long head biceps brachii
11) Tendon short head biceps brachii
12) Brachialis
13) Coracobrachialis
14) Triceps brachii--long head
15) Axillary nerve
16) Radial nerve--travels posteriorly to humerus
17) Musculocutaneous nerve
18) Median nerve
19) Ulnar nerve
20) Lateral antebrachial cutaneous nerve

CROSS SECTION OF ARM

A FASCIAL SEPTUM divides the arm into anterior and posterior compartments. The medial aspect of the arm containing the median nerve and brachial artery is sometimes referred to as the neurovascular compartment.

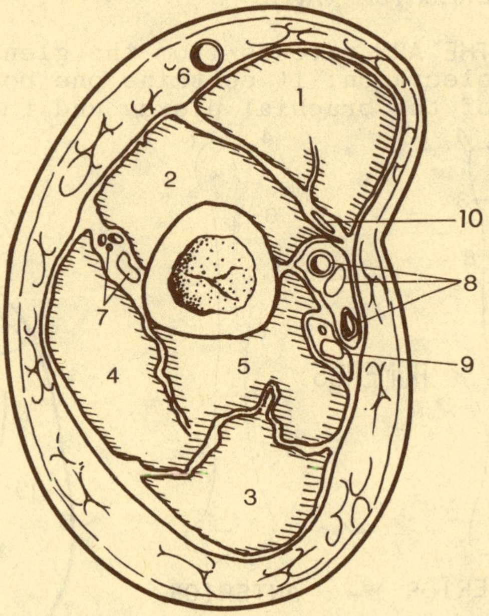

CAUDAL VIEW

ANTERIOR COMPARTMENT:
1) Biceps
2) Brachialis

POSTERIOR COMPARTMENT:
3) Long head triceps
4) Lateral head triceps
5) Medial head triceps

OTHER STRUCTURES:
6) Cephalic vein
7) Radial nerve, profunda brachii artery and vein
8) Median nerve, brachial artery and vein
9) Ulnar nerve
10) Musculocutaneous nerve

ANTERIOR COMPARTMENT:

11) BICEPS BRACHII
- O-Short head--coracoid process
  Long head--supraglenoid tubercle
- I-Radial tuberosity, bicipital aponeurosis
- A-Flex and supinate forearm, weakly flex arm
- N-Musculocutaneous n.
- B-Brachial artery

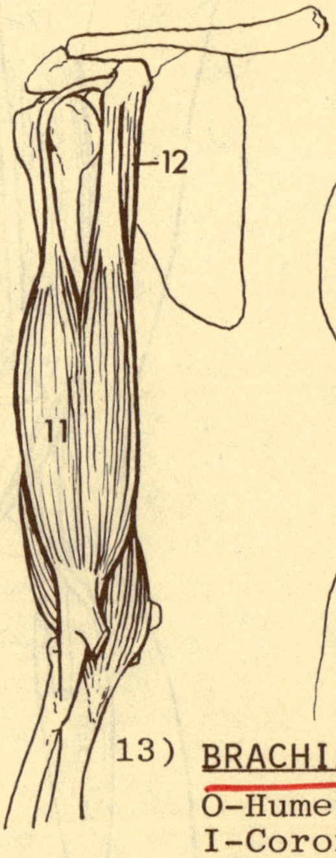

12) CORACOBRACHIALIS
- O-Coracoid process
- I-Shaft of humerus
- A-Flexion, adduction and medial rotation of the arm
- N-Musculocutaneous n.
- B-Brachial artery

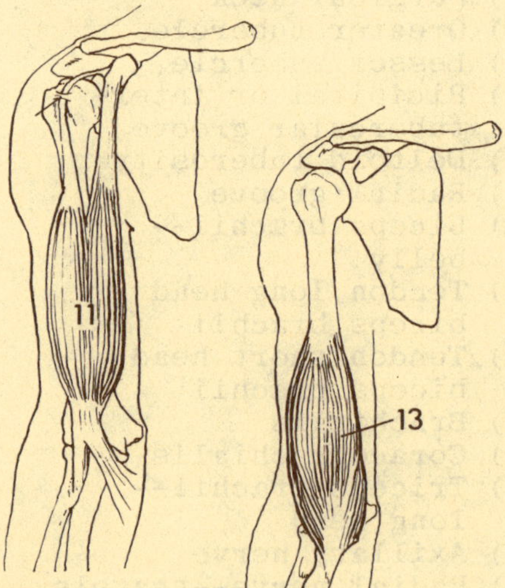

13) BRACHIALIS
- O-Humerus
- I-Coronoid process ulna
- A-Flexion at elbow
- N-Musculocutaneous n.
- B-Brachial artery

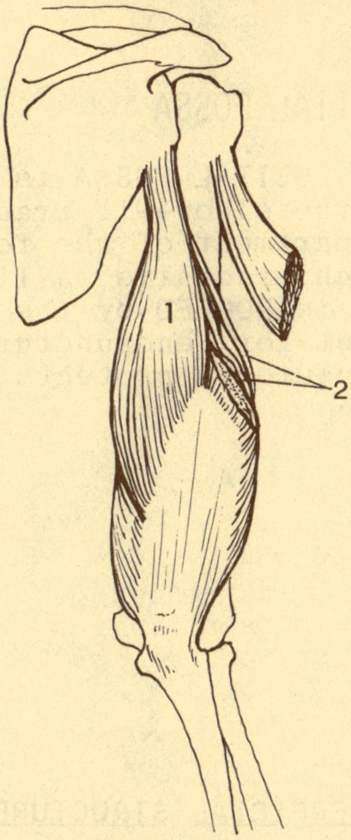

POSTERIOR COMPARTMENT:

1) TRICEPS BRACHII
O-Long head--infraglenoid tubercle
Lateral head--humerus above radial groove
A-Extension at elbow
N-Radial nerve
B-Profunda brachii a.

OTHER STRUCTURES:

2) Radial nerve and profunda brachii a. in radial--spiral--groove

THE OLECRANON (ELBOW) JOINT

THE ELBOW JOINT is complex and constitutes the articulation between arm and forearm--antebrachium--as well as the proximal articulation between the ulna and radius. The actions at the joint include flexion and extension of the forearm as well as pronation and supination of the forearm-- it is a hinge joint as well as a pivot joint.

STRUCTURES:

1) Ulnar collateral ligament
2) Radial collateral ligament
3) Annular ligament--surrounds the radial head, attaches only to the ulna
4) Area of contact of the ulnar nerve

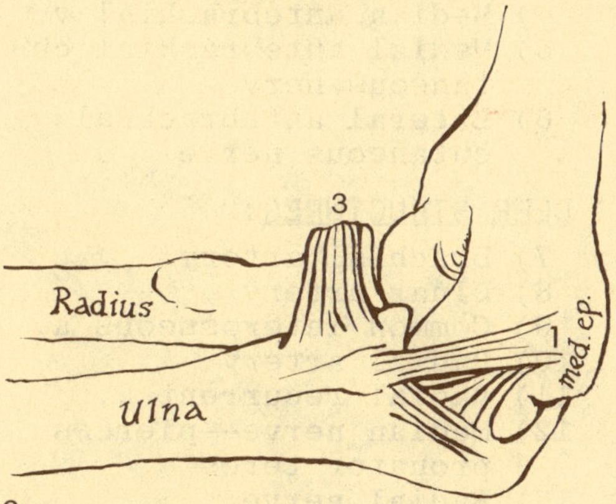

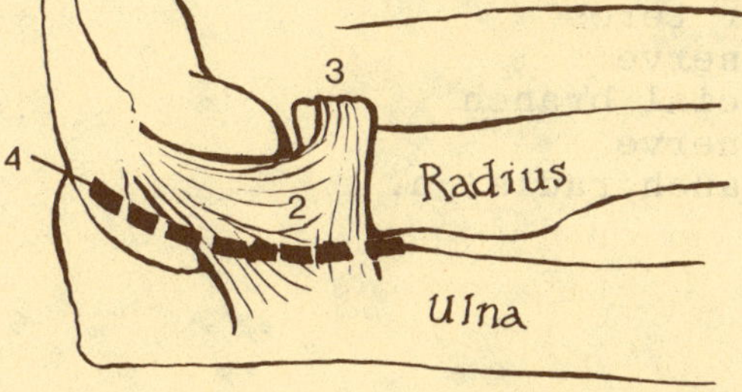

IV. CUBITAL FOSSA

THE CUBITAL FOSSA is a triangular intermuscular space anterior to the elbow--it transmits nerves and vessels to the anterior compartment of the forearm. IT IS BOUNDED by the pronator teres, brachioradialis, a line connecting the epicondyles of the humerus, and is ROOFED by the bicipital aponeurosis. It is a favorite point for venepuncture of the median cubital vein--the bicipital aponeurosis protects the deeper structures.

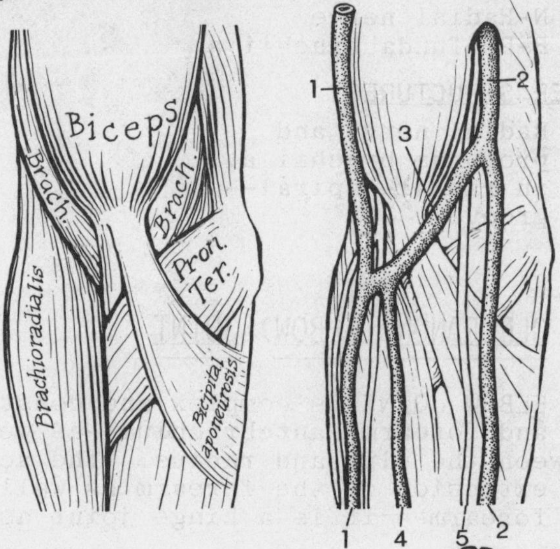

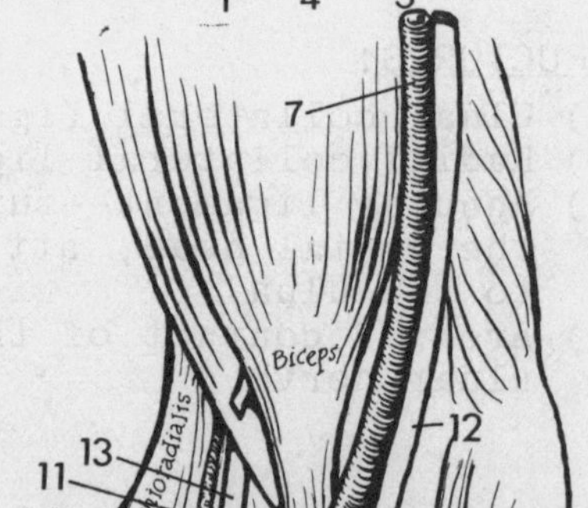

SUPERFICIAL STRUCTURES:

Subject to much variation

1) Cephalic vein
2) Basilic vein
3) Median cubital vein
4) Median antebrachial v.
5) Medial antebrachial cutaneous nerve
6) Lateral antebrachial cutaneous nerve

DEEP STRUCTURES:

7) Brachial artery
8) Ulnar artery
9) Common interosseous a.
10) Radial artery
11) Radial recurrent a.
12) Median nerve--pierces pronator teres
13) Radial nerve
14) Superficial branch radial nerve
15) Deep branch radial n.

COMMENTS

1. The anatomical neck is just distal to the head of the humerus while the surgical neck is at a narrowing of the shaft of the humerus distal to the tubercles. The surgical neck is at the level of the humeral circumflex arteries and is a common place for fractures of the humerus.
2. If a supracondylar process is present, the median nerve may be injured by it. The ulnar nerve--funny bone--is commonly injured by blows to the medial aspect of the elbow.
3. The intercostobrachial nerve is a branch of an upper intercostal nerve--T 1-2--that innervates the medial aspect of the arm. This nerve is at the same spinal cord level as the entrance of sensory fibers from the heart; heart pain may be referred to the medial aspect of the left arm.
4. 'Tennis elbow' is tendonitis of the common extensor tendon at the lateral epicondyle. 'Students elbow' is bursitis of the superficial olecranon bursa from rubbing of the elbows on a desk top or from other trauma. 'Nursemaids elbow' or 'pulled elbow' results when a sudden jerk or pull on a pronated forearm tears the annular ligament of the radius and allows part of the head of the radius to slip out of place.

FRACTURES

Fractures or breaks in the bone of the upper extremity are common. Here is a list of definitions of terms concerning fractures:

CLOSED--not open to skin
COMMINUTED--bone broken into several pieces
COMPLICATED--significant injury to nearby tissues
COMPOUND--a wound in the skin, open to the fracture; the fracture may protrude through the skin
GREENSTICK--bending of bone with break on the convex side of the bend, but not the concave
IMPACTED--one of the fractured pieces is driven into cancellous bone of the other piece
INCOMPLETE--bone does not fracture completely into 2 or more pieces
LONGITUDINAL--fractures along longitudinal axis of bone
OBLIQUE--fractures obliquely across the bone
SPIRAL--crack spirals up the long axis of the bone
TRANSVERSE--fractures run transversely to the long axis of the bone

V. FOREARM, WRIST AND HAND

THE FOREARM, WRIST AND HAND WILL BE DISCUSSED AS A UNIT, since many of the long muscles of the forearm insert into structures of the wrist and hand.

OSTEOLOGY:

FOREARM

1) Ulna
2) Olecranon process
3) Coronoid process
4) Trochlear notch
5) Radius
6) Radial tuberosity
7) Styloid process
8) Interosseous crest
9) Radial tubercle of Lister
10) Radial notch

CARPAL (WRIST) BONES

11) Scaphoid
12) Lunate
13) Triquetrum
14) Pisiform
15) Trapezium
16) Trapezoid
17) Capitate
18) Hamate
19) Metacarpals

PHALANGES

20) Proximal
21) Middle
22) Distal

JOINTS

23) Metacarpo-phalangeal--MP
24) Proximal interphalangeal--PIP
25) Distal interphalangeal--DIP
26) Interphalangeal --IP

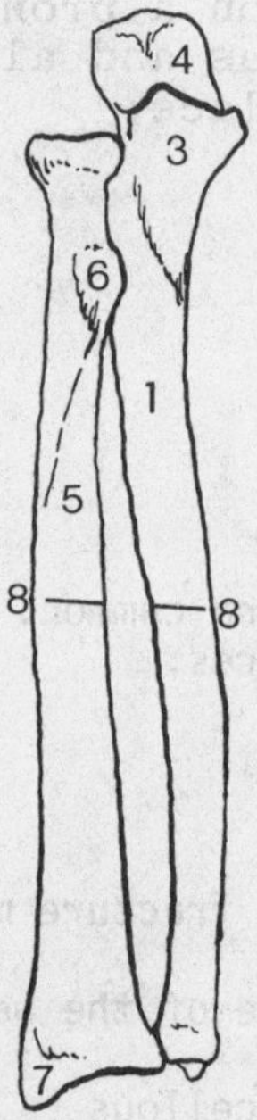

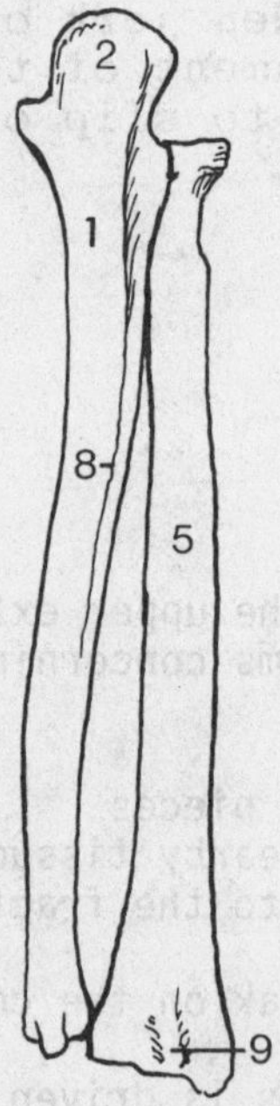

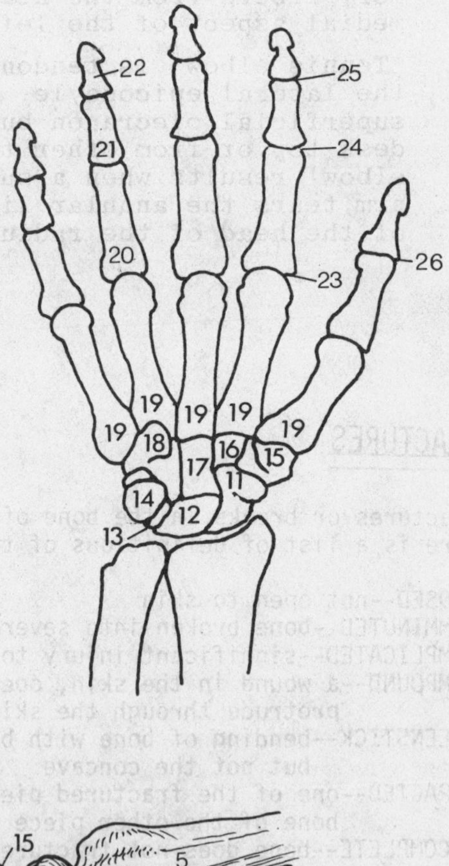

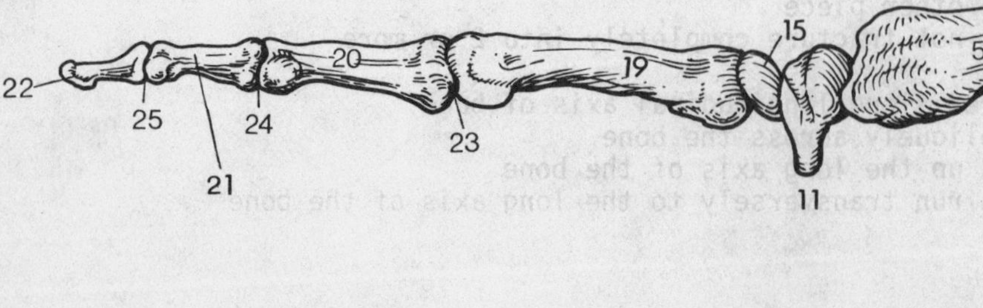

CROSS SECTION OF THE FOREARM

CAUDAL VIEW

FASCIAL SEPTAE divide the forearm into anterior--flexor--and posterior--extensor compartments as in the arm. The interosseous membrane also contributes to the division. These compartments do not communicate with those of the arm, but do continue into the hand. The posterior compartment contains one flexor, the brachioradialis muscle.

STRUCTURES:

1) Ulna
2) Radius
3) Interosseous membrane
4) Radial n. and a.
5) Median nerve
6) Ulnar n. and a.
7) Anterior compartment
8) Posterior compartment
9) Cephalic vein
10) Basilic vein

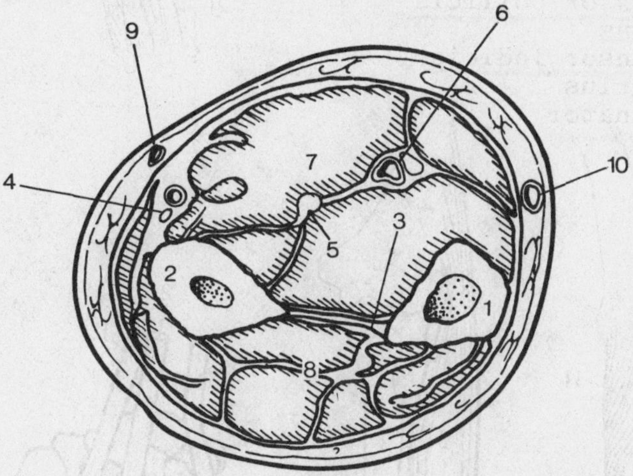

POSTERIOR COMPARTMENT:

SUPERFICIAL GROUP:

1) Extensor carpi radialis longus
2) Extensor carpi radialis brevis
3) Extensor digitorum communis
4) Extensor digiti minimi
5) Extensor carpi ulnaris
6) Brachioradialis--actually a weak flexor, functions to hold the forearm in the position of function, half-way between supination and pronation

DEEP GROUP

7) Abductor pollicis longus
8) Extensor pollicis brevis
9) Extensor pollicis longus
10) Extensor indicis proprius
11) Supinator

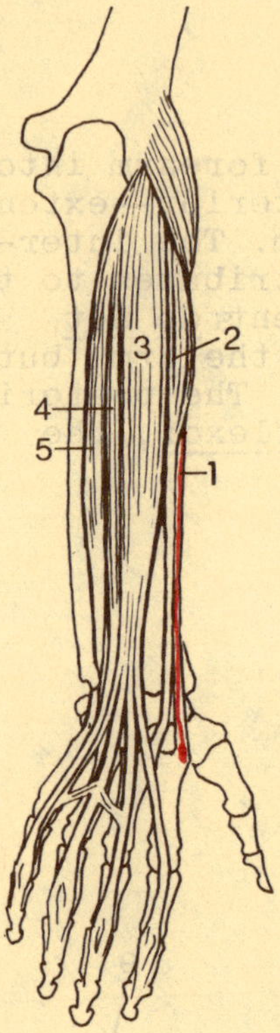

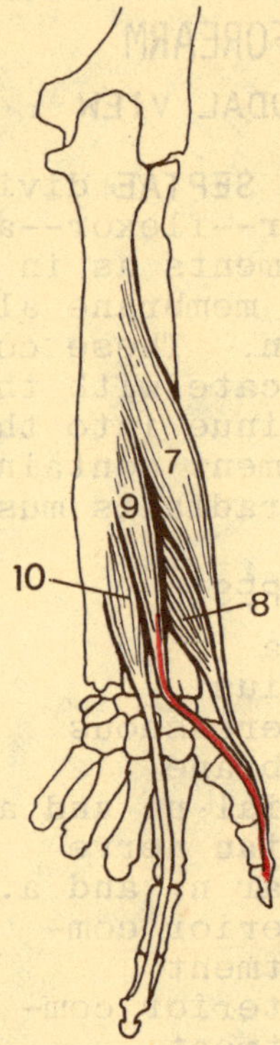

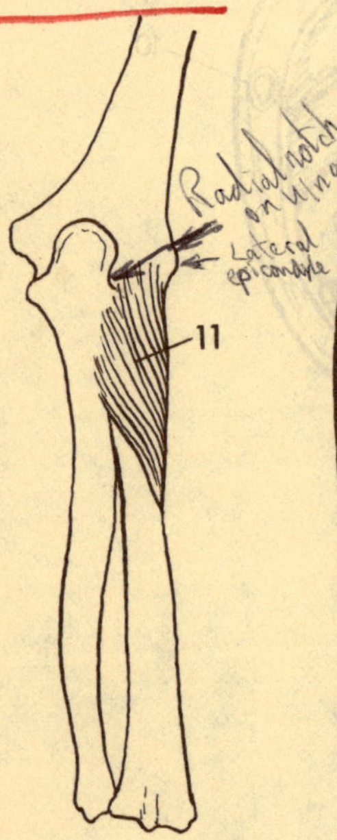

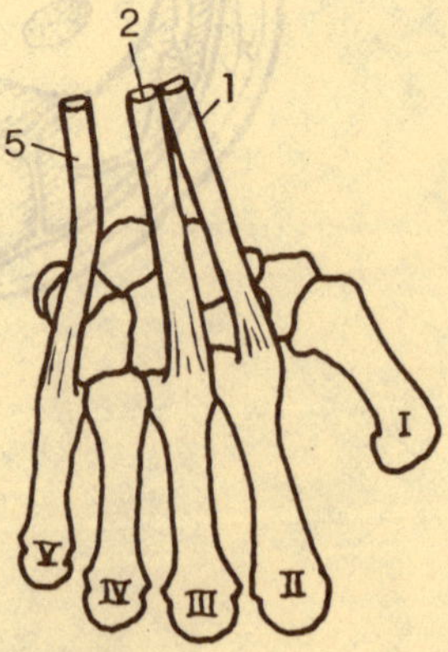

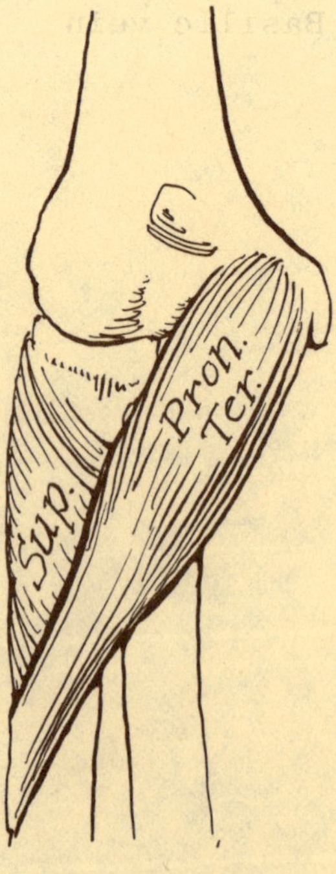

FOREARM

POSTERIOR COMPARTMENT

	ORIGIN	INSERTION	ACTION	INNERVATION
BRACHIORADIALIS	LSR	RADIUS	FLEX FOREARM	RADIAL
E CARPI RADIALIS LONGUS	LSR	METACARPAL 2	EXTEND, ABDUCT HAND	RADIAL
E CARPI RADIALIS BREVIS	LE	METACARPAL 3	EXTEND, ABDUCT HAND	DEEP RADIAL
E DIGITORUM	LE	E E	EXTEND PROXIMAL PHALANX	DEEP RADIAL
E DIGITI MINIMI	LE	E E	EXTEND PROXIMAL PHALANX	DEEP RADIAL
E CARPI ULNARIS	LE, ULNA	METACARPAL 5	EXTEND, ADDUCT HAND	DEEP RADIAL
E INDICIS PROPRIUS	ULNA, IM	E E	EXTEND PROXIMAL PHALANX	POSTERIOR INTER-OSSEOUS
E POLLICIS BREVIS	RADIUS, IM	PROX PHALANX THUMB	EXTEND PROXIMAL PHALANX	POSTERIOR INTER-OSSEOUS
E POLLICIS LONGUS	ULNA, IM	DISTAL PHALANX THUMB	EXTEND DISTAL PHALANX	POSTERIOR INTER-OSSEOUS
ABDUCTOR POLLICIS LONGUS	RADIUS, ULNA	METACARPAL 1, TRAPEZIUM	ABDUCT METACARPAL 1	POSTERIOR INTER-OSSEOUS
SUPINATOR	LE, ULNA, ANN-ULAR LIGAMENT	RADIUS	SUPINATION	DEEP RADIAL
ANCONEUS	LE	OLECRANON	EXTENSION FORE-ARM	RADIAL

SUMMARY

CONTAINS EXTENSORS OR SUPINATORS
INNERVATED BY RADIAL NERVE OR ITS BRANCHES
SUPPLIED BY BRANCHES OF RADIAL ARTERY AND POSTERIOR INTEROSSEOUS ARTERY

LSR = LATERAL SUPRACONDYLAR RIDGE   IM = INTEROSSEOUS MEMBRANE

E = EXTENSOR   EE = EXTENSOR EXPANSION

LE = LATERAL EPICONDYLE

EXTENSOR EXPANSION:

AN EXTENSOR EXPANSION OR APONEUROSIS is formed by the aponeurosis of the extensor tendons as they reach the phalanges. One central and two lateral bands serve as common insertion for the extensors, interossei and lumbrical muscles.

STRUCTURES:

12) Extensor tendon
13) Expansion
14) Central band
15) Lateral band
16) MP joint
17) PIP joint
18) DIP joint
19) Flexor profundus tendon
20) Flexor superficialis tendon--note how it splits and allows the deep flexor to access the distal phalanx
21) Vincula longus
22) Vincula brevis--the vinculae transmit neurovascular elements

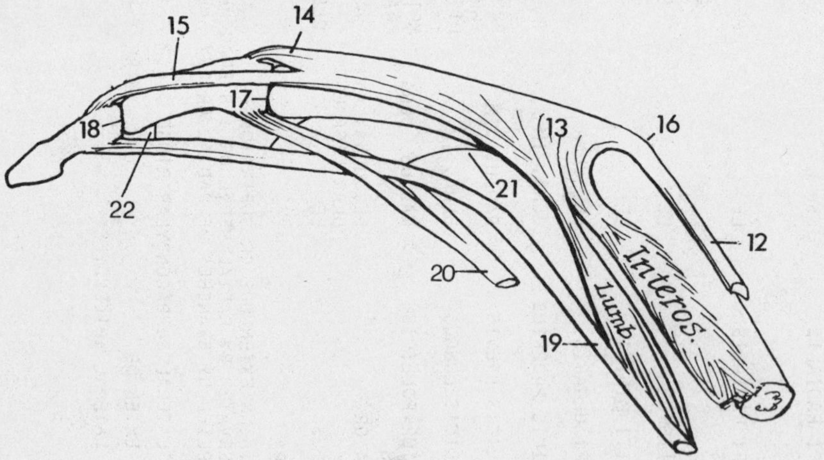

ANTERIOR COMPARTMENT:

SUPERFICIAL:

1) Pronator teres
2) Flexor carpi radialis
3) Palmaris longus
4) Flexor carpi ulnaris--has two heads

INTERMEDIATE:

5) Flexor superficialis or sublimis

DEEP:

6) Flexor profundus--usually has several heads for independent flexion
7) Flexor pollicis longus
8) Pronator quadratus
9) Supinator--from posterior compartment, insertion shown

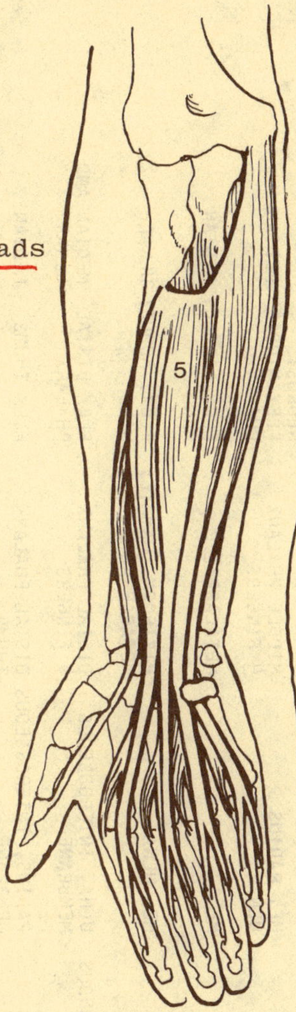

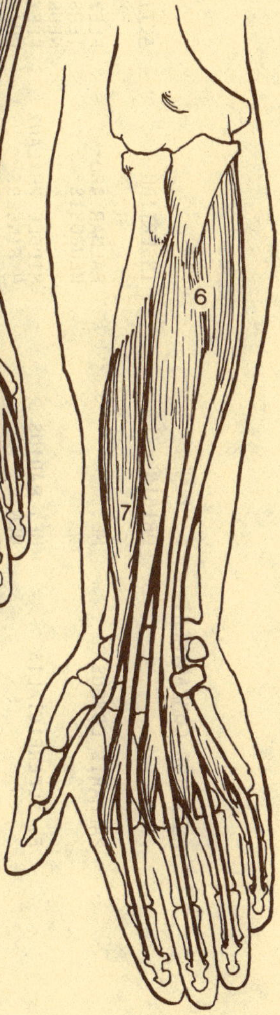

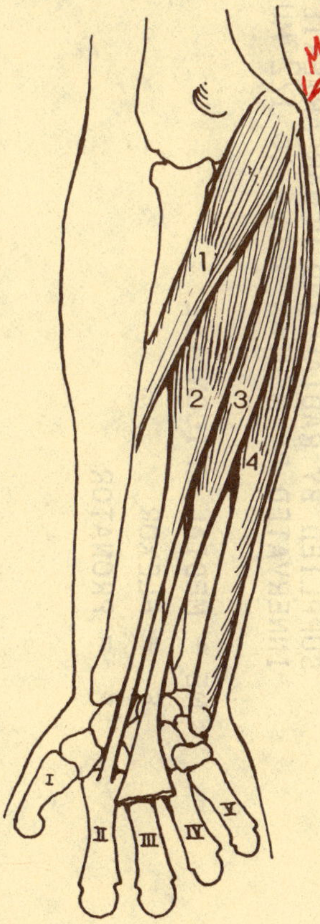

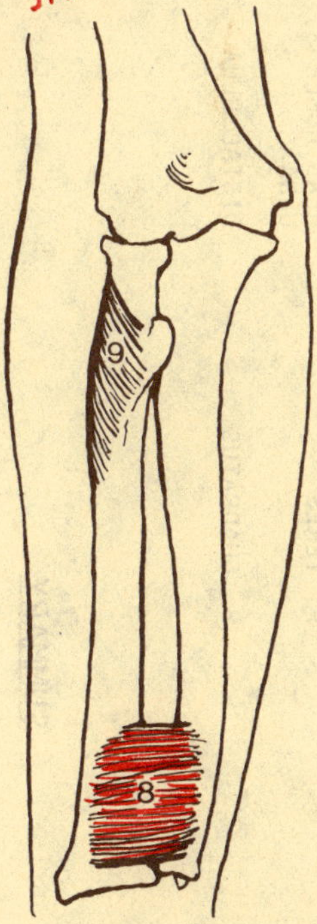

FOREARM

ANTERIOR COMPARTMENT

	ORIGIN	INSERTION	ACTION	INNERVATION
PALMARIS LONGUS	ME	PALMAR APO-NEUROSIS	FLEX HAND, TENSE APO-NEUROSIS	MEDIAN
F. SUPERFICIALIS	ME, RADIUS ULNA	MIDDLE PHALANX 4 FINGERS	FLEX MIDDLE PHALANX	MEDIAN
F. CARPI RADIALIS	ME	METACARPAL 2	FLEX, ABDUCT HAND	MEDIAN
F. CARPI ULNARIS	ME, ULNA	PISIFORM, HAMATE, METACARPAL 5	FLEX, ADDUCT HAND	ULNAR
F. DIGITORUM PROFUNDUS	ULNA, INTEROSSEOUS MEMBRANE	DISTAL PHALANX 4 FINGERS	FLEX DISTAL PHALANX	MEDIAN AND ULNAR
F. POLLICIS LONGUS	RADIUS, INTEROSSEOUS MEMBRANE	DISTAL PHALANX THUMB	FLEX THUMB	MEDIAN
P. TERES	HEAD HUMERUS, ME, ULNA	RADIUS	PRONATION	MEDIAN
P. QUADRATUS	DISTAL ULNA	DISTAL RADIUS	PRONATION	MEDIAN

SUMMARY

CONTAINS FLEXORS OR PRONATORS
SUPPLIED BY RADIAL AND ULNAR ARTERIES AND THEIR BRANCHES
INNERVATED BY MEDIAN NERVE (6 1/2 MUSCLES) AND ULNAR NERVE (1 1/2 MUSCLES)

ME = MEDIAL EPICONDYLE
F = FLEXOR
P = PRONATOR

WRIST:

DENSE BANDS OF FASCIA, called the EXTENSOR AND FLEXOR RETINACULUM, surround the carpal bones and prevent 'bowstringing' of the tendons during flexion and extension. On the anterior surface, the flexor retinaculum forms a roof over the concavity of carpal bones, creating the CARPAL TUNNEL.

STRUCTURES:

1) Pisiform
2) Hamate
3) Capitate
4) Trapezoid
5) Trapezium
6) Palmaris longus
7) Ulnar a. and n.
8) Flexor carpi radialis
9) Median n.
10) Flexor pollicis longus
11) Flexor superficialis
12) Flexor profundus
13) Extensor carpi ulnaris
14) Extensor digiti minimi
15) Extensor digitorum
16) Extensor indicis
17) Extensor carpi radialis brevis
18) Extensor carpi radialis longus
19) Extensor pollicis longus
20) Radial artery
21) Extensor pollicis brevis
22) Abductor pollicis longus

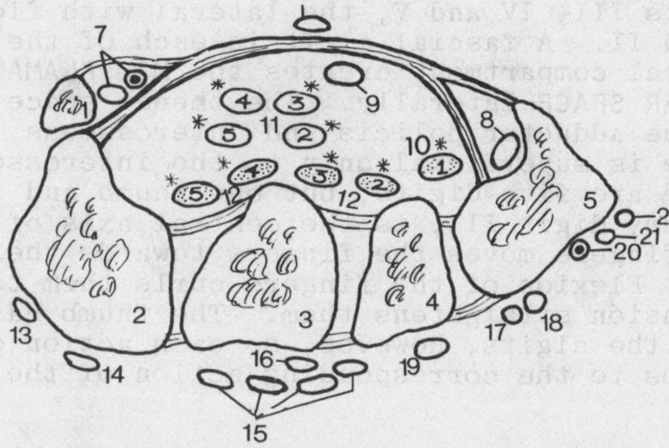

(*) Number within tendon indicates digit number

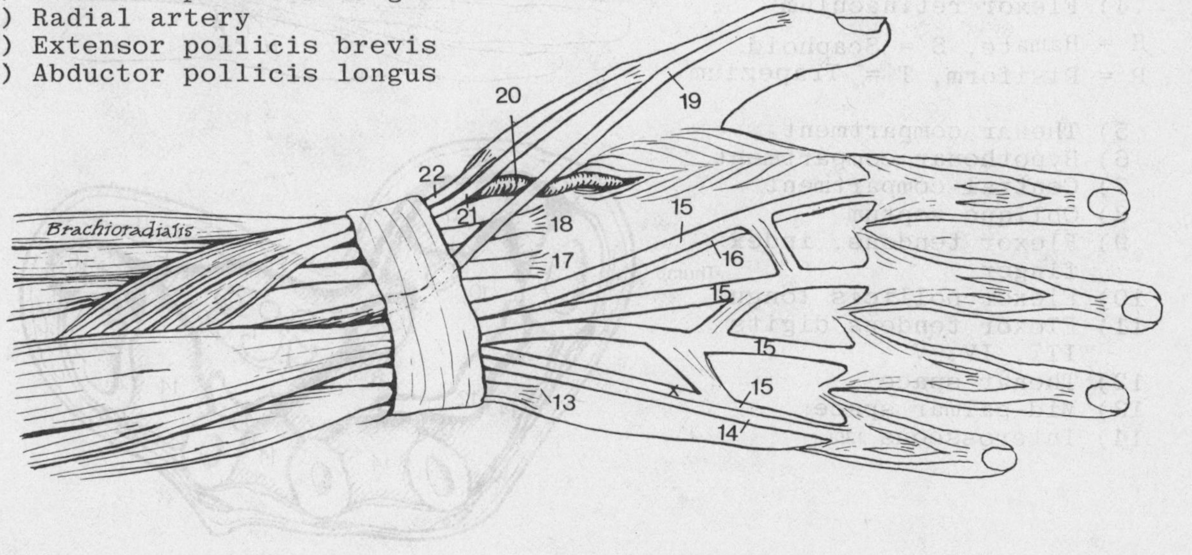

HAND:

THE PALMAR APONEUROSIS is the dense fascia of the palm and the insertion of the palmaris longus. Extensions of the fascia divide the palm into different compartments. The THENAR compartment con-contains muscles of the thumb, the HYPOTHENAR compartment contains muscles of the little finger, and the CENTRAL compartment contains tendons of the long flexors. An OBLIQUE SEPTUM divides the central compartment into two areas, the medial with flexor tendons for digits III, IV and V, the lateral with flexor tendons for digits I and II. A fascial cleft in each of the two divisions of the central compartment creates the MID-PLAMAR SPACE medially and the THENAR SPACE laterally. The thenar space intervenes superficial to the adductor policis and interosseous, while the mid-palmar space is superficial only to the interosseous muscles.
There are five digits, but one thumb and four fingers. The second finger, digit III, is the central axis of the hand. Adduction of the fingers moves the fingers towards the middle digit, abduction away. Flexion of the fingers curls them towards the palm, while extension straightens them. The thumb has rotated 90° out of phase with the digits, however, so each action of the thumb is at right angles to the corresponding action of the fingers.

STRUCTURES:

1) Palmar aponeurosis
2) Superficial transverse metacarpal ligaments
3) Palmaris longus
4) Flexor retinaculum

H = Hamate, S = Scaphoid
P = Pisiform, T = Trapezium

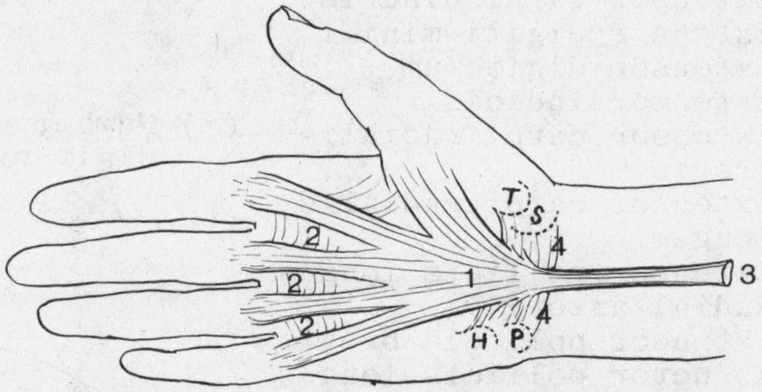

5) Thenar compartment
6) Hypothenar compartment
7) Central compartment
8) Oblique septum
9) Flexor tendons, index finger
10) Flexor pollicis longus
11) Flexor tendons digits III, IV, V
12) Thenar space
13) Mid-palmar space
14) Interosseous mm.

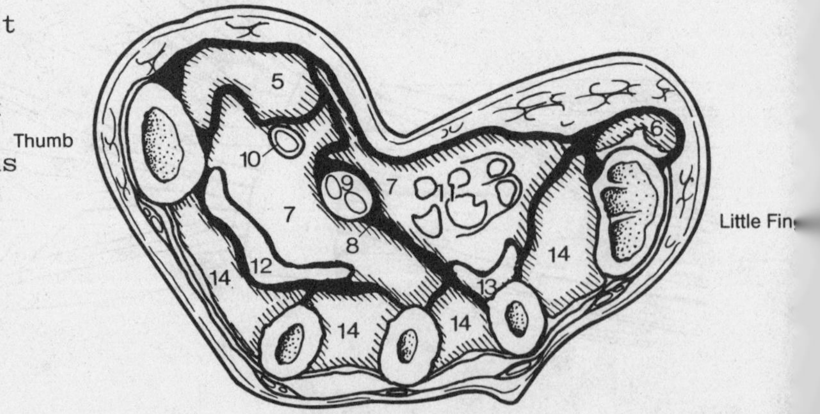

THENAR EMINENCE:

1) Abductor pollicis brevis
2) Flexor pollicis brevis
3) Opponens pollicis
4) Adductor pollicis actually deep to the thenar eminence

HYPOTHENAR EMINENCE:

5) Abductor digit V
6) Flexor digit V
7) Opponens digit V

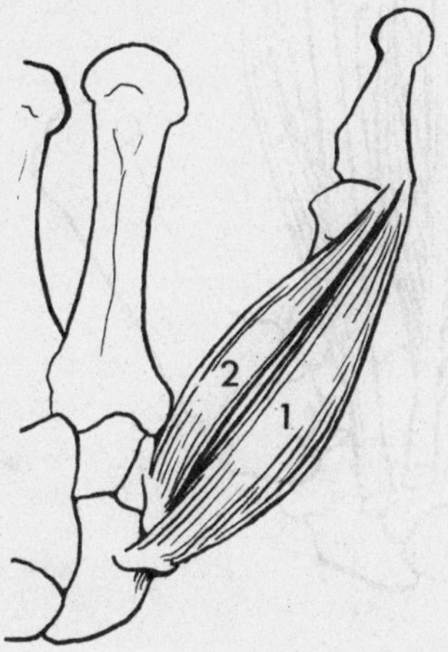

SUPERFICIAL THENAR MUSCLES*

REMEMBER:
ABDUCTORS-OUTSIDE
FLEXORS-INSIDE
OPPONENS-DEEP

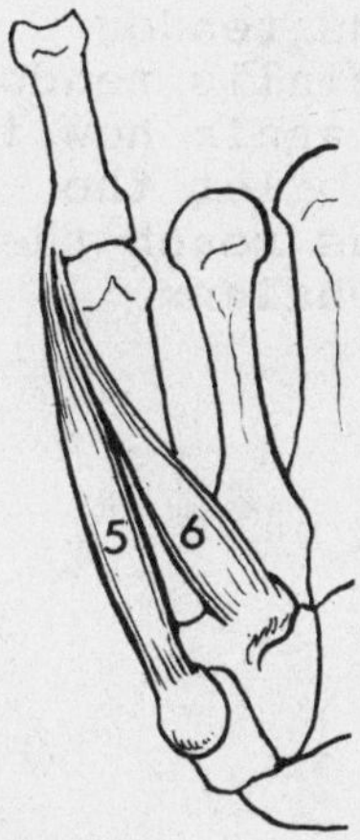

SUPERFICIAL HYPOTHENAR* MUSCLES

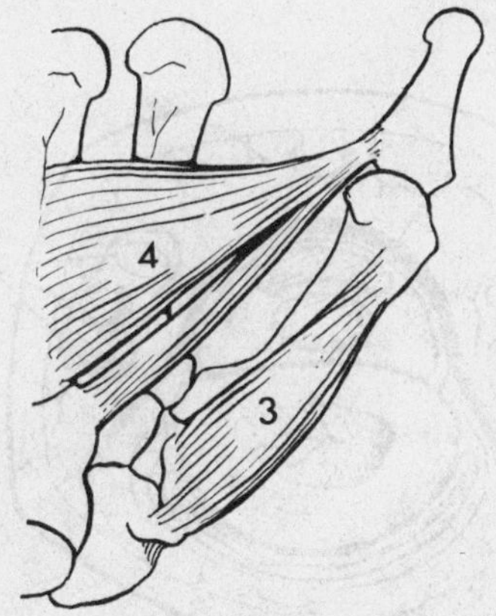

DEEP THENAR MUSCLES*

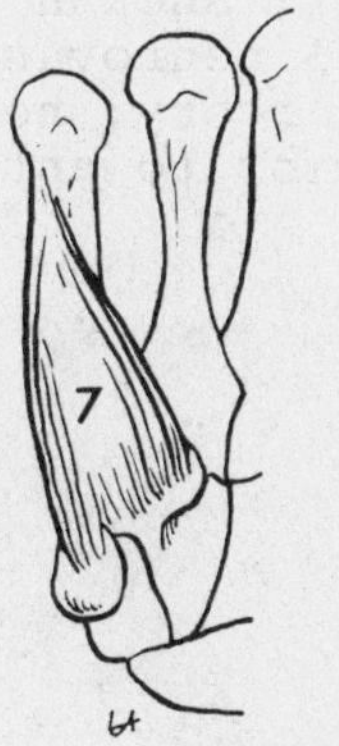

DEEP HYPOTHENAR MUSCLES*

*Only bony origins are indicated

CENTRAL COMPARTMENT:

1) Lumbricals--arise from profundus tendon and insert on radial side of extensor expansion
2) Profundus tendon
3) Superficialis tendon--observe again how it splits to let the profundus reach the distal phalanx

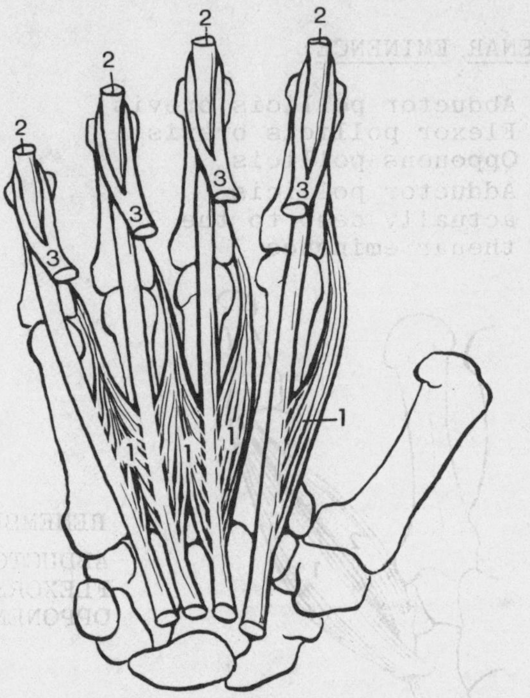

CROSS SECTION MIDDLE PHALANX:

4) Extensor expansion
5) Synovial sheath flexor tendons
6) Digital neurovascular bundle--vein, artery, nerve from posterior to anterior
7) Bone

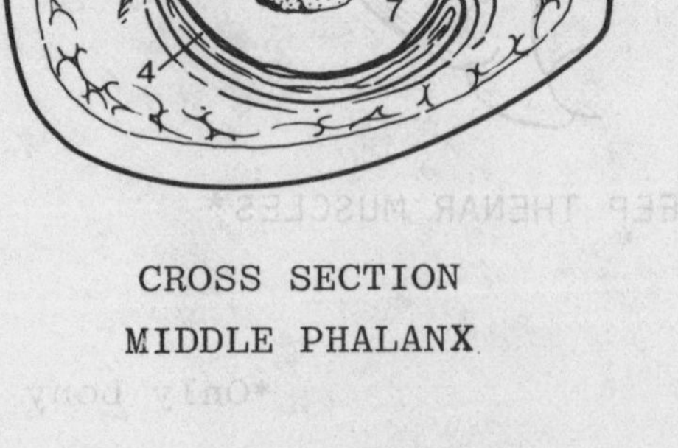

CROSS SECTION
MIDDLE PHALANX

DEEP COMPARTMENT:

THE DEEP COMPARTMENT contains 8 interosseous muscles and the adductor pollicis muscle. The interosseous muscles are divided into 4 palmar and 4 dorsal muscles.

PALMAR INTEROSSEOUS MUSCLES
ADDUCT--MNEMONIC PAD
Previously the first palmar interosseous has been referred to as the deep head of the flexor pollicis brevis

DORSAL INTEROSSEOUS MUSCLES
ABDUCT--MNEMONIC DAB

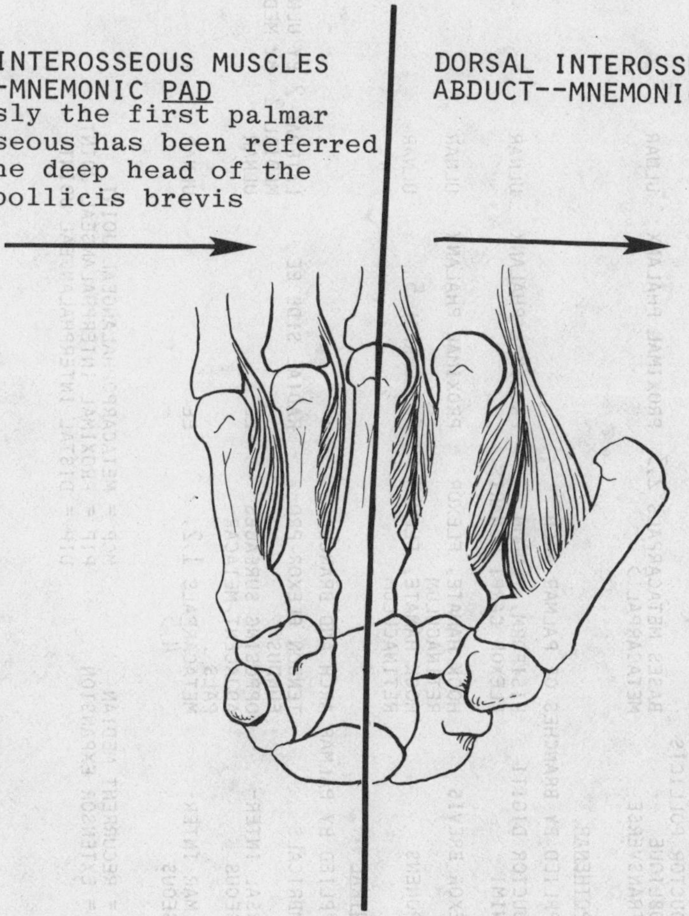

VENTRAL VIEW INTEROSSEI
COMPOSITE

HAND

	ORIGIN	INSERTION	INNERVATION	ACTION
THENAR				
SUPPLIED BY PRINCEPS AND DORSALIS POLLICIS ARTERIES; BRANCHES OF PALMAR ARCHES				
ABDUCTOR POLLICIS BREVIS	FLEXOR RETINACULUM, SCAPHOID, TRAPEZIUM	PROXIMAL PHALANX THUMB	RM	
FLEXOR POLLICIS BREVIS	FLEXOR RETINACULUM, TRAPEZIUM	PROXIMAL PHALANX THUMB	RM	
OPPONENS POLLICIS	FLEXOR RETINACULUM, TRAPEZIUM	METACARPAL 1	RM	
ADDUCTOR POLLICIS				
OBLIQUE	BASES METACARPALS 2,3	PROXIMAL PHALANX	ULNAR	
TRANSVERSE	METACARPAL 3			
HYPOTHENAR				
SUPPLIED BY BRANCHES OF PALMAR ARCH				
ABDUCTOR DIGITI MINIMI	PISIFORM, TENDON FLEXOR CARPI ULNARIS	PROXIMAL PHALANX	ULNAR	
FLEXOR BREVIS	HOOK HAMATE, FLEXOR RETINACULUM	PROXIMAL PHALANX	ULNAR	
OPPONENS	HOOK HAMATE, FLEXOR RETINACULUM	METACARPAL 5	ULNAR	
CENTRAL				
SUPPLIED BY PALMAR ARCH AND BRANCHES				
LUMBRICALS	TENDON FLEXOR PROFUNDUS	RADIAL SIDE EE	LATERAL 2 BY ULNAR MEDIAL 2 BY MEDIAN	FLEX MCP, EXTEND PIP, DIP
DORSAL INTEROSSEOUS	OPPOSING SURFACES ADJACENT METACARPALS	EE	ULNAR	ABDUCT
PALMAR INTEROSSEOUS	METACARPALS 1,2, 4,5	EE	ULNAR	ADDUCT

RM = RECURRENT MEDIAN
EE = EXTENSOR EXPANSION

MCP = METACARPOPHALANGEAL JOINT
PIP = PROXIMAL INTERPHALANGEAL JOINT
DIP = DISTAL INTERPHALANGEAL JOINT

ARTERIES OF HAND:

1) Ulnar
2) Deep branch of the ulnar
3) Superficial branch of the ulnar
4) Radial
5) Superficial branch radial
6) Superficial palmar arch
7) Deep palmar arch
8) Princeps pollicis
9) Radialis indicis
10) Common digital branch
11) Proper digital branches

A dorsal network of arteries is made up of branches from the radial and ulnar arteries. The network gives perforating branches to the deep arch and branches extend distally along the digits.

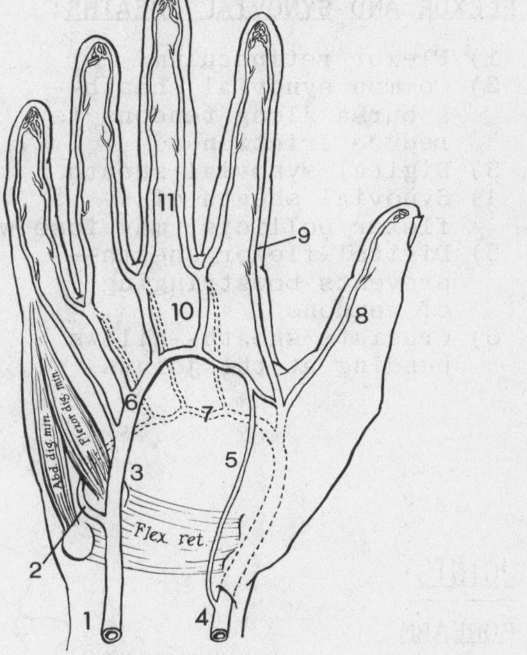

NERVES OF HAND:

1) Median
2) Recurrent--to thenar muscles
3) Digital branches
4) Ulnar
5) Superficial branch
6) Digital branches
7) Deep branch

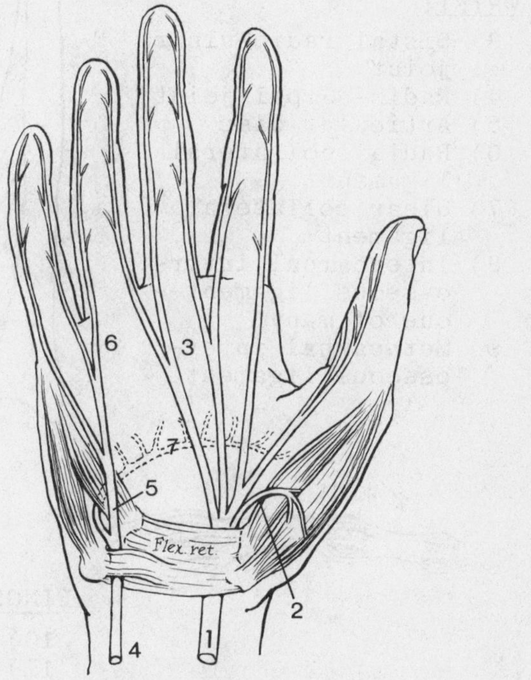

CUTANEOUS INNERVATION

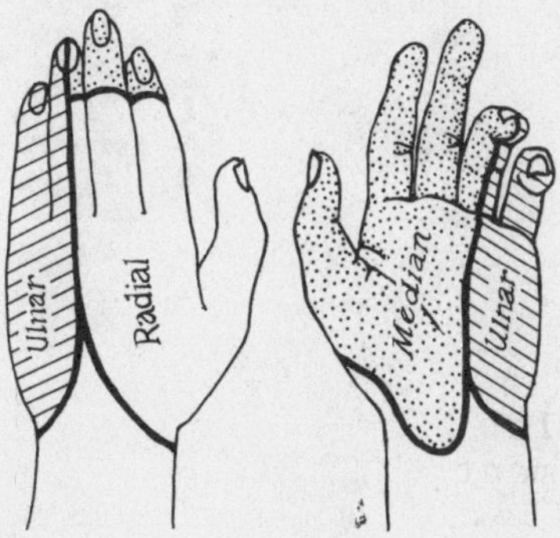

FLEXOR AND SYNOVIAL SHEATHS:

1) Flexor retinaculum
2) Common synovial sheath--a bursa along tendons to reduce friction
3) Digital synovial sheath
4) Synovial sheath of flexor pollicis, may fuse with (2)
5) Digital flexor sheath--prevents bowstringing of tendons
6) Cruciate sheath--allows bending at the joints

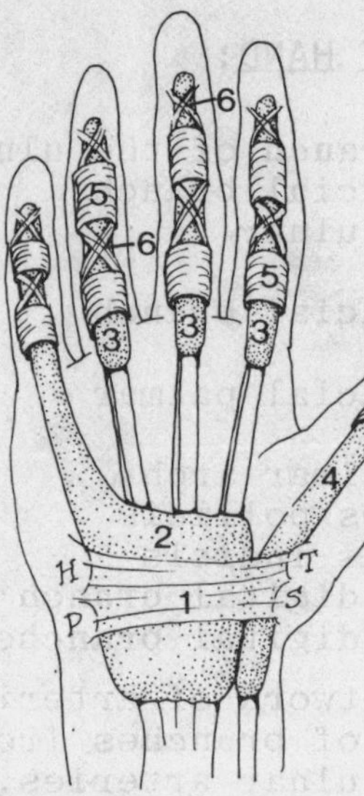

JOINTS:

FOREARM:

1) Interosseous ligament
2) Arterial foramen

WRIST:

3) Distal radio-ulnar joint
4) Radio-carpal joint
5) Articular disc
6) Radial collateral ligament
7) Ulnar collateral ligament
8) Intercarpal interosseous ligament--one of many
9) Metacarpal interosseous ligament

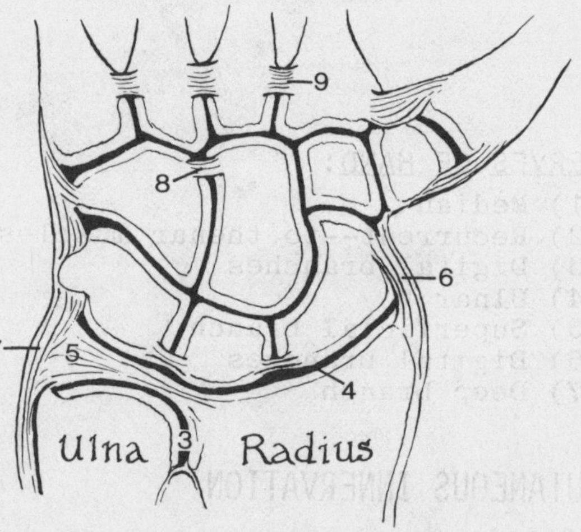

FINGER:

10) Articular capsule
11) Collateral ligament

COMMENTS

1. A whitlow or felon is an infection in the pad of the distal phalanx which causes the pad to swell. Due to the direction of the retinaculae cutis, the infection is often directed deeply to the bone.

2. Mallet finger results when the distal phalanx is hyperflexed rapidly and the most distal attachment of the extensor expansion is torn. The distal phalanx presents a flexed appearance. It is common in all such tendon tears to avulse--tear loose--bone where the tendon was attached.

3. Fibers of the interosseous membrane act as a shock absorber by transmission of energy from the wrist to the radius, from the radius to the ulna, and from the ulna to the humerus.

4. Carpal Tunnel Syndrome is a neurovascular compression of structures deep to the flexor retinaculum and is diagnosed by noting impairment of the median nerve and the muscles it innervates. The thumb is the most affected. Division of the retinaculum to release the structures is recommended.

5. Dupuytren's Contracture is a shortening of fibers of the palmar aponeurosis such that the fingers curl into flexion with the third and forth usually affected the first and worst. Absence of the palmaris longus is not rare.

5\. Due to the number of lacerations around the wrist, it is important to know the order of tendons and neurovascular structures. It would be embarrassing to reconnect the extensor digitorum tendon proximally to the indicis proprius distally.

6\. The scaphoid is the most commonly fractured bone of the wrist. The lunate is easily dislocated during hyperextension. Rarely, the lunate and triquetrum may fuse.

7\. The long flexor tendon sheaths are known as 'No man's land'--an area where surgery is difficult and blood supply, which travels along the vinculae, is easily jeopardized. Arrangement of the common sheath in the palm and the usual connection to the sheath around the 4th finger explains why an infection in the distal 4th finger--but no others--can spread to the forearm. The long flexor of the thumb has its own long sheath and infections may also spread from the thumb to the forearm.

REVIEW ARTERIES OF UPPER EXTREMITY-- COLLATERAL CIRCULATION:

1) Dorsal scapular
2) Suprascapular
3) Subscapular
4) Scapular anastomoses
5) Brachial
6) Humeral circumflex
7) Profunda
8) Posterior ulnar recurrent
9) Anterior ulnar recurrent
10) Radial recurrent
11) Interosseous recurrent
12) Ulnar
13) Radial
14) Common interosseous
15) Posterior interosseous
16) Anterior interosseous
17) Dorsal carpal
18) Volar carpal
19) Deep palmar arch
20) Superficial palmar arch
21) Pisiform bone
22) Epicondyles

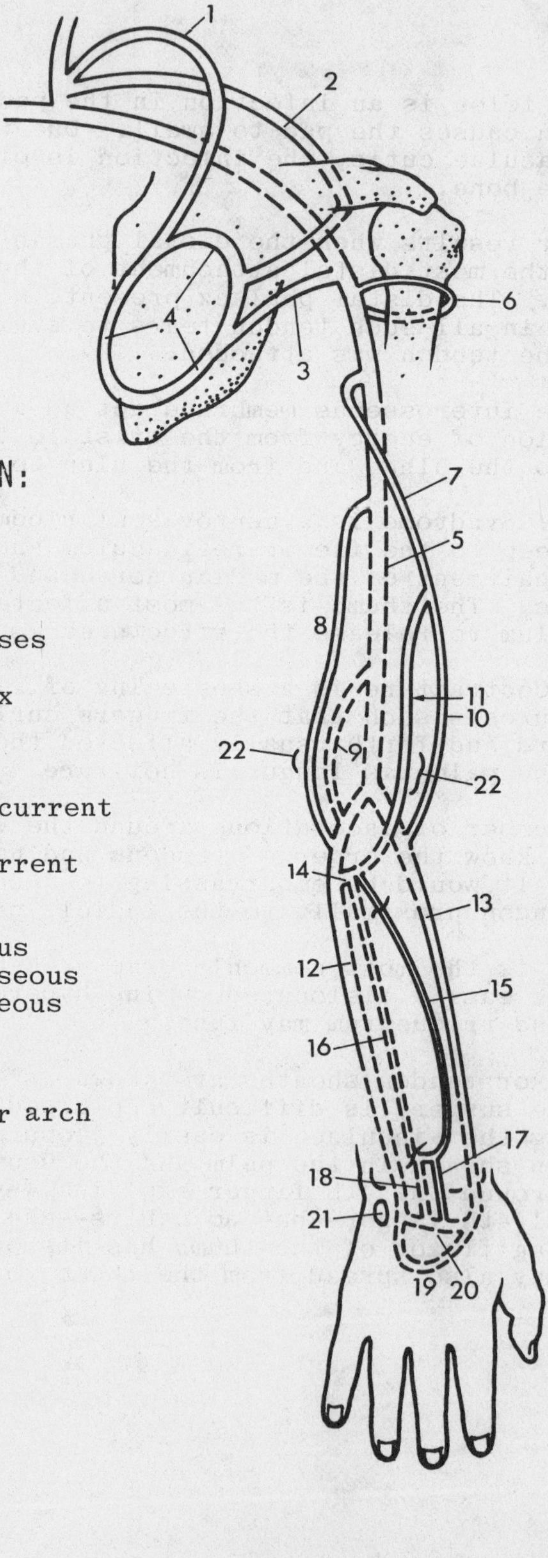

REVIEW NERVES OF UPPER EXTREMITY:

1) Axillary
2) Musculocutaneous
3) Radial
4) Deep radial
5) Superficial radial
6) Median
7) Anterior interosseous branch of median
8) Recurrent median
9) Ulnar
10) Deep branch ulnar

The ulnar is called the 'Musician's' nerve because of its control of fine movements of the digits.

The median is called the 'Carpenter's' nerve because of its control of coarser movements of the forearm.

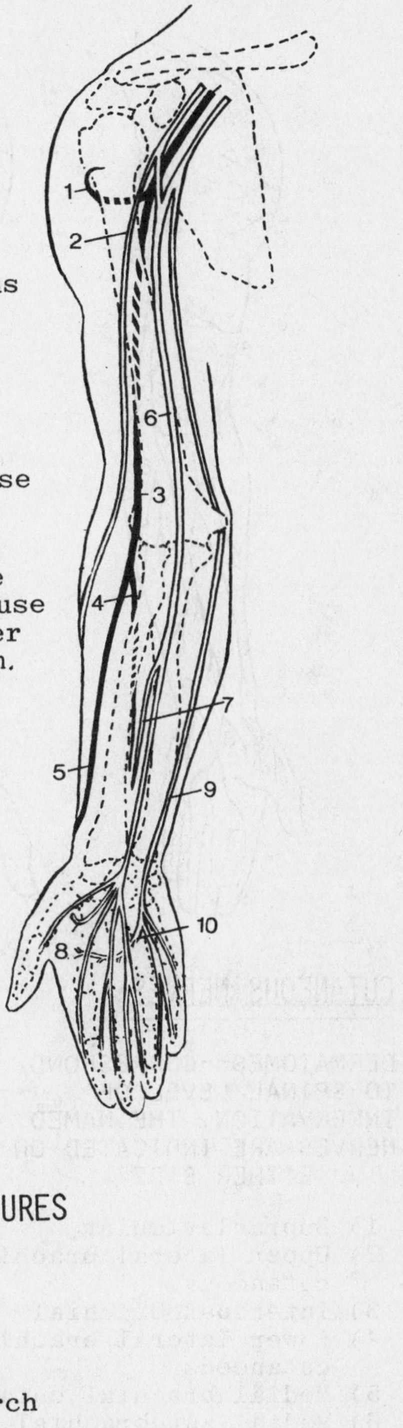

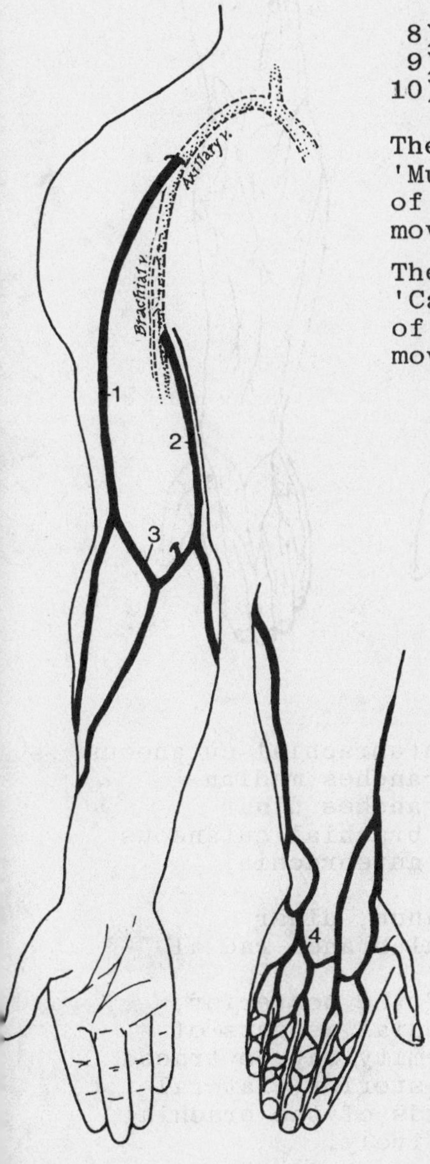

VI. CUTANEOUS STRUCTURES

SUPERFICIAL VEINS:

1) Cephalic
2) Basilic
3) Median cubital
4) Dorsal venous arch or network

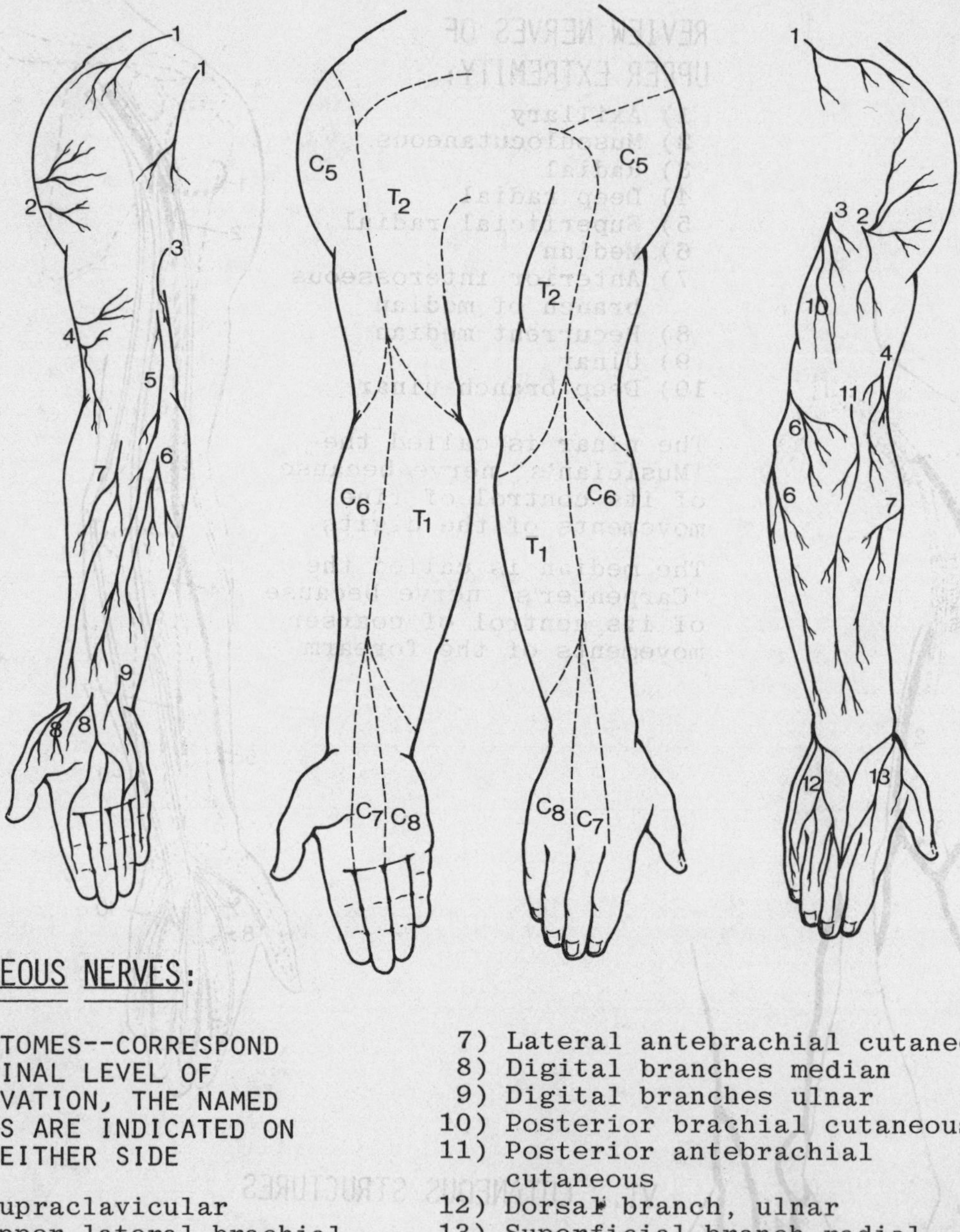

CUTANEOUS NERVES:

DERMATOMES--CORRESPOND TO SPINAL LEVEL OF INNERVATION, THE NAMED NERVES ARE INDICATED ON EITHER SIDE

1) Supraclavicular
2) Upper lateral brachial cutaneous
3) Intercostobrachial
4) Lower lateral brachial cutaneous
5) Medial brachial cutaneous
6) Medial antebrachial cutaneous
7) Lateral antebrachial cutaneous
8) Digital branches median
9) Digital branches ulnar
10) Posterior brachial cutaneous
11) Posterior antebrachial cutaneous
12) Dorsal branch, ulnar
13) Superficial branch radial

Innervation of the posterior, lateral and medial aspects of the upper extremity may be traced back to the posterior, lateral and medial cords of the brachial plexus respectively.

VII. LESIONS OF THE BRACHIAL PLEXUS

The upper roots of the brachial plexus supply the more proximal muscles; the lower roots supply the more distal. The segmental innervation of the skin does not correlate with the supply of the muscles beneath.

LESION OF THE UPPER ROOTS (C 5&6)

Erb-Duchenne Palsy--'Waiter's Tip'--most common type of nerve injury.
Nerves affected--Suprascapular, musculocutaneous, axillary
Muscles paralyzed--All of rotator cuff except subscapularis, biceps, brachialis, coracobrachialis, deltoid
Effect--arm hangs limply, medially rotated by unopposed subscapularis, pronated due to loss of biceps, loss of abduction

LESION OF THE LOWER ROOTS (C 8 & T 1)

Klumpke's Palsy

Nerves affected--Ulnar and median
Muscles paralyzed--all intrinsic muscles of hand
Effect--Clawed hand, due to hyperextension of MP joints and flexion of digits by forearm flexors unopposed by interossei

OTHER SYNDROMES:

Wrist drop--injuries to radial nerve, extension of hand and MP joints is impossible. PIP and DIP extension is normal.
Ape Hand--Median nerve injury. Since the extensors of the thumb are unopposed, the thumb rotates to lie in the same plane as the fingers.
Ulnar Claw Hand--Ulnar nerve injury at the wrist. Fingers are flexed at DIP and PIP joints, extended at MP due to unopposed actions of extensors. Most pronounced at the little finger, gradually less towards the index finger.
True Claw Hand--Median and ulnar nerve injury at the wrist. All digits are flexed at IP joints, extended at MP joint. In addition, the thumb is rotated into the same plane as the fingers.

STUDY QUESTIONS: UPPER EXTREMITY

1. Nerve supply to the pectoralis major is/are the _____ nerve(s).	1. medial and lateral pectoral
2. The pectoralis minor inserts on the _____ and is innervated by the _____ nerve(s).	2. coracoid process; medial pectoral
3. The deltoid muscle, supplied by the _____ nerve, may abduct the arm approximately from _____ to _____ degrees.	3. axillary; 15-20°; 90° (initiated by supraspinatus muscle)
4. Which vascular structure travels between the deltoid and pectoralis muscles?	4. cephalic vein (It pierces the clavipectoral fascia about 1 inch below the clavicle.)
5. The latissimus dorsi is supplied by the _____ nerve, and its main action is _____ of the upper extremity.	5. thoracodorsal; adduction and medial rotation
6. Teres major inserts on the _____ of the humerus.	6. medial lip of the bicipital groove
7. Which muscle is a powerful protractor of UE?	7. serratus anterior muscle

Matching 8-11

Match the following structures with their respective border of the axilla.

A. anterior
B. medial
C. lateral
D. posterior
E. apex
F. base

8. pectoralis muscles _____	8. A
9. humerus _____	9. C
10. serratus anterior _____	10. B
11. midpoint of clavicle and first rib	11. E

Matching 12-18

Match the following structures or statements to the appropriate vascular structure.

A. axillary artery
B. subclavian artery
C. supreme thoracic artery
D. thoracodorsal artery
E. lateral thoracic artery
F. subscapular artery
G. anterior humoral circumflex
H. post. humoral circumflex
I. profundus brachii artery

12. travels with axillary nerve	12. H
13. origin of internal thoracic artery	13. B
14. supplies extensor compartment of arm	14. I
15. thyrocervical trunk	15. B
16. supplies latissimus dorsi muscle	16. D
17. breast arterial supply	17. E
18. branches supply pectoralis minor	18. E

19. The elbow is an example of a _____ joint.

19. hinge

20. The most common direction for a dislocation of the humerus is _____.

20. anterior and inferior

21. Which muscles comprise the rotator cuff?

21. the SITS muscles: supraspinatus; infraspinatus; teres minor; subscapularis

22. Which muscles comprise the flexor compartment of UE?

22. coracobrachialis; biceps brachii; brachioradialis

23. The triceps insert into the _____.

23. olecranon of ulna

24. What nerve supplies the lateral head of triceps?

24. radial nerve

25. The brachioradialis muscle is a/an _____ of the wrist, even though it lies in the _____ compartment.

25. flexor; posterior or extensor

26. Which nerve supplies the supinator muscle?

26. radial nerve

27. The flexor carpi radialis inserts _____.

27. base of II & III metacarpal

28. Describe the nerve supply to the flexor digitorium superficialis.

28. The ulnar aspect is supplied by the ulnar nerve, while the radial aspect is supplied by the median nerve.

29. What is the nerve just beneath the flexor retinaculum?

29. median nerve

30. Which nerve pierces the pronator teres in the cubital fossa?

30. median nerve

31. Which structure is involved in Dupuytren's contracture?

31. palmar aponeurosis

32. Is the ulnar nerve superficial or deep to the flexor retinaculum?

32. superficial

33. Name the two general actions of the lumbricals.

33. flex M/P joint; extend I/P joint

34. List the 3 muscles of the thenar eminence and their innervation.

34. abductor pollicis brevis, flexor pollicis brevis, and opponens pollicis; median nerve

35. What is the innervation of adductor pollicis?

35. ulnar nerve

36. Palmar interossei insert on which aspect of each metacarpal?

36. medial, i.e., on the side closest to the middle finger (midline of hand)

37. Describe the attachments of the levator scapula.

37. Origin C 1-4 transverse process. Insert: medial border of scapula above spine

38. Which muscle is an antagonist of serratus anterior?

38. rhomboid major

39. Which muscle is the chief lateral rotator of the arm?

39. infraspinatus

40. Which are the MOST COMMON lymph nodes involved in metastatic cancer of the breast?

40. axillary

41. Which blood vessel supplies the clavicle?

41. the clavicular branch of the thoracoacromial artery

42. A lesion of the medial pectoral nerve would affect which muscle(s)?

42. pectoralis major and pectoralis minor

43. Which is the only articulation of the upper extremity with the axial skeleton?

43. the sternoclavicular joint

44. Winging of the scapula is seen with lesion of which nerve?

44. long thoracic nerve

45. Ape hand (thenar atrophy) is commonly due to lesion of the _____ nerve at the _____.

45. median; wrist

46. Erb's palsy (waiter's tip) is due to lesion of the _____ trunk.

46. upper

47. Wrist drop is due to lesion of which nerve?

47. radial nerve

48. Motor innervation to flexor compartment of the arm is via _____

48. musculocutaneous nerve

49. Sensation to anterior-lateral aspect of forearm is via which nerve?

49. musculocutaneous nerve

50. Innervation of hypothenar muscles is via which nerve(s)?

50. ulnar and median nerves

51. Most of the deep palmar arch is from which artery?

51. radial artery

52. The thumb and index finger have cutaneous innervation from which ventral cervical ramus?

52. C-6

53. Suppination of the forearm is controlled by _____ nerve while pronation is controlled by _____ nerve.

53. radial; median

54. What muscles comprise the borders of the triangular space?

54. long head of triceps, teres major and teres minor

55. The axillary sheath is an extension of which neck fascial layer?

55. prevertebral fascia

56. Which muscle abducts the fingers?

56. dorsal interossei

THORAX

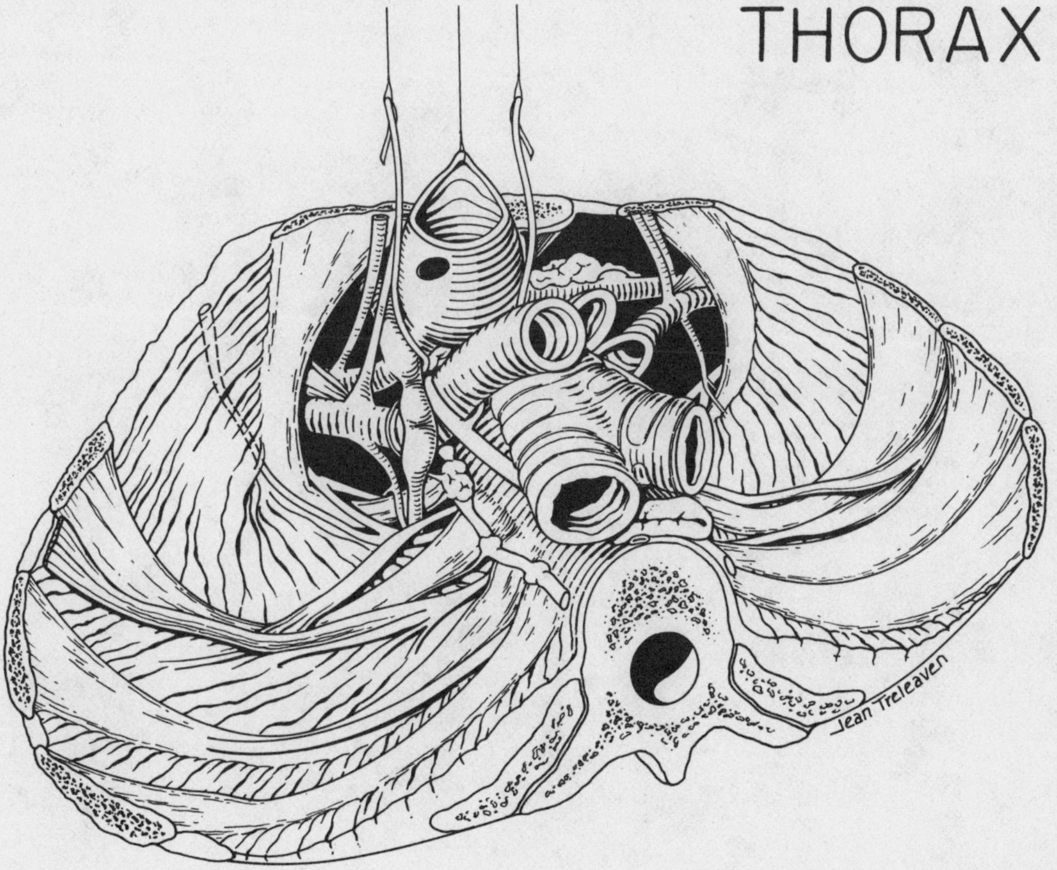

THORAX

I. SURFACE CORRELATIONS

By vertebral levels

T1---Highest point of apex of lungs
T2---Jugular notch
T3---Base of spine of scapula
T4---Sternal angle, aortic arch begins and ends, 2nd costal cartilage, trachea ends and bronchi begin, right and left pleurae meet retrosternally
T5-7-Thoracic duct crosses from inferior right to superior left, posterior to the esophagus
T8---IVC passes through the diaphragm
T9---Xiphosternal junction
T10--Esophageal hiatus of the diaphragm
T12--Aortic hiatus of the diaphragm

II. THORACIC WALL

STERNUM

1) Manubrium
2) Gladiolus or body
3) Xiphoid process
4) Angle of Louis or manubrio-sternal joint
5) Clavicle
6) Sterno-clavicular joint
7) Costal cartilage attachments
8) Jugular notch
9) Esophagus
10) Trachea
11) Aortic arch
12) Pulmonary artery
13) Common carotid artery
14) Subclavian artery

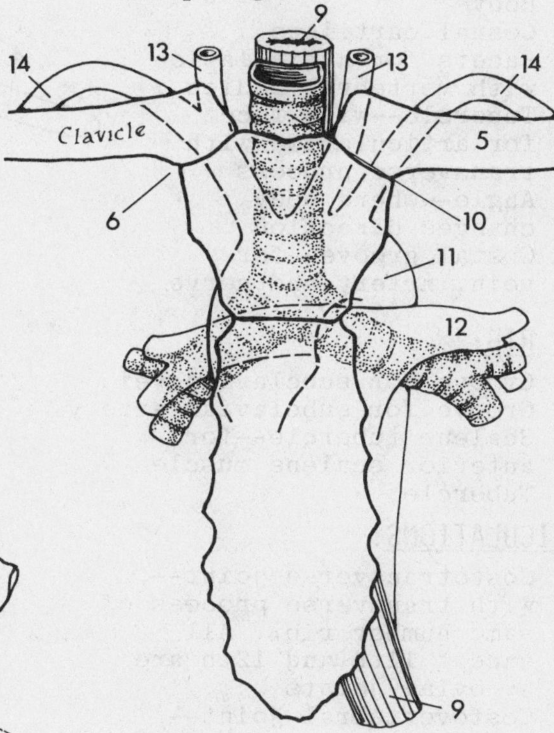

PROJECTION OF MEDIASTINAL STRUCTURES

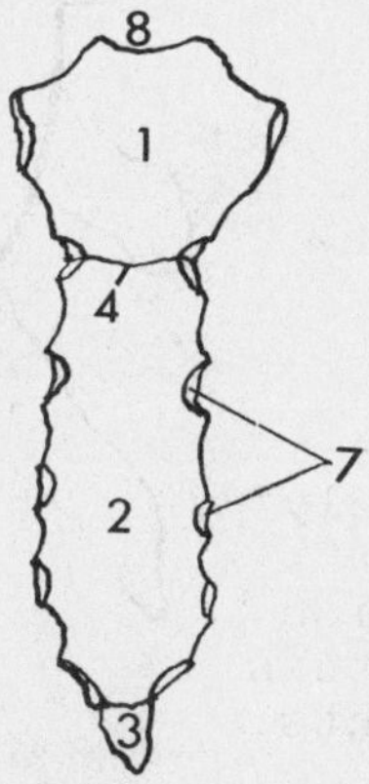

STERNUM
ANTERIOR VIEW

RIBS

There are 12 pairs of ribs:

- 1-7 are TRUE RIBS--the costal cartilages articulate directly with the sternum
- 8-10 are FALSE RIBS--the costal cartilages articulate with superior costal cartilages, form a costal arc or margin
- 11-12 are FLOATING RIBS--there are no articulations of the body of the rib

TYPICAL RIB:

Ribs 3-10

1) Head
2) Neck
3) Body
4) Costal cartilage
5) Facets for articulation with vertebral bodies
6) Tubercle--with facet for articulation with transverse process
7) Angle--where body changes direction
8) Costal groove--for vein, artery and nerve

1ST RIB:

9) Groove for subclavian vein
10) Groove for subclavian artery
11) Scalene tubercle--for anterior scalene muscle
12) Tubercle

ARTICULATIONS:

13) Costotransverse joint--with transverse process of same number rib. All except 11th and 12th are synovial joints
14) Costovertebral joint--with body of vertebra of same number and the body immediately above--via demifacets--except for ribs 1, 10, 11 and 12 which have only 1 articulation at this point. All are synovial joints.

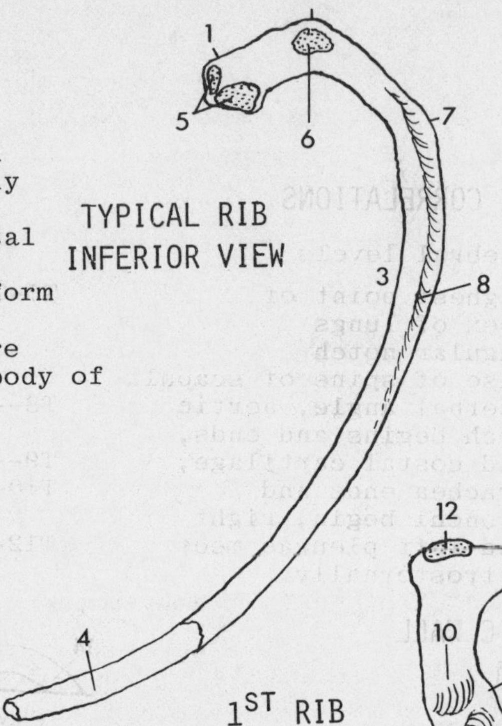

TYPICAL RIB
INFERIOR VIEW

1ST RIB
SUPERIOR VIEW

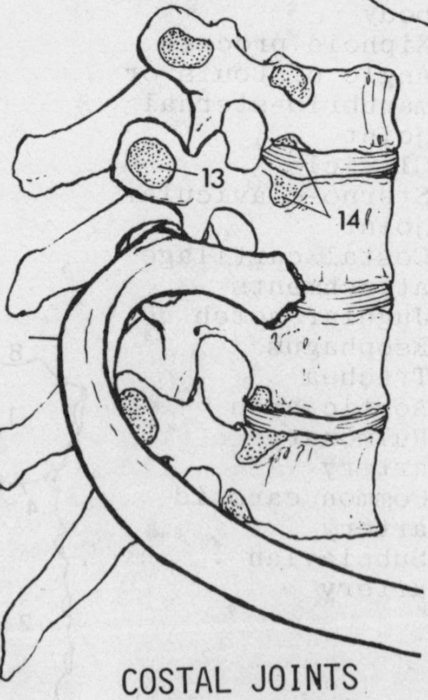

COSTAL JOINTS
LATERAL VIEW

COMMENTS

1. The sternum has a marrow cavity accessible via a sternal puncture for aspiration of red marrow--useful for diagnostics.

2. Sternal fractures from automobile accidents are common. The gladiolus may separate at the Angle of Louis and be driven posteriorly, damaging or rupturing the aorta or its branches.

3. The sternal angle marks the articulation of the 2nd rib. The articulation of the 1st rib is not readily palpable. Counting of the ribs for various purposes is a necessity and is easily done using the sternal angle as a starting point.

4. "Flail chest" results from a rib broken in two or more places; the corresponding section of chest wall moves outwards in expiration and inwards during inspiration due to pressure changes in the thoracic cavity--this is a paradoxical movement. Flail chest may also occur if the sternum is broken from the ribs on both sides.

5. A separated rib refers to a disjuncture of the body of a rib from its costal cartilage; costal cartilages in young people are less rigid than in the adult, providing a "shock absorber" so that crush and direct injuries to the ribs are less common in the young. Crush injuries usually fracture the rib just anterior to the angle of the rib, while direct injuries will fracture the rib at the point(s) of impact. In a slipping rib, the distal portion of a costal cartilage from a false rib separates from the costal margin and overrides the superior costal cartilages.

6. Rib fractures and resultant bone fragments may puncture the lungs and cause a pneumothorax--air in the pleural cavity--or a hemothorax--blood in the pleural cavity. Pain from a fractured rib may be relieved by anesthetizing the corresponding intercostal nerve.

7. Cervical ribs may be associated with cervical vertebrae, usually the 7th; they may compress the brachial plexus and subclavian artery causing a neurovascular syndrome--coldness, numbness, tingling--in the upper extremity. Lumbar ribs associated with lumbar vertebrae may be mistaken for fractured transverse processes or cause a mistake in ascertaining vertebral levels.

8. The middle ribs are most commonly fractured; they are longest and most securely attached anteriorly and posteriorly. The lateral diameter of the chest is greatest at rib 8.

SCHEMATIC CROSS-SECTION
THROUGH ICS

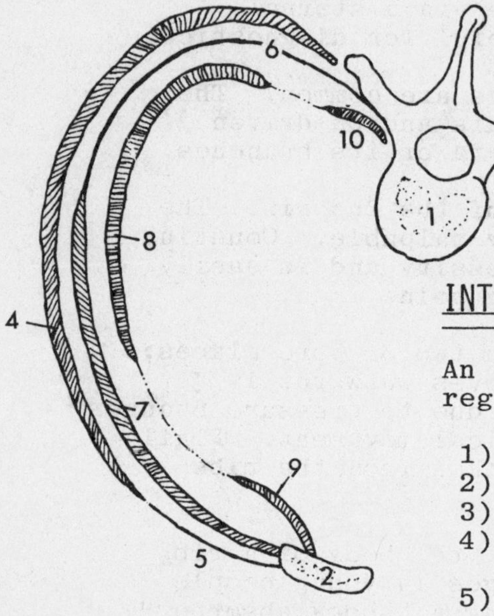

INTERCOSTAL SPACES (ICS)

An intercostal space is the region bounded by two ribs.

1) Vertebra
2) Sternum
3) Rib
4) External intercostal muscle
5) External intercostal membrane
6) Internal intercostal membrane
7) Internal intercostal muscle
8) Innermost intercostal muscle
9) Transversus thoracis muscle
10) Subcostalis muscle
11) Intercostal neurovascular bundle in costal groove
12) Costotransverse ligament
13) Anterior longitudnal ligament
14) Manubrium
15) Clavicle
16) Internal thoracic artery and vein

SAGITTAL SECTION
THROUGH ICS

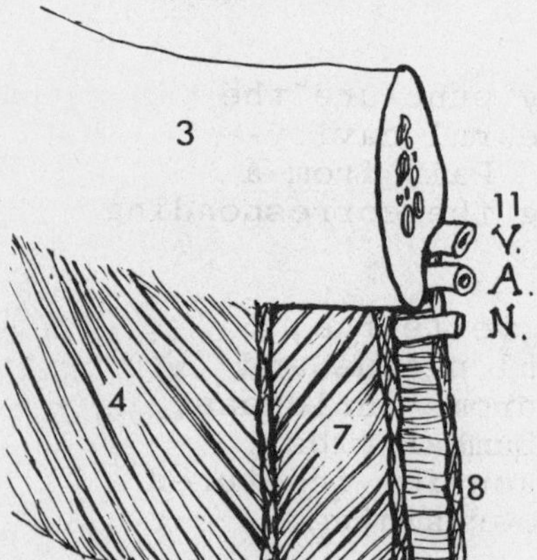

POSTERIOR SURFACE
OF THE STERNUM

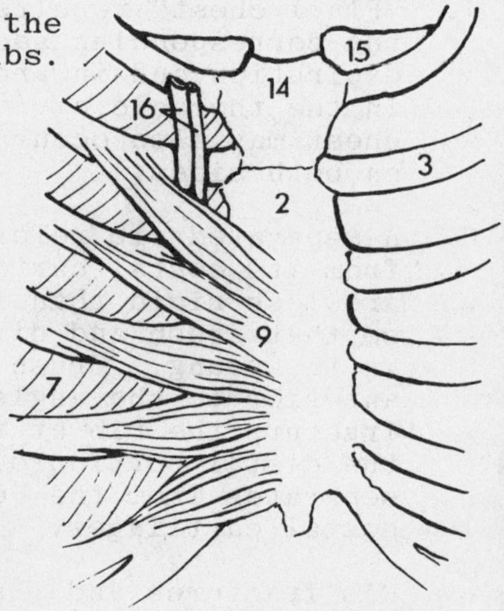

POSTERIOR
THORACIC WALL

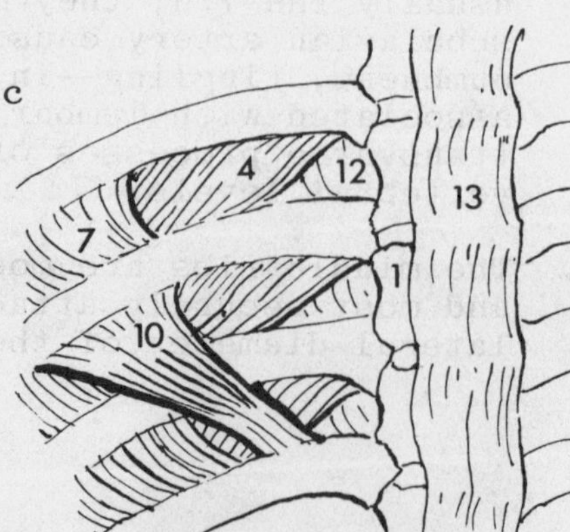

GENERAL DISCUSSION

1. INNERVATION is from the spinal nerves which form a pattern common for all spinal nerves. Innervation of the thoracic wall is segmental in nature, but to completely anesthetize any dermatome or segment, the nerve above and below must be deadened also. This is the phenomenon of overlap. The intercostal nerves from lower ICS continue diagonally inferiorly to innervate skin of the abdominal wall.

2. ARTERIAL BLOOD is derived from the aorta via intercostal arteries and from internal thoracic arteries through direct end-to-end anastomoses. However, the aorta does not extend superiorly to T4, even though the thorax does. Intercostal branches to upper ICS vary greatly. The supreme intercostal artery from the costocervical trunk and supreme thoracic artery from the axillary artery supply the upper intercostal spaces.

3. VENOUS DRAINAGE from the ICS generally finds its way into the azygous system and then to the SVC or may go into the brachiocephalic vessels. These vessels will be discussed with the posterior mediastinum.

COMMENTS

1. Thoracentesis or pleuracentesis is the introduction of a needle or similar instrument into the thorax. This should be performed along the superior border of the rib in order not to jeopardize the intercostal neurovascular bundle. When thoracentesis is performed below the 7th ICS care must be taken not to penetrate the diaphragm.

2. Coarctation or narrowing of the aorta just distal to the origin of the left subclavian artery is a common birth defect. A route for collateral circulation is through the internal thoracic artery to the intercostal arteries and then to the aorta. The intercostal arteries enlarge and will produce an erosion or 'notching' of the inferior border of the ribs, visible in X-rays.

3. Abdominal viscera including the kidneys, spleen, stomach and liver are protected by the ribs.

4. In shingles (Herpes zoster) a viral infection of the dorsal root ganglion is followed by an eruption of vesicles along the course of the nerves, mapping the dermatome.

III. THE THORACIC CAVITY

BOUNDARIES:

1) The thoracic inlet--a term referring to the superior aperture of the bony thorax. It is closed by soft tissues in life and is discussed below as the apex of the thorax
2) The thoracic outlet--a term referring to the inferior aperture of the bony thorax. It is closed by the diaphragm in life and is discussed below as the base of the thorax. The excursion of the diaphragm in inspiration (dashed line) and expiration (solid line) is indicated
3) Anterior--the sternum and ribs
4) Posterior--thoracic vertebrae and ribs
5) Lateral--ribs

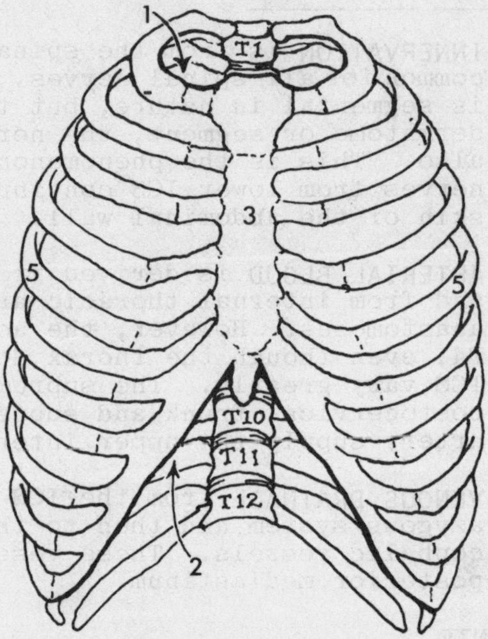

ANTERIOR VIEW
THORAX

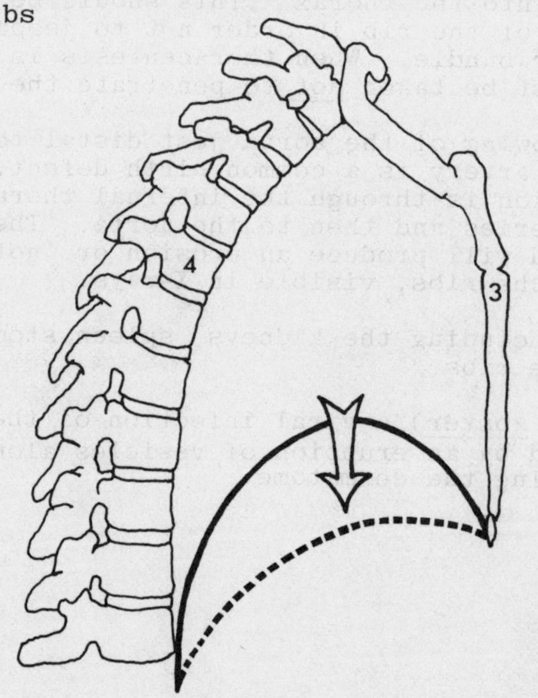

LATERAL VIEW
THORAX

THORACIC APEX:

1) Vagus
2) Recurrent laryngeal n.
3) Ansa subclavia
4) Stellate or inferior cervical ganglion
5) Internal thoracic vessels
6) Left common carotid a.
7) Left subclavian a.
8) Left brachiocephalic v.
9) Right brachiocephalic v.
10) Superior vena cava--SVC
11) Ostium of azygos v.
12) Sympathetic trunk
13) Brachiocephalic a.
14) Trachea--observe the bifurcation
15) Esophagus
16) Thoracic duct
17) Intercostal n.
18) Right subclavian a.
19) Phrenic n.
20) Scalenus anterior
21) Lower roots of the brachial plexus

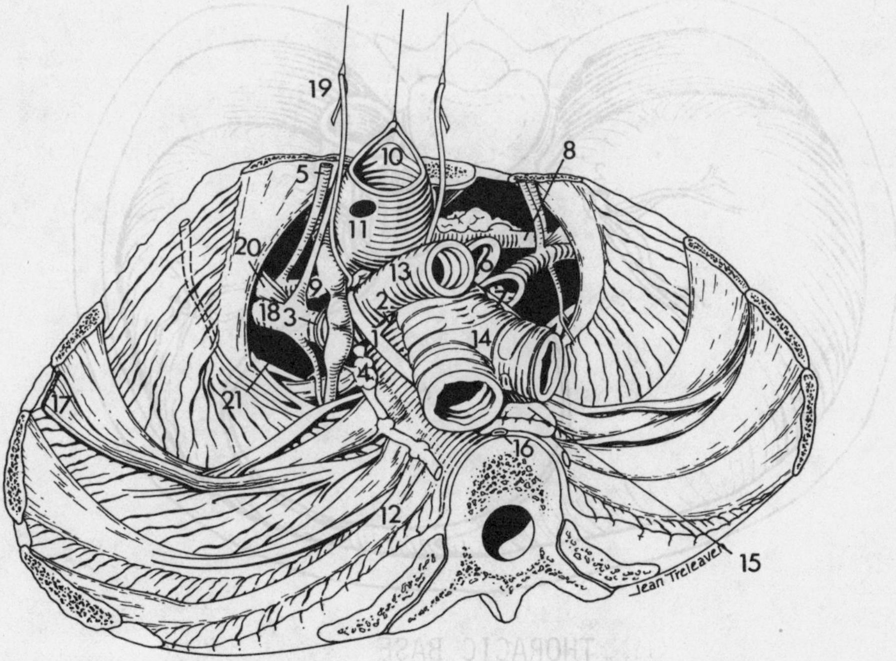

THORACIC APEX

THORACIC BASE:

1) Esophagus
2) Phrenic neurovascular bundle
3) Inferior vena cava--IVC
4) Aorta
5) Central tendon of the diaphragm
6) Pericardium

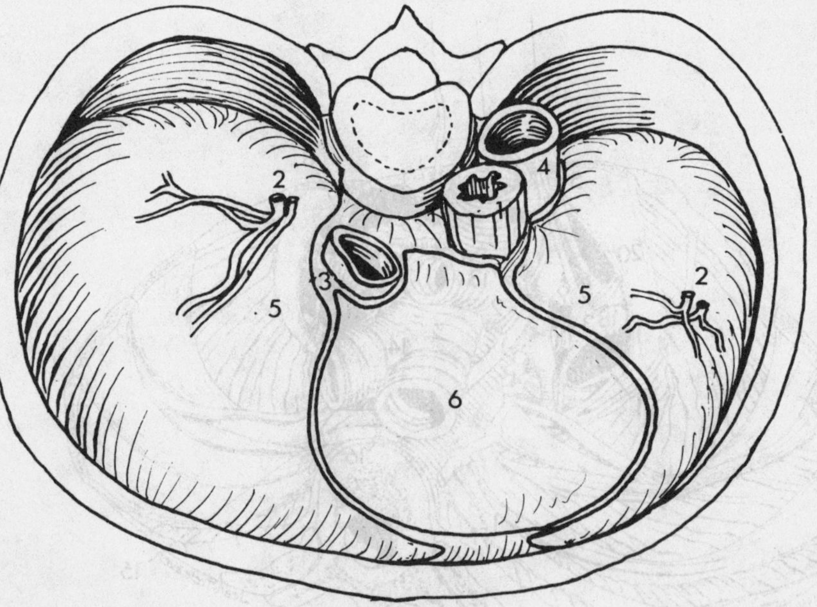

THORACIC BASE

CONTENTS OF THE THORACIC CAVITY

The thoracic cavity is divided into 2 pleural cavities and the mediastinum, a region between the pleural cavities. The pleural cavity is a space bounded by the pleural membranes. Although we commonly say the lungs are 'in' the pleural cavities, they actually are not.

THE PLEURAE

Each pleural sac is a complete envelope. The right lung is completely separated from the left lung. During development, the lung bud pushes into the cavity lining; this cavity lining will be the pleurae. There are thus two continuous pleural linings each subclassified by anatomical location. The visceral pleura is directly applied to the lungs, The parietal pleura is applied to the thoracic wall by areolar connective tissue termed the endothoracic fascia (over the diaphragm it is the phrenico-pleural fascia). In addition, the parietal pleura is subdivided according to its location. The costal pleura is adjacent to the ribs, diaphragmatic pleura covers the diaphragm, and mediastinal pleura lies against the mediastinum. A small area of thoracic cavity protrudes above the first rib and is lined by cervical pleura. The parietal and visceral pleurae are continuous with each other and are normally separated by a small, well lubricated pleural cavity. Pleural recesses are potential spaces created by a folding or reflection of the parietal pleura. The lungs may expand into the recesses during inspiration.

PARIETAL PLEURA:

1) Costal
2) Mediastinal
3) Cervical--forms the cupola
4) Diaphragmatic
6) Location of Costo-diaphragmatic recess
7) Location of Costo-mediastinal recess

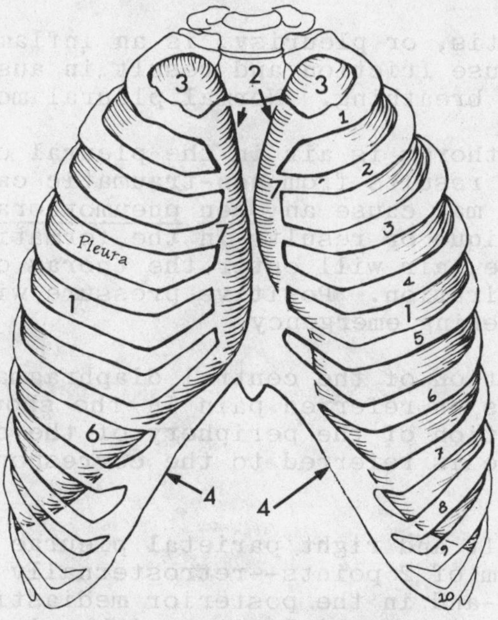

ANTERIOR PROJECTION
OF THE PARIETAL PLEURA

INNERVATION OF THE PLEURAE:
Parietal--sensitive to pain and temperature, innervated by the intercostal nerves and phrenic nerve
Visceral--not sensitive to pain and temperature, but some sensory fibers travel with autonomic nerves that innervate the lungs

VASCULATURE:
Parietal--from intercostal, internal thoracic, pericardiacophrenic and anterior mediastinal arteries and veins
Visceral--from bronchial vessels and pulmonary trunks

LYMPHATICS from the parietal pleura drain with those of the thoracic wall, while those of the visceral pleura drain with lymphatics from the lung

COMMENTS

1. The pleural cavity is normally a potential space, with little fluid. Pyothorax is pus in the thoracic cavity, a qualification of the general term empyema. A chyle thorax is lymphatic fluid in the pleural cavity.

2. Pleuritis, or pleurisy, is an inflammation of the pleurae which may cause friction and result in auscultation of the pleurae during breathing. Normal pleural movements are non-auscultatory.

3. Pneumothorax is air in the pleural cavity. A spontaneous pneumothorax results from non-traumatic causes, while penetrating wounds may cause an open pneumothorax. If the penetrating wound is oblique or results in the formation of a flap which can act as a valve, air will enter the thorax on inspiration but not leave on expiration. Positive pressure will build, creating a life threatening emergency.

4. Irritation of the central diaphragmatic and mediastinal pleurae results in referred pain in the shoulder and lower neck. Irritation of the periphery of the diaphragmatic and costal pleurae is referred to the corresponding region of the thoracic wall.

5. The left and right parietal pleurae contact each other at a minimum of 2 points--retrosternally from T4 to the cardiac notch--and in the posterior mediastinum in Truesdale's Triangle. These are potential sites of herniation and/or inadvertent puncture of the contralateral pleural cavity during a surgical procedure.

THE LUNGS

The lungs are highly elastic, and unfixed tissue removed from the thorax quickly becomes shapeless; however, fixation hardens the lungs so that relations of the lungs may be studied. The right lung has 3 lobes while the left lung has 2 lobes with a long lingular process.

The trachea is an air passageway originating at the lower end of the larynx. Although the trachea exists in the superior mediastinum, it will be discussed at this point due to its functional association with the lungs. A series of "C" shaped cartilaginous structures maintain the open nature of the tube. At the level of T4 in the cadaver (Angle of Louis), but lower in the living, the trachea bifurcates into primary bronchi. An internal ridge known as the carina separates the openings of the primary bronchi. The right bronchus is larger and leaves the trachea at a 25° angle and divides into secondary bronchi before entering the lung. The left primary bronchus departs the trachea at a 45° angle and does not divide into secondary bronchi until it reaches the lung. Each secondary bronchus divides into tertiary bronchi. The number and position of the tertiary or segmental bronchi are fairly constant. Each segmental bronchus serves an anatomically distinct portion of lung tissue called a bronchopulmonary segment. This is a classic description of bronchopulmonary segments with 10 right segments and 8 left segments. One should be aware, however, of the distinct possibility of 10 left segments. A concept of bronchopulmonary segments can be crucial for treatments such as bronchoscopy.

LUNGS AND BRONCHOPULMONARY SEGMENTS

ANTERIOR VIEW

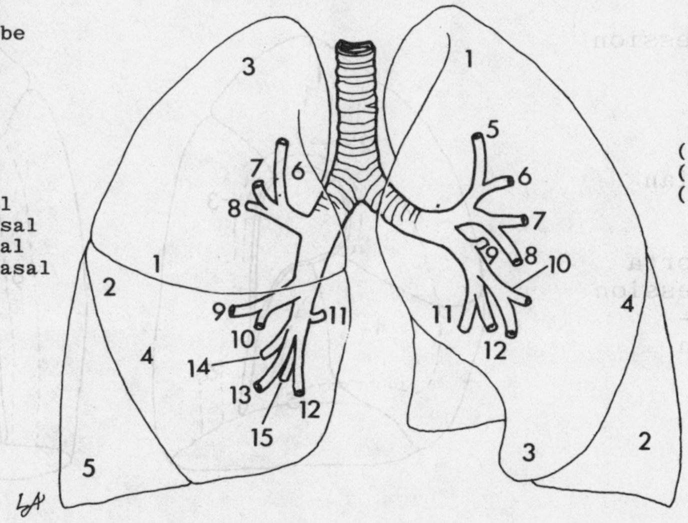

RIGHT LUNG:
(1) Horizontal fissure
(2) Oblique fissure
(3) Superior lobe
(4) Middle lobe
(5) Inferior lobe

SEGMENTS
(6) Apical
(7) Anterior
(8) Posterior
(9) Lateral
(10) Medial
(11) Superior
(12) Medial Basal
(13) Anterior Basal
(14) Lateral Basal
(15) Posterior Basal

LEFT LUNG:
(1) Upper lobe
(2) Lower lobe
(3) Lingulum
(4) Oblique fissure

SEGMENTS
(5) Apicoposterior
(6) Anterior
(7) Superior Lingular
(8) Inferior Lingular
(9) Superior
(10) Anteromedial Basal
(11) Posterior Basal
(12) Lateral Basal

ROOT STRUCTURES

Each lung is suspended by its root and pulmonary ligament. The root of the lung includes the bronchi, pulmonary vessels, lymphatics and nerves, which enter the lung through a wedge shaped area termed the hilum. The pulmonary ligament is a fold of pleura formed as the visceral pleura drapes around the root of the lung and fuses inferiorly. The right hilum is round, with the pulmonary artery adjacent to the bronchus, while the left hilum is elongated with the pulmonary artery above the bronchus.

ROOT STRUCTURES AT THE HILUM:

1) Pulmonary ligament
2) Bronchial artery
3) Bronchus
4) Pulmonary artery
5) Pulmonary veins --these are most anterior and inferior.

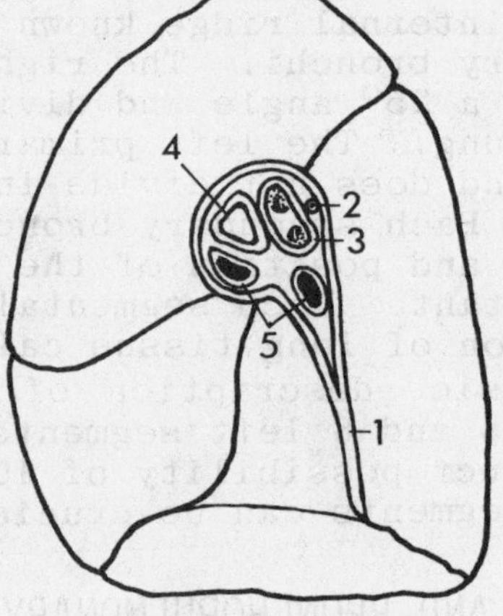

MEDIAL VIEW
RIGHT LUNG

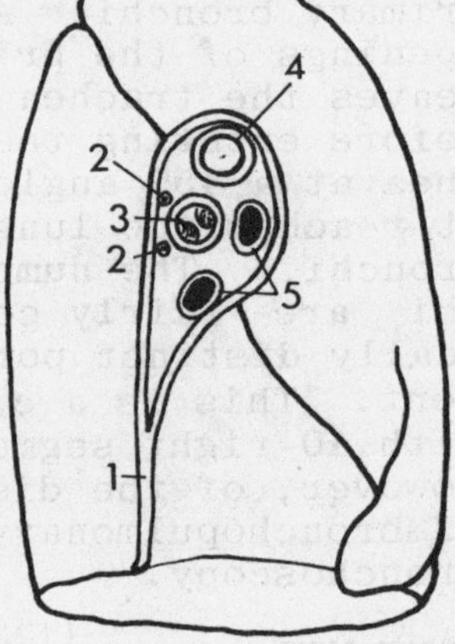

MEDIAL VIEW
LEFT LUNG

LUNG IMPRESSIONS AND RELATIONS:

RIGHT LUNG:

1) Right subclavian artery
2) SVC
3) Azygos vein
4) Cardiac impression
5) IVC
6) Esophagus

LEFT LUNG:

7) Left subclavian artery
8) Aortic arch
9) Descending aorta
10) Cardiac impression
11) Left brachiocephalic vein

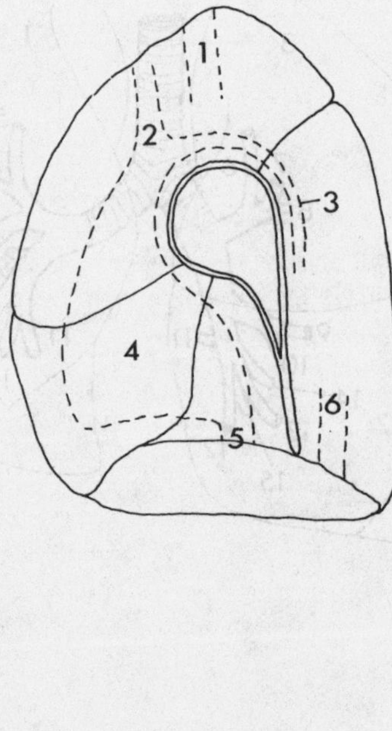

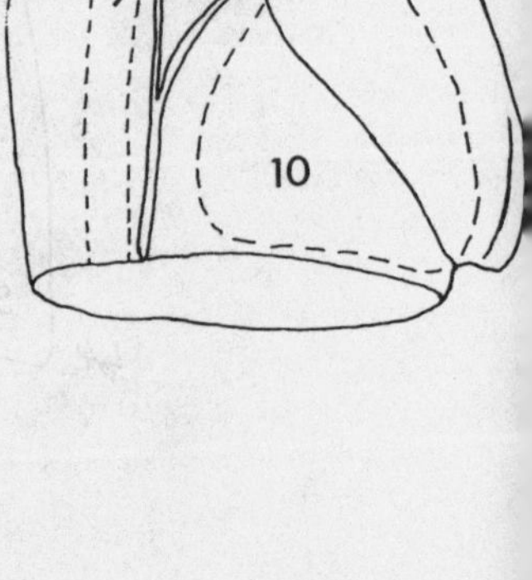

VASCULATURE

PULMONARY:

Pulmonary arteries carry blood to the lungs for oxygenation. These vessels are central in each bronchopulmonary segment and branch in a manner similar to the segmental bronchi. They are considered segmental in nature.

Pulmonary veins, however, are at the periphery of each segment, and are considered intersegmental.

BRONCHIAL:

Nutritional supply to the tracheobronchial tree is via the bronchial vessels. There are 2 bronchial arteries on the left, 1 on the right. There is 1 bronchial vein on each side. The left bronchial vein drains to the accessory hemiazygos or superior intercostal vein, while the right bronchial vein drains to the azygos vein.

BRONCHIAL ARTERIES:

1) Left bronchial arteries
2) Intercostal arteries
3) Right bronchial artery--note common origin with the intercostal artery

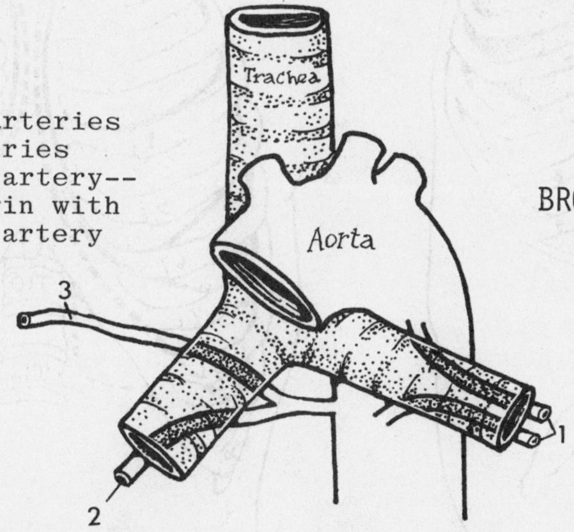

ANTERIOR VIEW
BRONCHIAL ARTERIES

INNERVATION:

Sensory fibers from the lungs and visceral pleura travel with the vagus nerve. Preganglionic parasympathetic fibers travel in the vagus nerves through the pulmonary plexus to the level of the tracheobronchial tree where they synapse with postganglionics. Sympathetic preganglionics from T1-6 synapse with postganglionics in the cervical and upper thoracic ganglia; postganglionics then descend to the cardiac and pulmonary plexuses. The PULMONARY PLEXUS lies on the anterior and posterior aspects of root structures of the lungs and is said to have anterior and posterior divisions. Both receive fibers from the vagus and cardiac plexus, while the posterior also receives from the thoracic sympathetic ganglia. Fibers then ramify to the lungs where parasympathetics are vasodilator, bronchoconstrictor and secretomotor in function, while sympathetics are vasoconstrictor, bronchodilator and decrease secretory function of mucous bronchial glands.

SURFACE PROJECTIONS OF THE LUNGS AND PLEURAL CAVITIES

It is important to learn the surface projections of the lungs and pleurae; remember a pneumothorax may occur if the pleural cavity is breached.

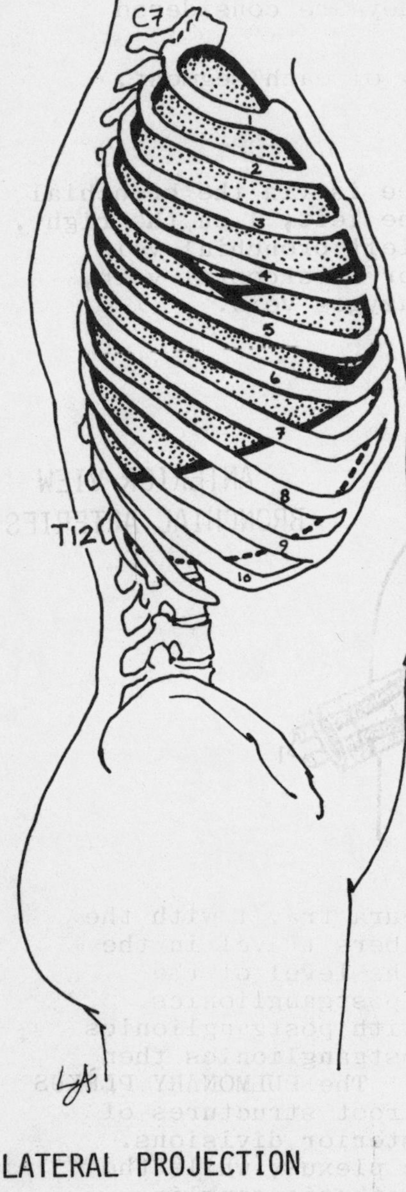

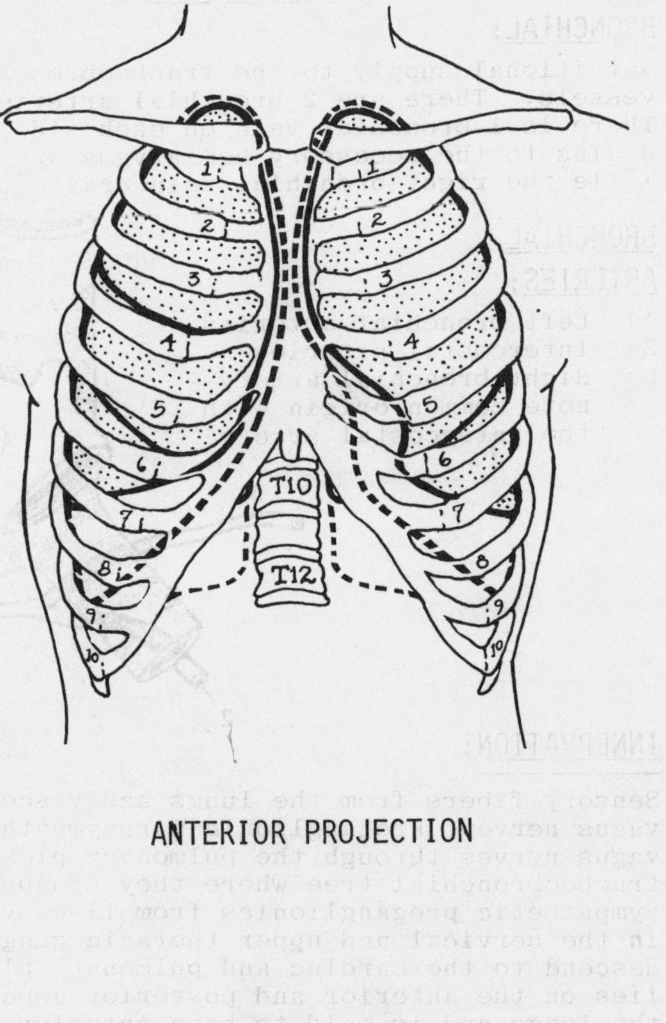

ANTERIOR PROJECTION

LATERAL PROJECTION

SURFACE PROJECTIONS OF THE LUNGS AND PARIETAL PLEURA

LUNGS:

Right Lung

Apex--1 inch above medial 1/3 of the clavicle
Anterior Border--slopes toward the midline, curves out between 4th and 6th costal cartilage
Inferior Border--6th costal cartilage at the lower end of the sternum, 6th rib at the midclavicular line, 8th rib at the midaxillary line, 10th thoracic vertebra at the paravertebral line.

Left Lung

Same as right, except the cardiac notch curves 1½ inches lateral to the sternum in the 5th ICS

Fissures

Oblique--extends from the spine of the scapula to the 6th costal cartilage
Horizontal--posterior to the 4th rib

PARIETAL PLEURA:

Anterior Border--essentially the same as the lungs
Inferior Border--6th costal cartilage at the lower end of the sternum, 8th rib at the midclavicular line, 10th rib at the midaxillary line, 12th thoracic vertebra at the paravertebral line. The pleura may be below the neck of the 12th rib, important to remember in incisions over the kidney.

LYMPHATICS

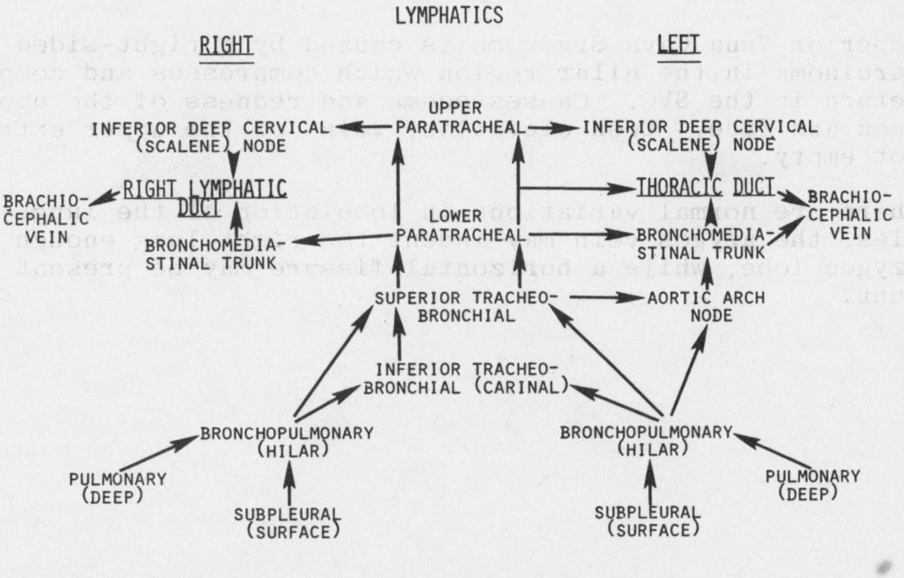

COMMENTS

1. The carina is an important landmark on X-rays for positioning of an endotracheal tube. It is the most distal point of the respiratory tree capable of the initiation of a cough reflex; indeed, the carina is one of the most sensitive areas in the respiratory tree.
2. The right primary bronchus is more directly in line with the trachea and is the most likely place for a foreign body to lodge. The left primary bronchus is closely related to the pulmonary trunk and aorta and therefore surgical procedures are more difficult on the left side.
3. Resection of a bronchopulmonary segment along the border of a segment offers the least bloody field; the pulmonary veins and occasionally pulmonary arteries have tributaries and branches which cross segmental boundaries.
4. Knowledge of the direction of the tertiary bronchioles and bronchopulmonary segments is important for proper positioning of the patient to encourage drainage of lung segments (postural drainage) or to predict the route of drainage. Blockage of a segment may result in segmental atelectasis or collapse.
5. Lymphshed is important in the lung owing to the incidence of cancer of the lung. The lymph from the left inferior lobe may normally cross to the right. Cancer of a bronchus, or bronchogenic carcinoma is common in men; a usual site of metastasis is the brain. The palpable supraclavicular nodes or 'sentinel' nodes are often enlarged in bronchogenic carcinoma.
6. Pancoast's Syndrome is usually caused by bronchogenic carcinoma of the cupola of the lung which compresses structures at the thoracic inlet (thoracic inlet tumor). It affects sympathetics to the head and compresses major vessels. It results in paresthesias and paresis of the arm and hand, pain, wasting and Horner's Syndrome.
7. Superior Vena Cava Syndrome is caused by a right-sided bronchogenic carcinoma in the hilar region which compresses and compromises venous return in the SVC. Causes edema and redness of the upper chest, neck and face. Upon elevation, veins of the upper extremity do not empty.
8. There are normal variations in lobulation of the lung. For examples, the azygos vein may indent the right lung enough to form an azygos lobe, while a horizontal fissure may be present on the left lung.

IV. THE MEDIASTINUM

The mediastinum is the portion of the thorax which lies between the two pleural sacs--in fact one original definition of mediastinum was septum. It is bounded laterally by the mediastinal pleurae. It extends from the inlet of the thorax superiorly to the diaphragm inferiorly, and from the sternum anteriorly to the twelve thoracic vertebrae posteriorly. It is further subdivided by an imaginary horizontal plane at the Angle of Louis.

SUMMARY OF MEDIASTINAL STRUCTURES

1) SUPERIOR MEDIASTINUM

Arch of the aorta
Brachiocephalic vessels
Common carotid arteries
Subclavian arteries
Ligamentum arteriosum
SVC--upper part
Thymus
Trachea and primary bronchi
Esophagus
Vagus nerve
Recurrent laryngeal nerve
Esophageal plexus
Phrenic nerve
Thoracic duct
Left highest intercostal vein
Cardiac nerves and sympathetic trunk

2) ANTERIOR MEDIASTINUM

Connective tissue
Lymph nodes and vessels
Inferior part of thymus

3) MIDDLE MEDIASTINUM

Heart
Pericardium
Great vessels
- Ascending aorta
- Pulmonary trunk
- Lower SVC

Root structures of the lung
Phrenic nerves
Pericardiacophrenic vessels

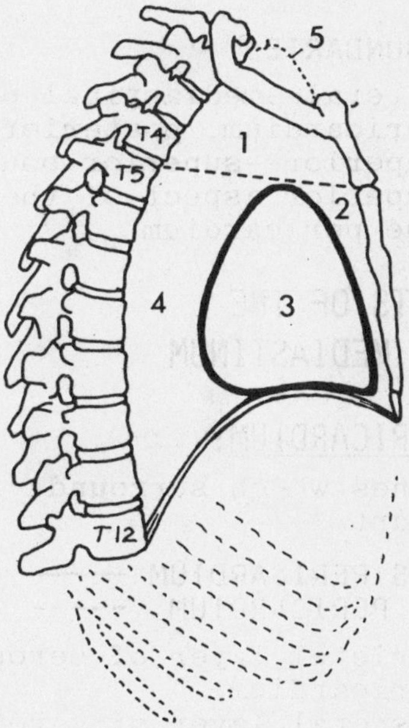

LATERAL PROJECTION OF THE MEDIASTINUM

4) POSTERIOR MEDIASTINUM

Descending aorta
Azygos vein
Hemiazygos vein
Esophagus
Vagus nerve
Sympathetic chain and ganglia
Splanchnic nerves
Thoracic duct
Lymph nodes and vessels

5) THORACIC INLET AND CUPOLA OF THE LUNG

ANTERIOR MEDIASTINUM

BOUNDARIES

Lateral--pleurae, anterior--sternum, posterior--pericardium, superior--horizontal plane at T4, inferior--superior aspect of the diaphragm

CONTENTS

Review previous page

MIDDLE MEDIASTINUM

BOUNDARIES

Lateral--mediastinal pleurae, anterior--anterior surface of the pericardium, posterior--posterior surface of the pericardium, superior--superior boundaries of the pericardium, inferior--superior aspect of the diaphragm in the region fused with the pericardium

CONTENTS OF THE MIDDLE MEDIASTINUM

THE PERICARDIUM:

Membranes which surround the heart

FIBROUS PERICARDIUM ———

SEROUS PERICARDIUM -----

1) Parietal layer of serous pericardium
2) Visceral layer of serous pericardium, also known as epicardium
3) Pericardial cavity--contains only enough fluid to moisten surfaces

PERICARDIAL SAC

Fibrous pericardium + parietal layer of serous pericardium

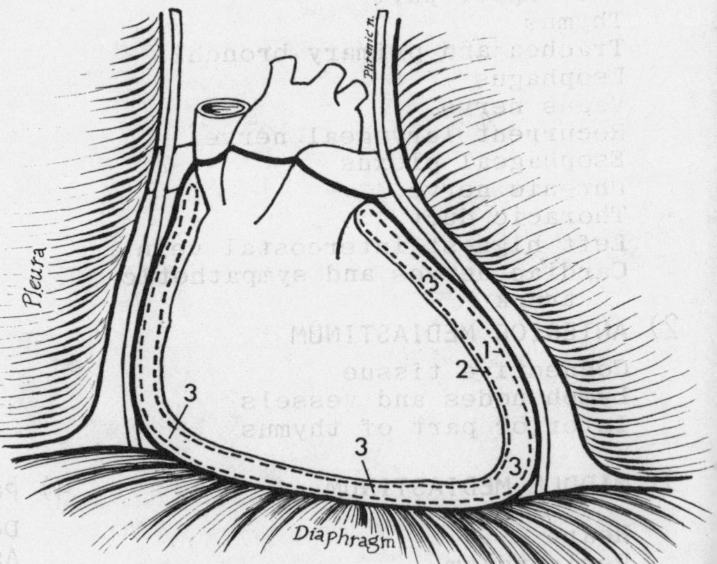

SCHEMATIC REPRESENTATION OF THE PERICARDIUM

COMMENTS

1. Pericarditis is an inflammation of the pericardial sac. Auscultation of movements of the pericardium is not normally possible; however a friction rub due to the contact of roughened, inflamed pericardial surfaces renders the movements audible. Pain is substernal or may be referred to the left shoulder and arm--a complication in diagnosis since this is the same area of referred pain from the heart.
2. Pericarditis may produce a large amount of fluid or pericardial effusion; effusion does not necessarily eliminate friction rub. The volume of effluent may become great enough to compress the heart within the pericardium and compromise its function; the result is a cardiac tamponade.
3. Hemopericardium may result from puncture wounds to the heart; cardiac tamponade may also occur; pericardiocentesis of the blood or other effluents may be performed.

THE HEART AND GREAT VESSELS:

SCHEME OF THE HEART AND BLOOD FLOW

1) Right atrium
2) SVC
3) IVC
4) Right ventricle
5) Tricuspid valve
6) Pulmonary trunk
7) Pulmonary valve
8) Left atrium
9) Pulmonary veins
10) Left ventricle
11) Mitral or bicuspid valve
12) Aortic valve
13) Aortic arch
14) Endocardium
15) Myocardium
16) Epicardium or visceral pericardium
17) Transverse sinus--this space represents the old separation between the venous and arterial ends of the primitive heart
18) Oblique sinus

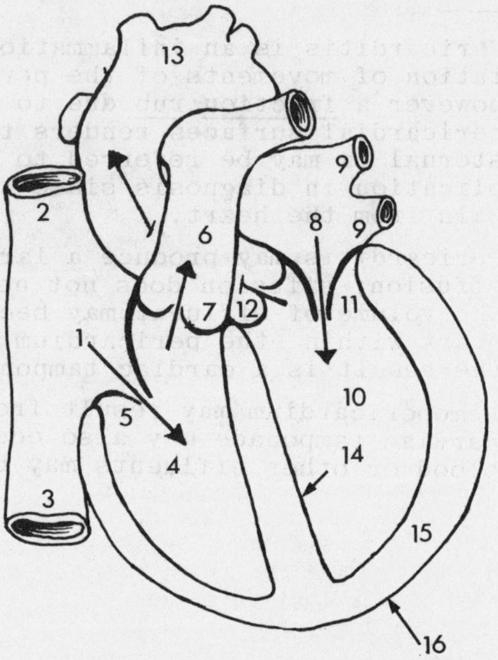

SCHEMATIC REPRESENTATION OF THE HEART

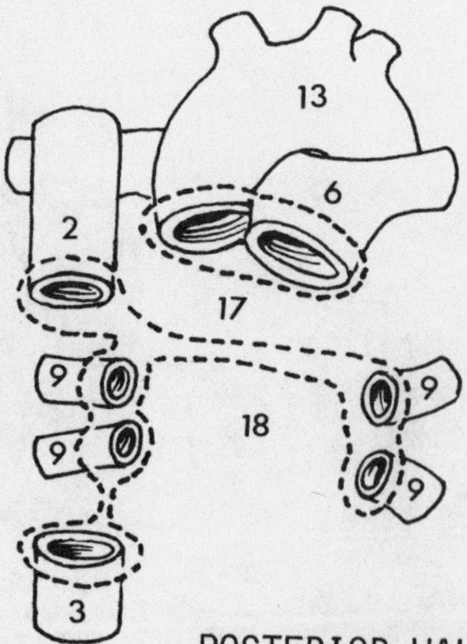

Dotted lines indicate the reflection of serous pericardium from the heart and great vessels onto the pericardial sac

POSTERIOR WALL OF THE PERICARDIAL SAC

SURFACE PROJECTIONS OF THE HEART

1) APEX--formed by left ventricle at 5th left intercostal space, 3½ inches from midline, point of strongest impulse
2) ACUTE OR RIGHT MARGIN--formed by right atrium, ½ inch from edge of sternum, extends from costal cartilages 3-6
3) OBTUSE OR LEFT MARGIN--formed by left ventricle, extends from left 2nd costal cartilage to apex
4) INFERIOR BORDER--formed by right and left ventricles, extends from 6th costal cartilage to apex

BASE-POSTERIOR ASPECT--formed by left atrium

DIAPHRAGMATIC SURFACE--formed by both of the ventricles, but mainly the left; at level of 6th costal cartilage, slightly concave, conforms to the diaphragm

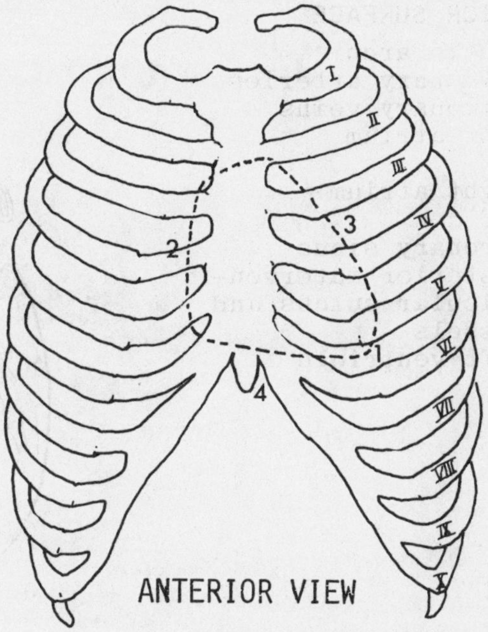

ANTERIOR VIEW

<u>SURFACES OF THE HEART</u>:

ANTERIOR SURFACE

1) Aortic arch
2) SVC
3) Pulmonary trunk
4) Right auricle
5) Right atrium
6) Coronary sulcus and vessels
7) Epicardial fat
8) IVC
9) Right ventricle
10) Ligamentum arteriosum
11) Left auricle
12) Left ventricle
13) Anterior interventricular sulcus and vessels

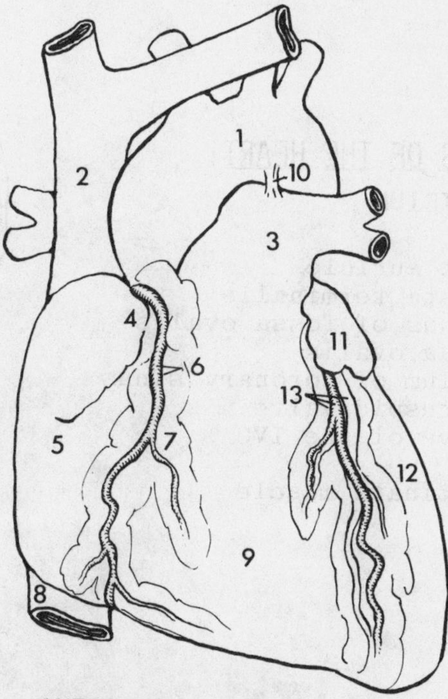

ANTERIOR SURFACE

POSTERIOR SURFACE

1) Aortic arch
2) Pulmonary arteries
3) Pulmonary veins
4) Left atrium
5) SVC
6) Right atrium
7) IVC
8) Coronary sinus
9) Posterior interventricular sulcus and vessels
10) Left ventricle

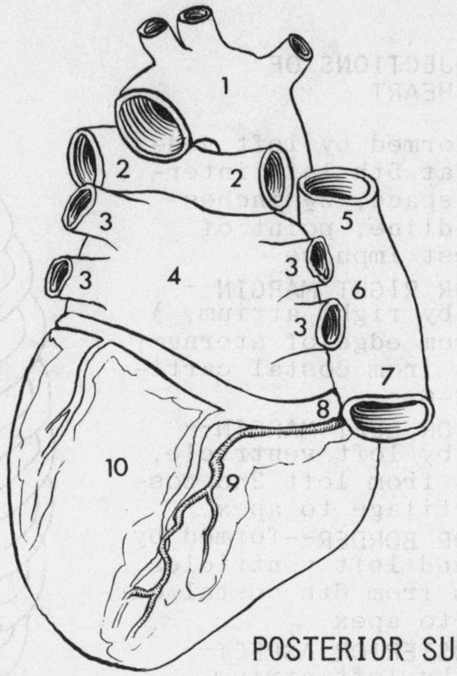

POSTERIOR SURFACE

CHAMBERS OF THE HEART:

RIGHT ATRIUM

1) SVC
2) Left auricle
3) Crista terminalis
4) Limbus of fossa ovalis
5) Fossa ovalis
6) Ostium of coronary sinus
7) Tricuspid valve
8) Valve of the IVC
9) IVC
10) Pectinate muscle

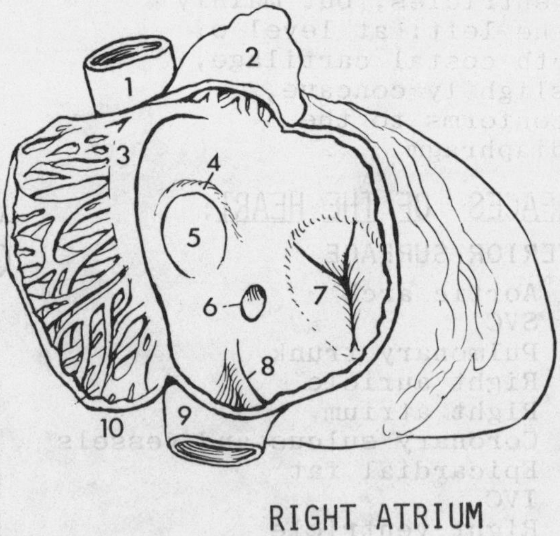

RIGHT ATRIUM

COMMENTS

1. The smooth walled portion of the atrium is derived from the sinus venosus while the trabeculated portion corresponds to the primitive right atrium. The auricle, properly termed the atrial appendage, is routinely used as a location for surgical entry.

2. Taenia sagittalis is a large pectinate muscle originating at the crista terminalis.

3. The Network of Chiari refers to a fenestrated IVC valve.

4. Probe patency of the foramen ovale is not infrequent.

RIGHT VENTRICLE
1) Pulmonary trunk
2) Infundibulum
3) Tricuspid valve
4) Chordae tendineae
5) Moderator band
Papillary Muscles:
6) Anterior
7) Posterior
8) Medial

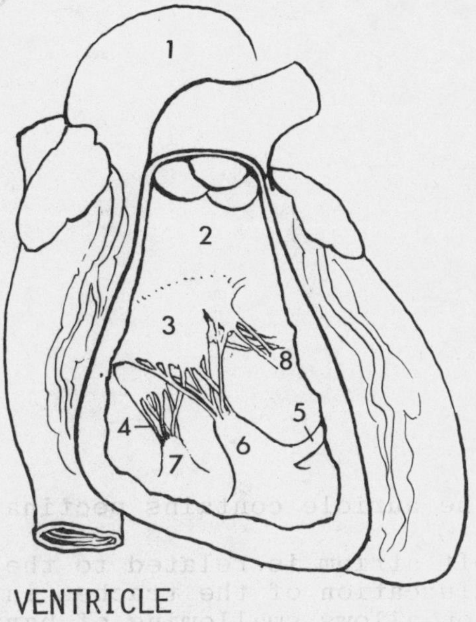

RIGHT VENTRICLE

COMMENTS

1. The supraventricular crest, septomarginal band and parietal band are muscular ridges which delimit an inflow tract containing the tricuspid valve and an outflow tract containing the infundibulum or conus arteriosus. The inflow tract contains many beam-like trabeculae carneae while the outflow tract has few trabeculae.

LEFT ATRIUM

1) Aortic arch
2) Left pulmonary artery
3) Right pulmonary artery
4) SVC
5) Pulmonary veins
6) Left atrium
7) Mitral or bicuspid valve
8) Right atrium
9) IVC
10) Coronary sinus

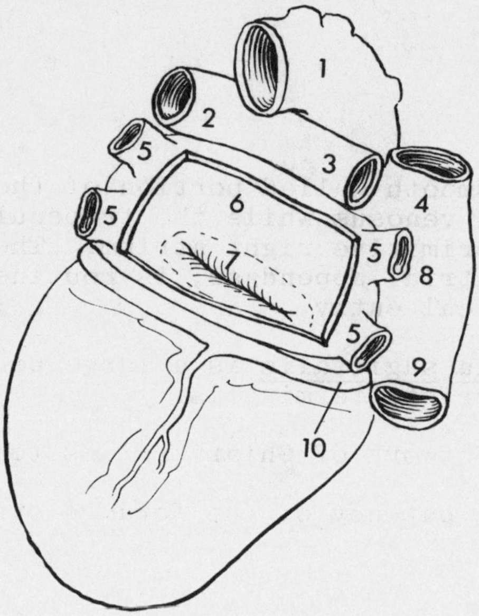

LEFT ATRIUM

COMMENTS

1. Only the auricle contains pectinate muscles.

2. The left atrium is related to the descending aorta, esophagus, and bifurcation of the trachea into primary bronchi. This relation allows swallowing of barium compounds to contrast against cardiac structures in radiography and allows microphones to aid in phonocardiography. In mitral stenosis the left atrium may enlarge enough to compress and deviate the esophagus and bronchi.

LEFT VENTRICLE-

1) Aorta
2) Aortic semilunar valves
3) Sinuses of Valsalva--between valves and aortic wall
4) Aortic vestibule
5) Mitral valve
6) Papillary muscles
7) Chordae tendineae

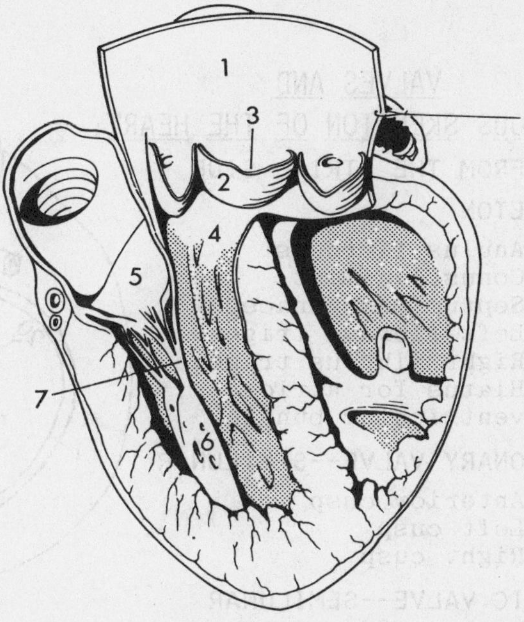

LEFT VENTRICLE

COMMENTS

1. The left ventricle is almost circular in transverse section. The septum protrudes into the right ventricle; the left wall is about three times thicker than the right.

2. The nonmuscular or membranous septum is that portion of the septum below the posterior and right aortic cusps. It is delimited by the limbus marginalis. Defects of the membranous portion of the septum --ventricular septal defects or VSD--are present in 50% of all congenital heart defects. Defects of the atrial septum--ASD--are also common, with a patent foramen ovale the most prevalent ASD.

VALVES AND FIBROUS SKELETON OF THE HEART:

FROM THE ATRIAL SIDE

SKELETON

1) Anulus fibrosus
2) Conus ligament
3) Septum membranaceum
4) Left fibrous trigone
5) Right fibrous trigone
6) Hiatus for atrio-ventricular bundle

PULMONARY VALVE--SEMILUNAR

7) Anterior cusp
8) Left cusp
9) Right cusp

AORTIC VALVE--SEMILUNAR

10) Left cusp
11) Right cusp
12) Posterior cusp
13) Left coronary artery
14) Right coronary artery

TRICUSPID VALVE

15) Septal cusp
16) Anterior cusp
17) Posterior cusp

MITRAL (BICUSPID) VALVE

18) Anterior cusp
19) Posterior cusp
20) Commissural cusps

A-V VALVE DETAIL

21) Cusp
22) Line of closure, with Nodules of Albini
23) Chordae tendineae
24) Papillary muscle

AORTIC SEMILUNAR VALVE DETAIL

25) Ostium of coronary artery
26) Cusp or valvule
27) Lunula
28) Nodule of Arantius
29) Sinus of Valsalva

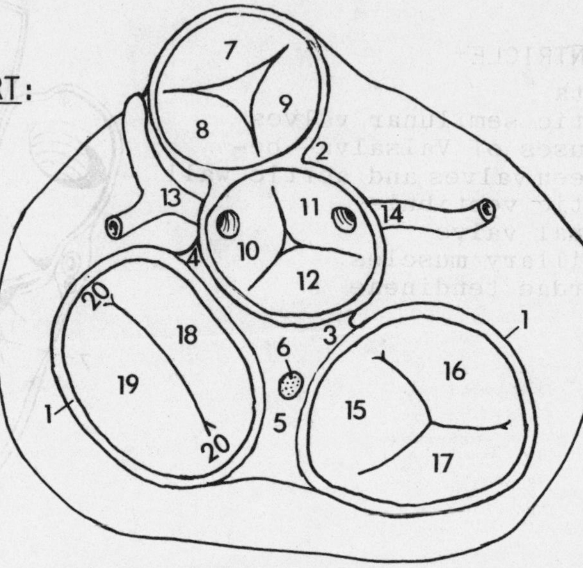

VALVES AND FIBROUS SKELETON OF THE HEART

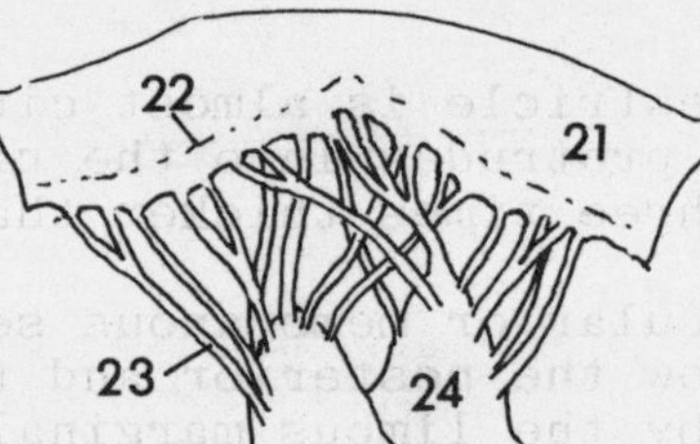

SEMILUNAR VALVE DETAIL
AORTIC

SURFACE PROJECTIONS AND AUSCULTATION POINTS FOR THE HEART VALVES

1) Pulmonary valve
2) Auscultation for pulmonary valve
3) Aortic valve
4) Auscultation for aortic valve
5) Mitral valve
6) Auscultation for mitral valve
7) Tricuspid valve
8) Auscultation for tricuspid valve

VALVE SOUNDS ARE HEARD WITH THE LEAST INTERFERENCE AT THE POINTS INDICATED, WHICH ARE DOWNSTREAM FROM THE VALVES.

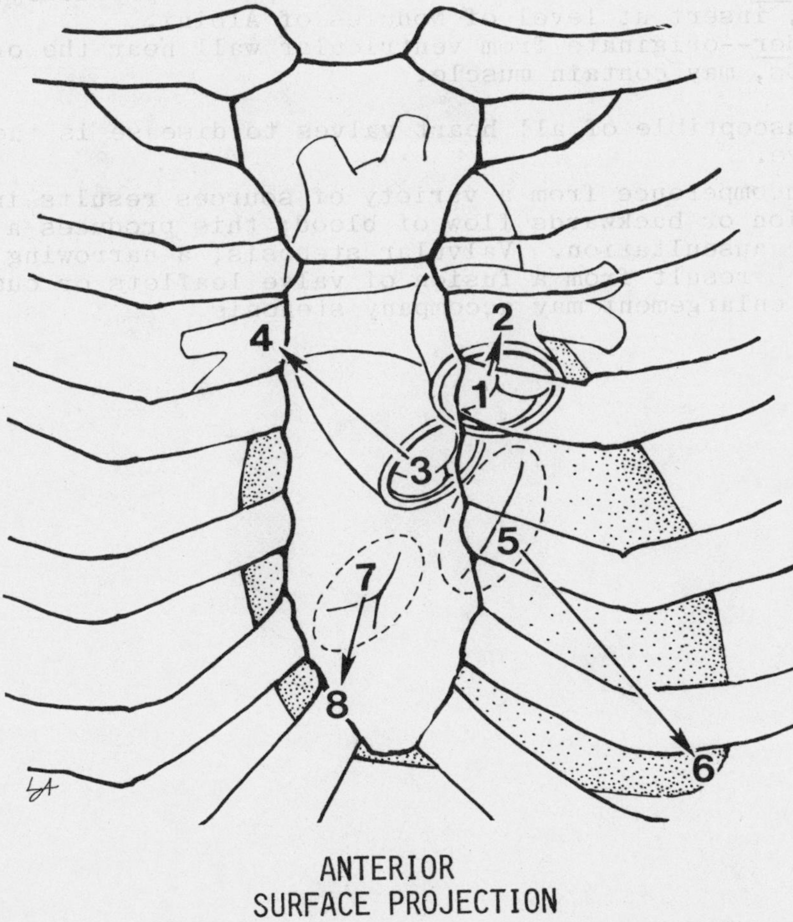

ANTERIOR
SURFACE PROJECTION

COMMENTS

1. Anuli fibrosi encircle the atrioventricular openings and provide an origin for atrial and ventricular musculature, an origin for the atrioventricular valves and electrical insulation between the atria and ventricles.

2. The origins of the pulmonary and aortic valves have also been described as anuli fibrosi; however clinicians describe the fibrous origin of the valves as scalloped, corresponding to the point of attachment to the vessel wall, with additional thickenings between the scallops.

3. Each papillary muscle sends chordae to more than 1 cusp. There are three orders of the chordae:
   1st Order--originate at or near the apex of papillary muscle, insert at free edge of cusp
   2nd Order--originate at or near the apex of papillary muscle, insert at level of Nodules of Albini.
   3rd Order--originate from ventricular wall near the origin of cusps, may contain muscle.

4. The most susceptible of all heart valves to disease is the mitral valve.

5. Valvular incompetence from a variety of sources results in a regurgitation or backwards flow of blood; this produces a murmur upon auscultation. Valvular stenosis, a narrowing process, may result from a fusion of valve leaflets or cusps. Myocardial enlargement may accompany stenosis.

CORONARY VASCULATURE:

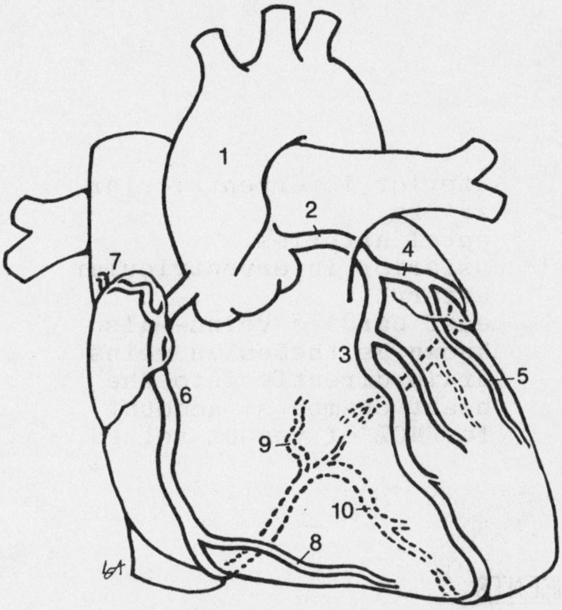

SCHEMATIC REPRESENTATION OF CORONARY ARTERIES

ARTERIES

In most individuals, 2 coronary arteries supply the heart and proximal portions of the great vessels.

1) Aorta
2) Left coronary artery
3) Anterior interventricular or anterior descending
4) Circumflex
5) Left marginal
6) Right coronary artery
7) Right atrial branch, to the sino-atrial node
8) Right marginal
9) Artery of the atrio-ventricular node
10) Posterior interventricular or posterior descending

VEINS

The coronary sinus receives most veins of the heart; however the Thebesian Veins empty directly into the ventricles.

11) Anterior cardiac veins
12) Coronary sinus
13) Small cardiac vein
14) Middle cardiac vein
15) Posterior vein of left ventricle
16) Great cardiac vein

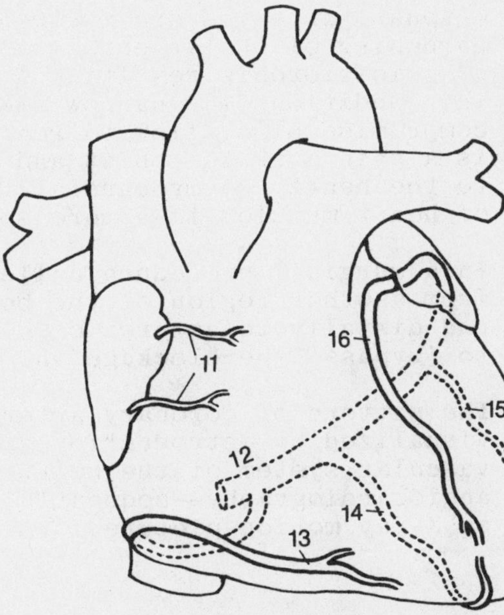

SCHEMATIC REPRESENTATION OF CORONARY VEINS

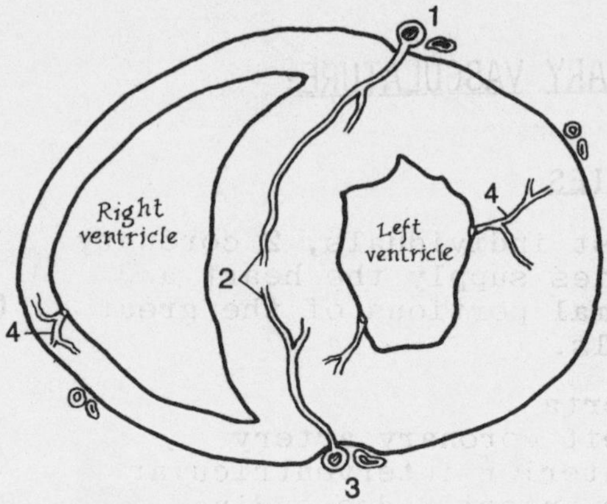

1) Anterior interventricular artery
2) Septal arteries
3) Posterior interventricular artery
4) Least cardiac veins--also known as thebesian veins, drain directly into the heart chambers, account for 40% of venous return

COMMENTS

1. Variations in the number and pattern of coronary arteries do occur; there may be as many as 3 or as few as 1. The right coronary artery is dominant--becomes the posterior interventricular--in about 70% of individuals.

2. Coronary arteries are frequent sites of arteriosclerosis--hardening of the arteries--and atherosclerosis, a form of arteriosclerosis resulting from deposits of lipid material. The conditions may narrow the lumen of the arteries and compromise blood flow to regions of the heart. Angina pectoris is a pain over the chest and left arm resulting from ischemia to the heart. A myocardial infarction--MI or death of a region of heart muscle--is a more severe result of ischemia.

3. In a surgical procedure called a coronary bypass, a vessel from another region of the body is connected both proximally and distally to an area of compromised coronary blood flow to 'bypass' the blockage and improve coronary circulation.

4. The pattern of coronary arteries and their patency may be visualized by introduction of a radio-opaque compound into the vascular system of the heart and subsequently viewed by angiocardiography--sequential X-rays--or by cineradiography--an X-ray motion picture.

INTRINSIC INNERVATION
of the heart is via a specialized conducting system of modified cardiac muscle fibers.

1) Sinoatrial--SA--node, in the wall of the right atrium where the crista terminalis meets the SVC; the node surrounds the nodal artery
2) Anterior internodal pathway, also gives rise to Bachmann's Bundle which courses to the left atrium
3) Middle internodal pathway or Wenckebach's Bundle
4) Posterior internodal pathway
5) Atrioventricular--AV--node, in the interatrial wall near the ostium of the coronary sinus
6) AV bundle of His
7) Moderator band
8) Right bundle branch
9) Terminal conducting fibers of Purkinje
10) Left bundle branch

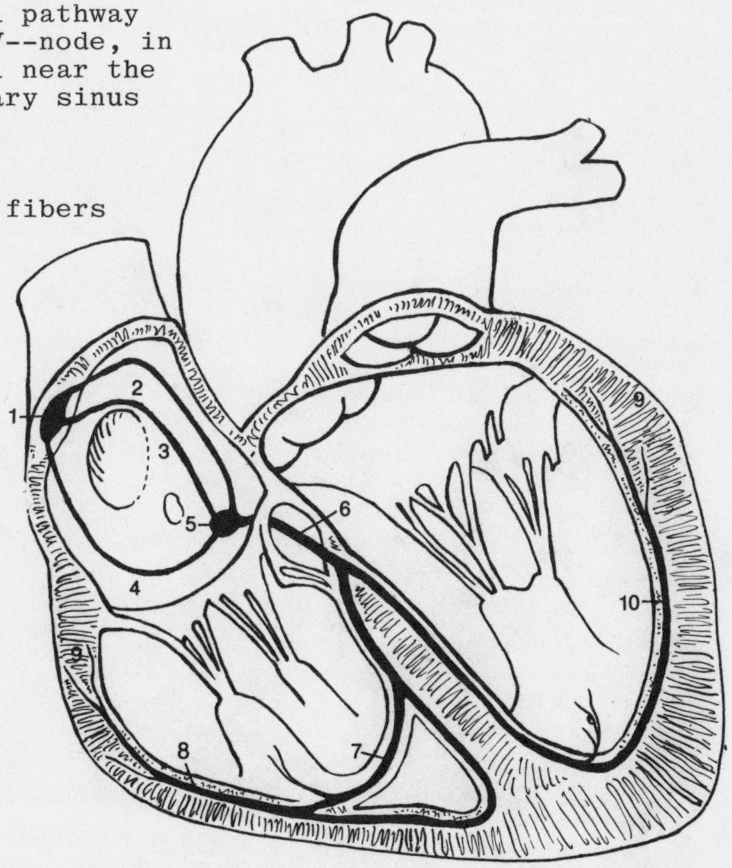

SCHEMATIC VIEW OF CONDUCTING SYSTEM OF THE HEART

COMMENTS

1. Cardiac muscle has the innate ability to contract; however it is the conducting system of the heart which regulates and controls heart contractions. At times the system fails to function properly--an infarction of the nodes for example--and the heart muscle may fibrillate or twitch irregularly and out of sequence. De-fibrillation by electric shock may result in a normalization of beat.
2. A cardiac pacemaker may be required to deliver a synchronizing impulse to cardiac muscle to compensate for a damaged conduction system.

EXTRINSIC INNERVATION: THE CARDIAC PLEXUS lies between the bifurcation of the trachea and the aortic arch; receives fibers from autonomic nerves in the neck.

PARASYMPATHETIC

Increases in parasympathetic activity cause slowing of heart rate and somewhat decrease force of contraction.

1) Vagus nerve
2) Recurrent laryngeal nerve
3) Superior cervical cardiac branch
4) Inferior cervical cardiac branch
5) Thoracic cervical cardiac branch
6) Cardiac plexus
7) Pulmonary plexus

SYMPATHETIC

Increases in sympathetic activity cause increases in rate and force of contraction.

8) Superior cervical ganglion
9) Middle cervical ganglion
10) Inferior cervical ganglion or cervicothoracic (stellate) ganglion if fused with T1
11) Cervical sympathetic cardiac nerves
12) Thoracic sympathetic cardiac nerves

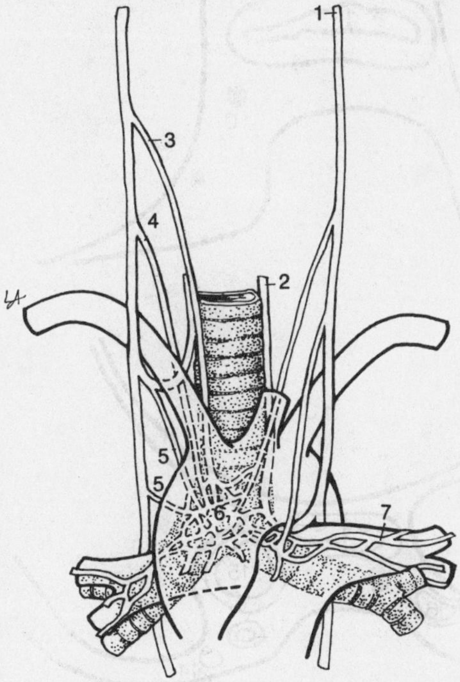

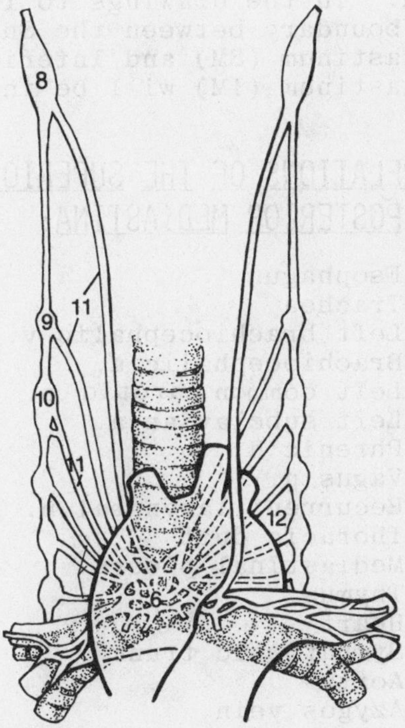

AFFERENT FIBERS from the heart travel in vagal and cervical sympathetic nerves. A Right and Left Coronary Plexus travels with the coronary arteries. They relay fibers from the cardiac plexus to the right and left sides of the heart.

SUPERIOR MEDIASTINUM

BOUNDED by superior aperture of the thorax, mediastinal pleurae, sternum, bodies T1-T4 and sternal angle (of Louis)

POSTERIOR MEDIASTINUM

BOUNDED by pericardium, bodies T5-T12, diaphragm, mediastinal pleurae, sternal angle (the boundary between superior and inferior mediastinum

THE SUPERIOR AND POSTERIOR MEDIASTINA will be discussed together since many structures pass through both. In the drawings to follow the boundary between the Superior Mediastinum (SM) and Inferior Mediastinum (IM) will be indicated.

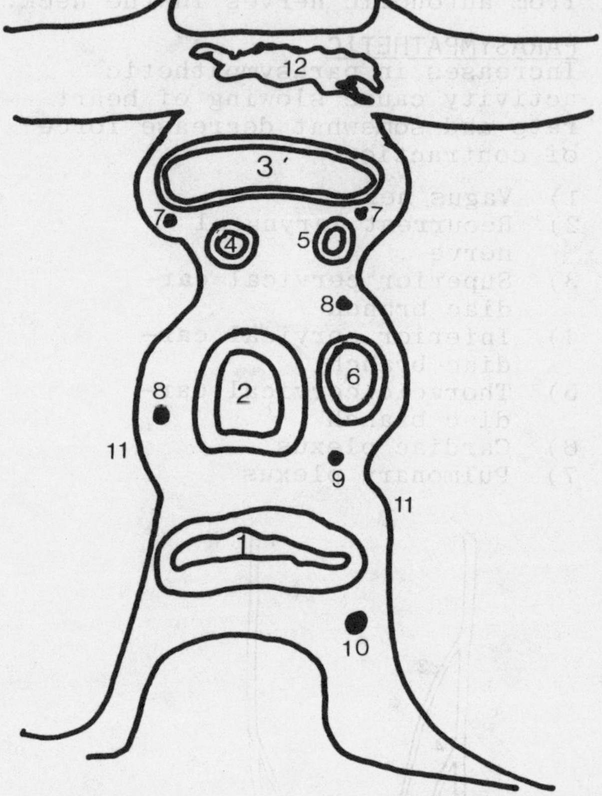

CORRELATIONS OF THE SUPERIOR AND POSTERIOR MEDIASTINA:

1) Esophagus
2) Trachea
3) Left brachiocephalic v.
4) Brachiocephalic a.
5) Left common carotid a.
6) Left subclavian a.
7) Phrenic n.
8) Vagus n.
9) Recurrent laryngeal n.
10) Thoracic duct
11) Mediastinal pleura
12) Thymus
13) Heart
14) Sympathetic trunk
15) Aorta
16) Azygos vein
17) Hemiazygos vein

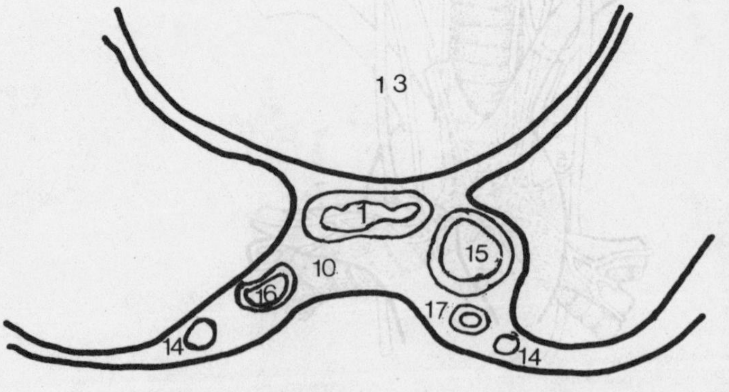

STRUCTURES OF THE SUPERIOR AND POSTERIOR MEDIASTINA:

1) Esophagus
2) Trachea
3) Aortic arch
4) Brachiocephalic a.
5) Right common carotid artery
6) Right subclavian a.
(7) Internal thoracic a.
8) Left common carotid a.
9) Left subclavian a.
10) SVC
11) Right brachiocephalic v.
12) Left brachiocephalic v.
13) Vagus nerves--note the left moves anteriorly on the esophagus while the right moves posteriorly
14) Phrenic nerve
15) Recurrent laryngeal nerve--note difference in origin between the right and left sides
16) Ligamentum arteriosum
17) Pulmonary trunk
18) Pulmonary artery
19) Primary bronchi
20) Pulmonary veins
21) Azygos vein
22) Intercostal vein
23) IVC
24) Pericardiacophrenic a.

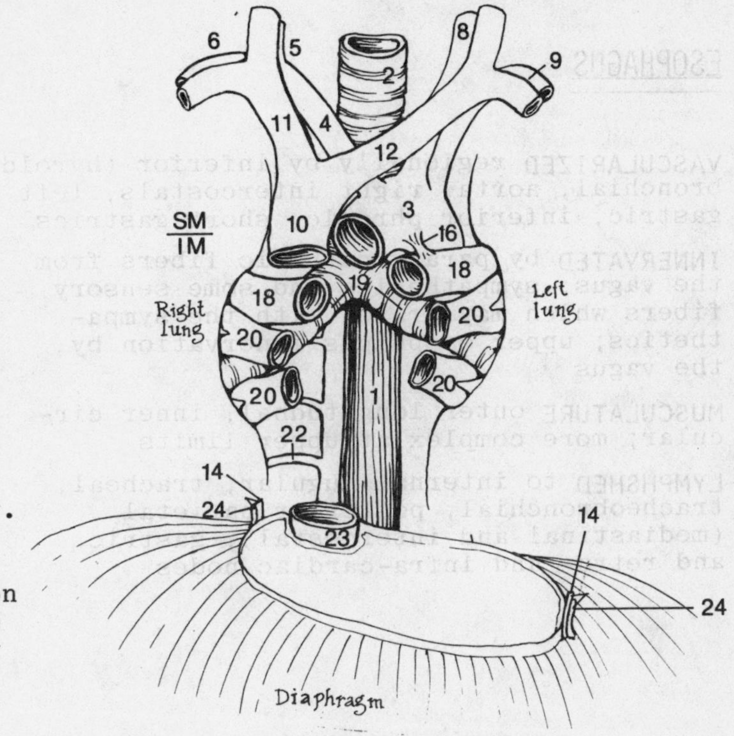

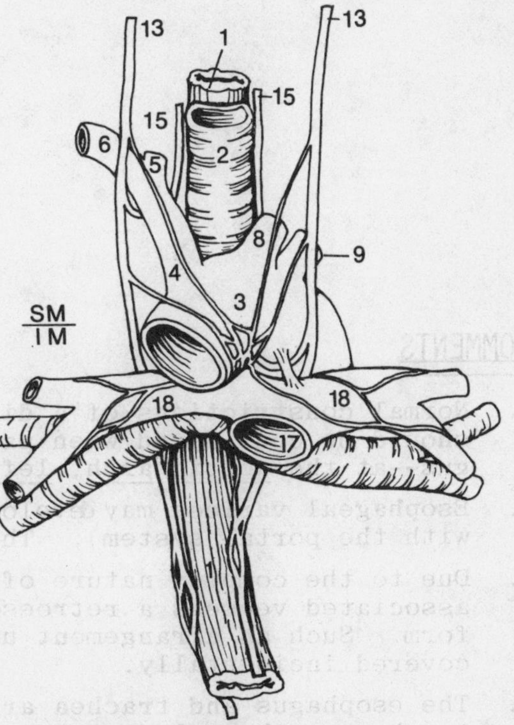

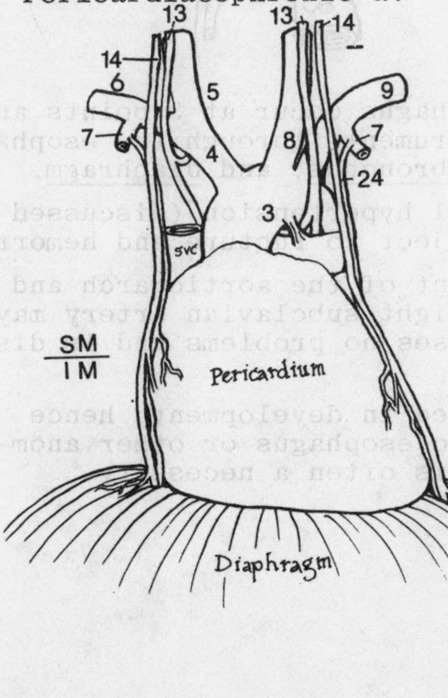

THREE ANTERIOR VIEWS
MEDIASTINAL STRUCTURES

ESOPHAGUS

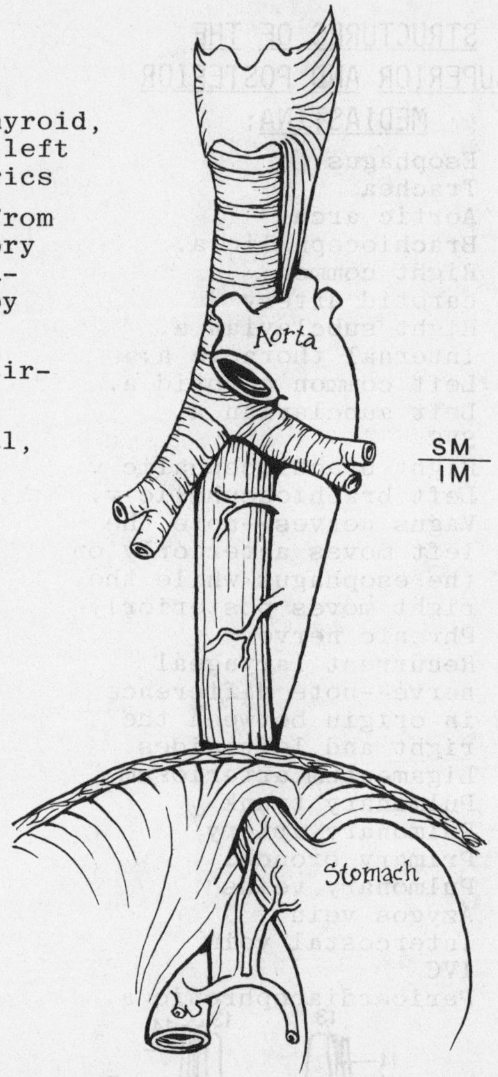

VASCULARIZED regionally by inferior thyroid, bronchial, aorta, right intercostals, left gastric, inferior phrenic, short gastrics

INNERVATED by parasympathetic fibers from the vagus; sympathetics and some sensory fibers which may travel with the sympathetics; upper esophagus innervation by the vagus

MUSCULATURE outer longitudnal, inner circular; more complex at upper limits

LYMPHSHED to internal jugular, tracheal, tracheobronchial, posterior parietal (mediastinal and intercostal), gastric and retro- and infra-cardiac nodes

COMMENTS

1. Normal constrictions of a dilated esophagus occur at 3 points and should be considered when passing instruments through the esophagus--at the aortic arch, left primary bronchus, and diaphragm.
2. Esophageal varices may develop in portal hypertension (discussed with the portal system). They are subject to rupture and hemorrhage
3. Due to the complex nature of development of the aortic arch and associated vessels a retroesophageal right subclavian artery may form. Such an arrangement usually causes no problems and is discovered incidentally.
4. The esophagus and trachea are associated in development; hence tracheo-esophageal fistulas, an atretic esophagus or other anomalies may occur. Surgical correction is often a necessity.

AZYGOS VEIN SYSTEM, LYMPHATICS AND SYMPATHETIC NERVOUS SYSTEM:

1) Right brachiocephalic v.
2) Left brachiocephalic v.
3) Left internal jugular v.
4) Left subclavian v.
5) SVC
6) Right superior intercostal v.
7) Azygos vein
8) Left superior intercostal v.
9) Accessory hemiazygos v.
10) Hemiazygos v.
11) Thoracic duct--enters left brachiocephalic vein at the confluens of the jugular and subclavian veins
12) Sympathetic ganglion and sympathetic trunk
13) Greater splanchnic nerve--communicates with coeliac ganglion, aorticorenal plexus and adrenal medulla
14) Lesser splanchnic nerve--communicates with aorticorenal plexus
15) Intercostal vein
16) Cisterna chyli

The Least splanchnic nerve is not illustrated

The splanchnic nerves have a variable origin and may not be bilaterally symmetrical

GREATER--T5 OR T6 TO T10
LESSER---T9 OR T10
LEAST----LAST THORACIC GANGLIA OR FROM THE LESSER SPLANCHNIC NERVE

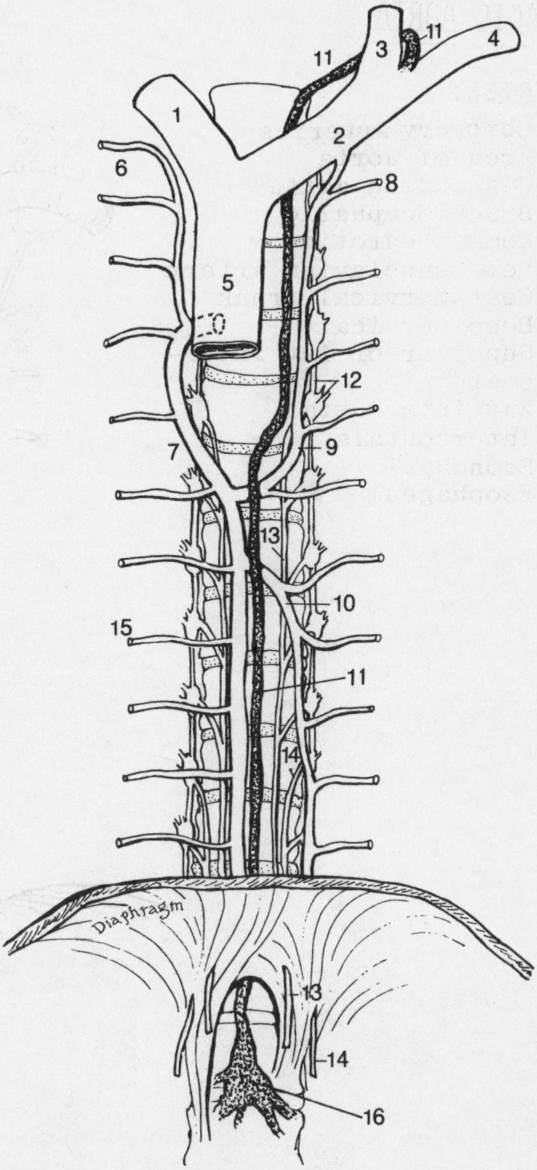

COMMENTS

1. The azygos venous system is connected to the vertebral venous plexus and the ascending lumbar venous system. These connections may be used for spread of infections from the thorax and/or pelvis to the vertebral venous plexus.

THORACIC AORTA

ARTERIES:

1) Coronary arteries
2) Arch of aorta
3) Descending aorta
4) Brachiocephalic
5) Common carotid
6) Left subclavian artery
7) Costocervical trunk
8) Deep cervical
9) Superior or 1st intercostal
10) 2nd intercostal
11) Intercostals
12) Bronchial
13) Esophageal

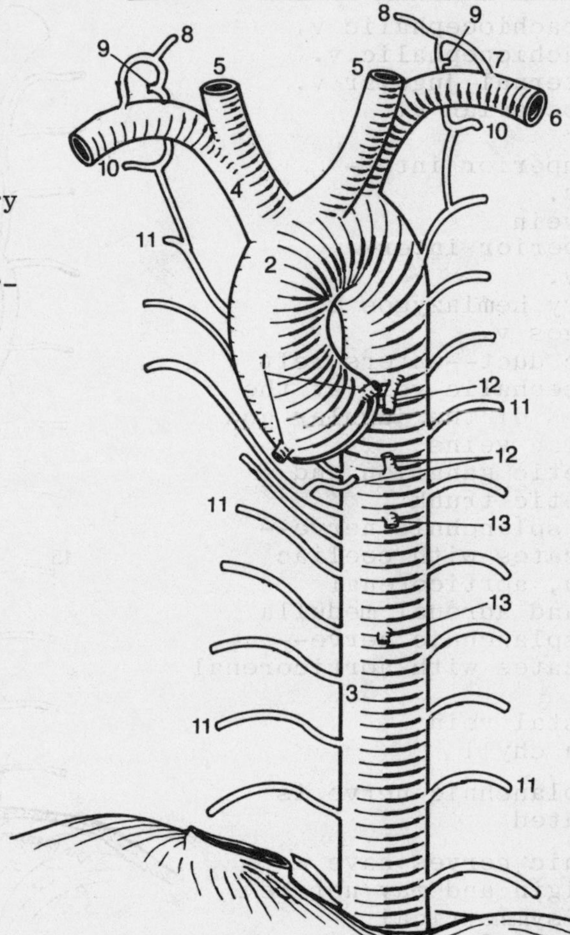

COMMENTS

1. Coarctation of the aorta is a narrowing of the aorta with a concomitant increase in size of the intercostal arteries to allow collateral flow to the aorta. The narrowed area may be proximal (preductal) or distal (postductal) to the ligamentum arteriosum. In preductal coarctation, common in infants, the ductus may be patent.

2. Developmental anomalies of the aorta may include transposition of the great vessels; a condition in which the aorta is connected to the right ventricle and the pulmonary artery to the left ventricle. In aortic atresia, the ascending aorta is markedly reduced in size, the ductus arteriosus is patent and there is usually an atrial septal defect.

3. Aneurysms of the aorta may be reinforced with Dacron grafts. Deceleration injuries, common in car wrecks, may tear or weaken the aorta. These may occur even in the absence of injuries to the bony thorax; contusions of the heart are common in such injuries.

4. In cystic medial necrosis of the aorta, a disease process results in loss of the integrity of elastic tissue in the aortic wall. The aorta increases in diameter. Dilatation of the aortic root can result in aortic valvular incompetence. Ruptures and aneurysms are common sequelae.

5. The tunica media of the thoracic aorta can be the site of damage in syphilis of the aorta. The aortic valves are commonly affected.

STUDY QUESTIONS: THORAX

1. The apex of the heart usually lies in the midclavicular line in the 5th ICS. T/F	1. True
2. The SVC is formed at the sternal angle. T/F	2. True
3. The manubrio-sternal joint is the point where the first rib articulates with the sternum. T/F	3. False - The manubrio-sternal joint is the sight for attachment of the 2nd rib.
4. Rib 10 is a false rib and a floating rib.	4. False - The 10th rib is a false rib, but the floating ribs are 11 & 12.
5. Within the intercostal space, the vessels lie between which 2 layers?	5. internal intercostal and innermost intercostal
6. Is the artery, vein or nerve most superior in the costal groove?	6. vein (vein, artery, nerve - VAN - superior to inferior).
7. The internal thoracic (mammary) vessels branch into the ___ and ___.	7. superior epigastric, musculophrenic
8. Posteriorly, the first two intercostal spaces are supplied by the _____, a branch of the _____.	8. highest thoracic artery, costocervical trunk
9. The azygos vein enters into the _____.	9. SVC
10. The left superior intercostal vein drains into the _____, while the right superior intercostal vein drains into the _____.	10. left brachiocephalic vein, azygos vein
11. The parietal pleura is subdivided into what three segments?	11. costal, mediastinal, and diaphragmatic
12. What is considered the lowest extent of the pleural sac?	12. costodiaphragmatic recess
13. Is the visceral pleura sensitive to pain and temperature?	13. No
14. Foreign bodies usually enter which bronchus?	14. the right bronchus, since it is wider and straighter than the left main bronchus
15. Does the right main bronchus divide into secondary bronchi before or after it enters the right lung?	15. Before, as opposed to the left main bronchus, which doesn't divide until it enters the lung.
16. The right bronchial artery is usually a branch from the _____.	16. 1st right intercostal artery
17. The superior and inferior mediastinum is separated by a horizontal plane at the _____ and _____.	17. sternal angle, T_4/T_5 intervertebral space
18. Name the layers of the pericardium.	18. parietal: fibrous outer layer and inner serous layer visceral: serous layer adherent to the myocardium (epicardium)
19. The right auricle is lined with roughened _____ muscles.	19. pectinate
20. The ____ is the border between the auricle and atrium proper.	20. crista terminalis
21. The moderator band is a muscle band between the septum and _____ in the _____ ventricle.	21. anterior papillary muscle, right

22. Name three functions of the anuli fibrosi.	22. (a) origin for AV valves; (b) origin of atrial and ventricular musculature; (c) electrical insulator between the atria and ventricles
23. Which is the non-coronary cusp of the aortic valve?	23. the posterior semilunar valve
For 24-27, Name the reference point for auscultation with respect to the sternum for the following:	
24. Aortic valve	24. 2nd ICS, 2 cm to right
25. Mitral valve	25. 5th ICS, 3½cm to left
26. Tricuspid valve	26. over xiphoid
27. Pulmonary valve	27. 2nd ICS, 2 cm to left
28. 75% of the time the _____ is the dominant artery.	28. right coronary - supplies the posterior interventricular artery
29. The posterior interventricular artery travels with which vein?	29. middle cardiac vein
30. The lower third of the esophagus drains into the _____ vein, which drains into the ____ vein, thus forming an important venous anastomosis.	30. left gastric, portal
31. The right thoracic duct drains what part of the body?	31. right half of the body above the diaphragm
32. The anterior esophageal plexus is from the _____ vagus.	32. left
33. The cupula of the lung extends about _____ inches into the neck, superior to the medial third of the clavicle.	33. 1½
34. What is the innervation of the transversus thoracus?	34. intercostal nerves
35. What is the origin of the nerve fibers of the phrenic nerve?	35. $C_{3,4,5}$ ("keeps your diaphragm alive")
36. What is the endothoracic fascia?	36. connective tissue which attaches the parietal pleura to the thoracic wall
37. What are the roughened inner muscles of the ventricles called?	37. trabeculae carneae
38. Is the SA node the normal pacemaker of the heart?	38. Yes
39. Describe the course of the esophagus through the thoracic cavity.	39. shallow reverse S curve
40. What is the ligamentum arteriosum?	40. fibrous bond between the left pulmonary artery and aorta in the adult which is the remnant of the fetal ductus arteriosus
41. The anterior surface of the heart, viewed in the anatomic position, is mostly composed of _____.	41. right ventricle

ABDOMEN

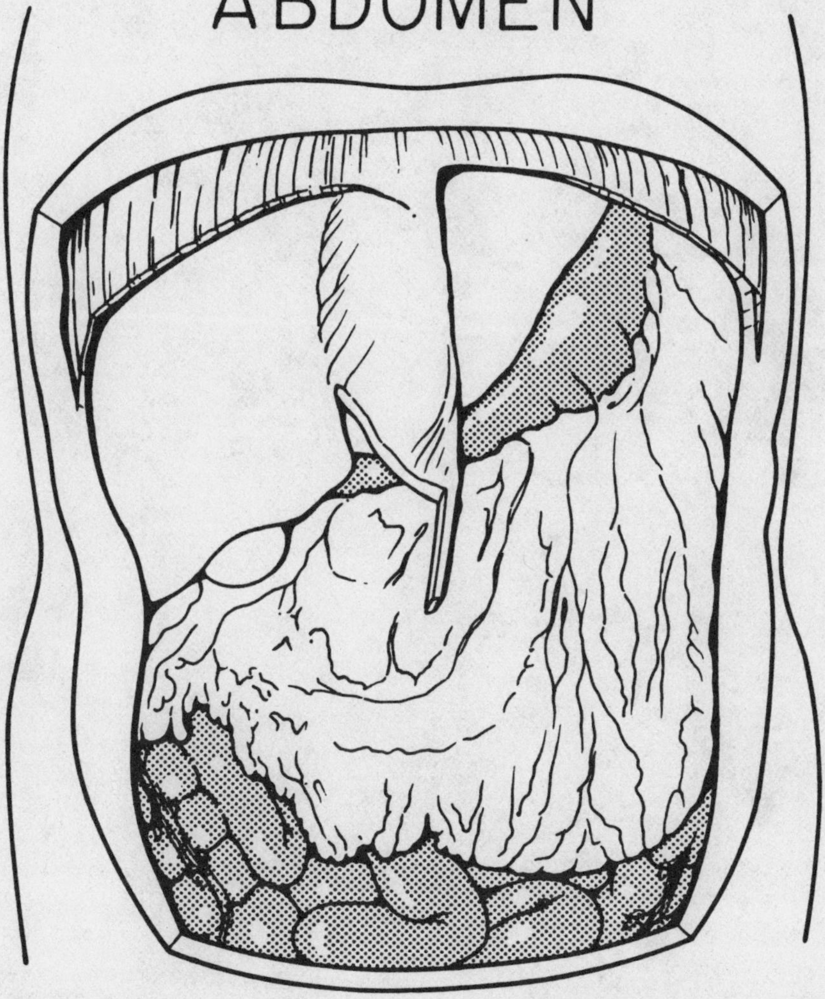

ABDOMEN

I. SURFACE ANATOMY

By vertebral levels

L1 Transpyloric Plane--hilum of kidneys (left just above, right just below), origin of superior mesenteric artery, coeliac artery above, renal artery below, pylorus, body of pancreas, duodenal-jejunal flexure inferior and below
L2 Azygos vein and thoracic duct begin, spinal cord ends in adult
L3 Subcostal Plane
L4 Umbilical Plane--umbilicus, aorta bifurcates to the left of midline, common iliac arteries begin, highest point of iliac crest
L5 IVC begins to right of midline, common iliac veins begin, iliac tubercles, arcuate line

SURFACE FEATURES

1) Sternal notch
2) Costal margin
3) Xiphoid process
4) Linea alba
5) Linea semilunaris
6) Inguinal ligament
7) Iliac crest
8) Tendinous intersection
9) Arcuate line
10) Umbilicus
11) Anterior superior iliac spine--ASIS
12) Pubic crest
13) Pubic tubercle

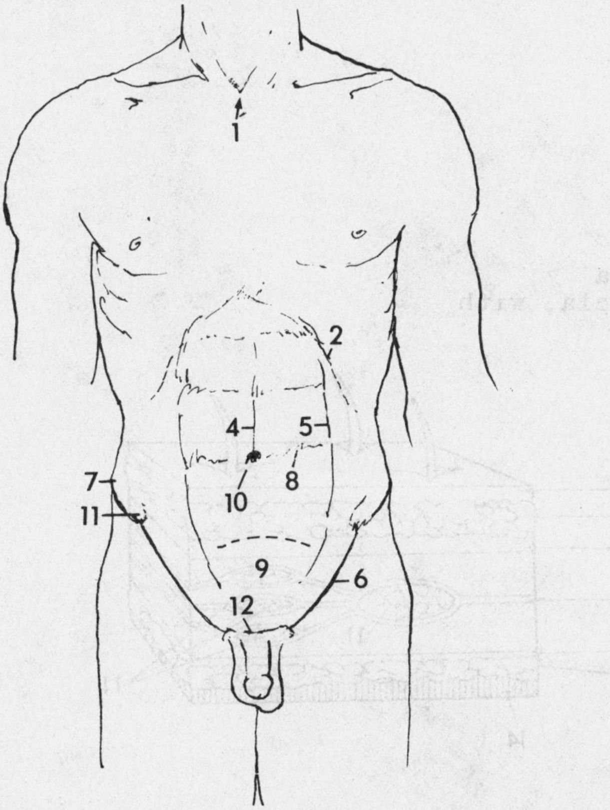

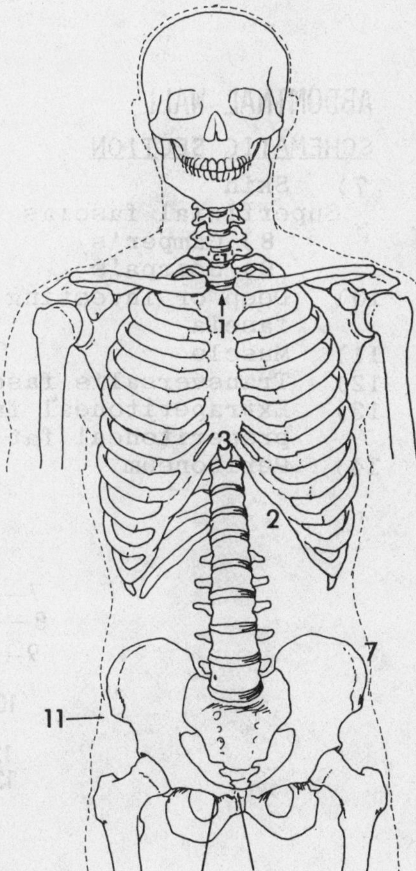

REGIONS AND QUADRANTS

This method uses the Transpyloric Plane and Intertubercular Plane for horizontal planes; some clinicians use Subcostal and Interspinous Planes. Midclavicular lines are used as vertical planes.

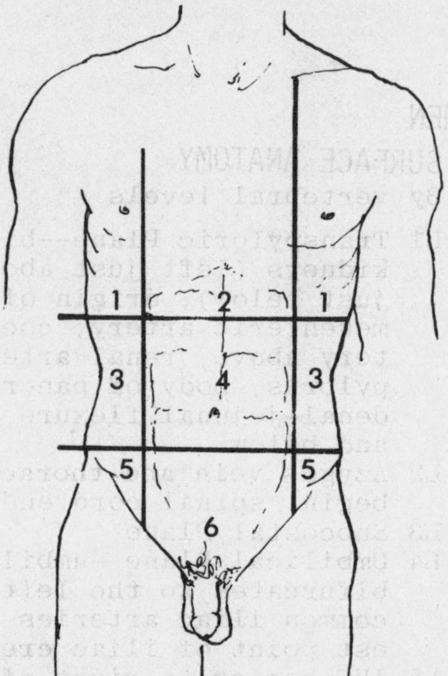

1)	Hypochondriac	4)	Umbilical
2)	Epigastric	5)	Inguinal
3)	Lateral	6)	Pubic

The abdomen is also described in quadrants with the umbilicus at the center. URQ = upper right quadrant, LRQ = lower right quadrant, ULQ = upper left quadrant, LLQ = lower left quadrant.

II. ABDOMINAL WALL

SCHEMATIC SECTION

7) Skin
Superficial fascias
8) Camper's
9) Scarpa's
10) Deep or investing fascia
11) Muscle
12) Transversalis fascia
13) Extraperitoneal fascia, with preperitoneal fat
14) Peritoneum

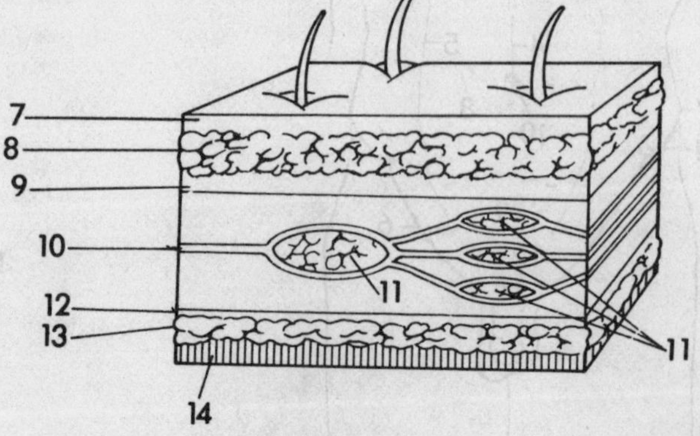

SKIN AND SUBCUTANEOUS TISSUES

1) Skin and Camper's Fascia
2) Scarpa's Fascia--membranous layer of superficial fascia
3) Spermatic cord
4) Iliohypogastric nerve T12-L1
5) Ilioinguinal nerve L1
6) Fundiform ligament--thickening of Scarpa's Fascia
7) Fascia lata of the thigh

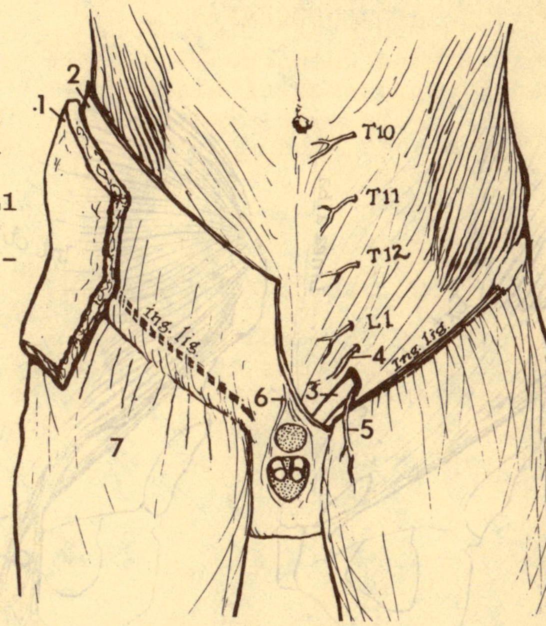

ANTERIOR ABDOMINAL WALL MUSCULATURE

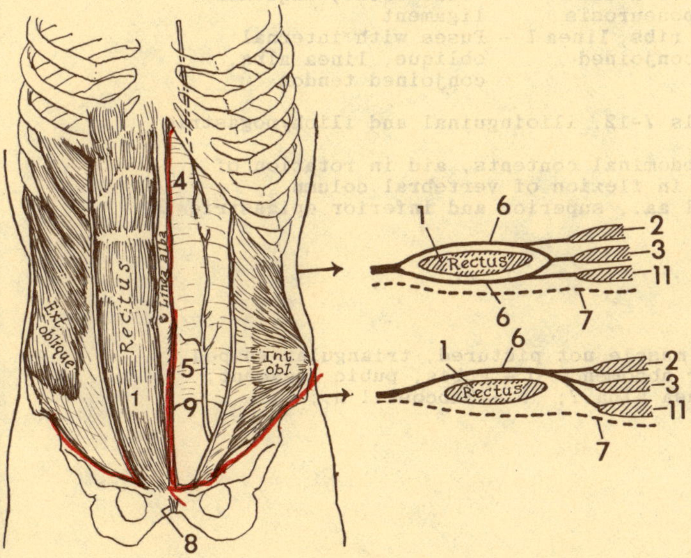

1) Rectus abdominus m.
2) External oblique m.
3) Internal oblique m.
4) Superior epigastric a.
5) Inferior epigastric a.
6) Rectus sheath--anterior and posterior laminae
7) Transversalis fascia
8) Pubis
9) Arcuate line
10) Transversus abdominus

OBSERVE THE DIFFERENCE IN COMPOSITION OF THE ANTERIOR AND POSTERIOR LAMINAE OF THE RECTUS SHEATH ABOVE AND BELOW THE ARCUATE LINE.

RECTUS ABDOMINUS

O - Pubic crest, pubic symphysis
I - Cartilages ribs 5-7
A - Flexes vertebral column, tenses abdominal wall, compresses abdominal contents
N - Intercostal nerves 7-12
B - Superior and inferior epigastric aa., some intercostal and subcostal aa.

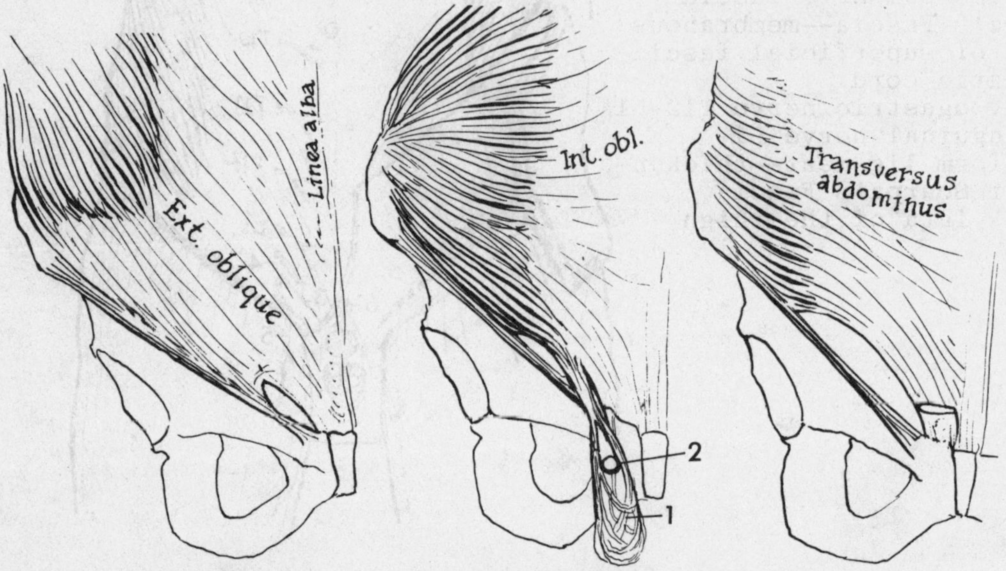

EXTERNAL OBLIQUE	INTERNAL OBLIQUE	TRANSVERSUS ABDOMINUS
O - Lower 8 ribs I - iliac crest, linea alba, thoracolumbar aponeurosis	O - Iliac crest, inguinal ligament, posterior abdominal aponeurosis I - Last 4 ribs, linea alba, conjoined tendon	O - Last 6 ribs, thoracolumbar aponeurosis, iliac crest, inguinal ligament I - Fuses with internal oblique, linea alba, conjoined tendon

ALL OF THE ABOVE--N - Intercostals 7-12, ilioinguinal and iliohypogastric nerves
A - Compress abdominal contents, aid in rotation of torso, aid in flexion of vertebral column
B - Intercostal aa., superior and inferior epigastric aa.

STRUCTURES:

1. Cremaster m.
2. Spermatic cord

PYRAMIDALIS is an inconstant muscle not pictured, triangular shaped within rectus sheath at lower abdomen. O - Pubis, pubic ligament; I - Linea alba; A - Tense linea alba ?; N - Subcostal n.; B - Inferior epigastric a.

THE INGUINAL CANAL

The inguinal canal is an oblique passageway through the abdominal wall for the spermatic cord or round ligament, and the ilioinguinal nerve. The deep inguinal ring is the internal orifice of the canal while the superficial or external ring is the superficial orifice.

THE INGUINAL LIGAMENT is the lower border of the aponeurosis of the external oblique, and extends from the ASIS to the pubic tubercle; supports the spermatic cord or round ligament.

1) **Inguinal ligament of Poupart**
2) **Lacunar ligament of Gumbernat--joins pectineal ligament, forms medial margin of femoral canal**
3) **Pectineal ligament of Cooper forms the deep surface of the femoral canal, continuation of (2)**

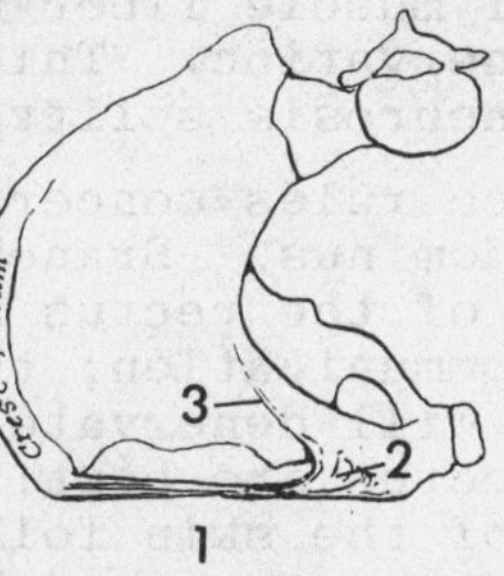

THE SPERMATIC CORD

LAYERS: (Skin and its fascias not shown)

4) External spermatic fascia from the external oblique muscle (5)
6) Cremasteric fascia and muscle from the internal oblique muscle (7)
8) Internal spermatic fascia from the transversalis fascia (9)
10) Tunica vaginalis

CONTENTS OF THE CORD:

11) Cremasteric branch of internal epigastric a.
12) Testicular artery
13) Deferential a., from superior vesical a.
14) Pampiniform plexus, from the testicular veins
15) Vas deferens
16) Genital branch of genitofemoral n.

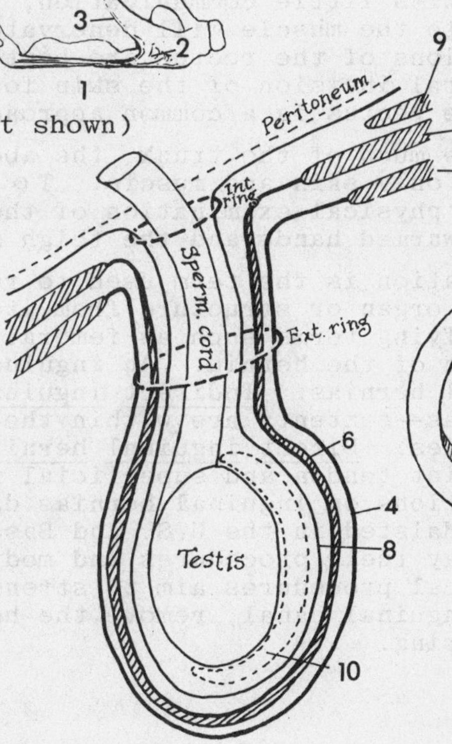

Also autonomic and sensory nn., lymphatics and deferential v.

THE CREMASTER MUSCLE IS A CONTINUATION OF THE INTERNAL OBLIQUE MUSCLE drawn into the scrotum as the testes migrate through the abdominal wall. The cremaster muscle draws the testes towards the inguinal rings; I - Genital branch of genitofemoral n. L1-2,;B - External spermatic artery (cremasteric) from the inferior epigastric.

COMMENTS

1 The abdominal wall is large in surface area and is used as an entry to the vast array of abdominal contents. It is frequently the site of trauma. In contemplating surgical entry as well as repair of trauma the cleavage lines of the skin, muscle fiber direction, pathway of blood supply and innervation must always be considered.

2 As a rule, surface incisions should be made in the direction of Langer's cleavage lines. Deeper incisions should be made in the direction of muscle fiber and not perpendicular to blood supply or innervation. This style incision is referred to as a muscle-aponeurosis splitting incision.

3 An exception to such rules concerns incisions around the midline and rectus abdominus. Branches of spinal nerves enter the lateral border of the rectus and proceed towards the midline with little communication; therefore an incision parallel to the muscle will denervate the muscle. Transverse incisions of the rectus are best. However, a paramedian sagittal incision of the skin followed by lateral reflection of the rectus is a common approach.

4 Unlike much of the trunk, the abdomen is not protected by bone, only skin and muscle. To prevent tensing of the muscles, physical examination of the abdomen should be performed with warmed hands and the thigh slightly flexed.

5 Herniation is the term used to refer to the abnormal extension of an organ or structure from its normal position in a cavity. Qualifying terms such as femoral or umbilical locate the position of the hernia. An inguinal hernia is the most common of all hernias. Indirect inguinal hernias: 75% of all inguinal hernias--contents are within the spermatic cord, most common in males. Direct inguinal hernias: exit directly through the conjoint tendon and superficial ring--most common in males. Operations on inguinal hernias date to the turn of the century with Halsted in the U.S. and Bassini in Italy. To the present day their procedures and modifications thereof are used. Surgical procedures aim to strengthen the posterior wall of the inguinal canal, remove the hernia sac and repair the deep ring.

THE INFERIOR ANTERIOR ABDOMINAL WALL

is important as an entry in appendectomy, cesarian deliveries, bladder surgeries and hernia repairs. A suprapubic incision may be made and the peritoneum dissected away to reach retroperitoneal structures.

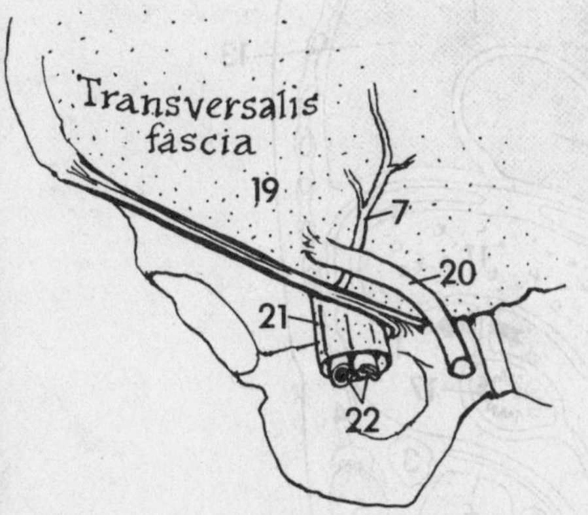

EXTERNAL VIEW

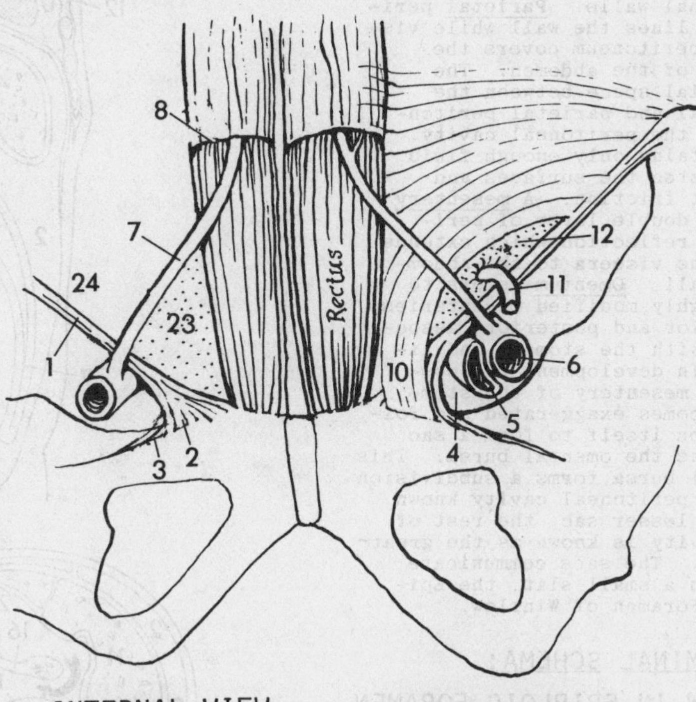

INTERNAL VIEW
PERITONEUM REMOVED

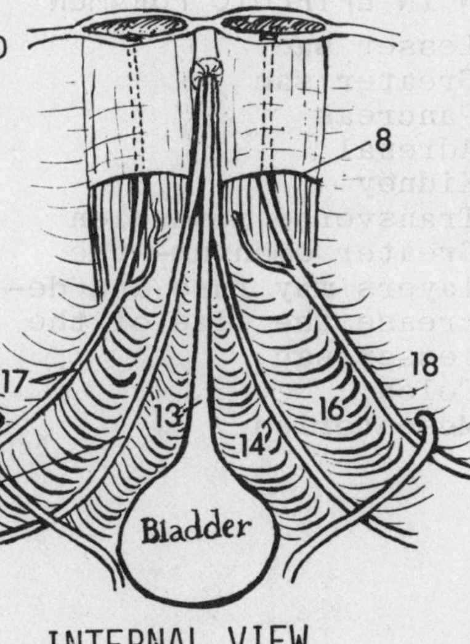

INTERNAL VIEW
PERITONEUM IN SITU

LEGEND:

1) Inguinal ligament
2) Lacunar ligament
3) Pectineal ligament
4) Femoral canal
5) External iliac v.
6) External iliac a.
7) Inferior epigastric a.
8) Arcuate line
10) Conjoint tendon
11) Ductus deferens
12) Deep inguinal ring
13) Median umbilical fold --urachus
14) Supravesicular fossa
15) Medial umbilical fold
16) Medial inguinal fossa
17) Lateral umbilical fold --over (7)
18) Lateral inguinal fossa
19) Transversalis fascia
20) Spermatic cord
21) Femoral sheath
22) Femoral vessels
23) Hesselbach's Triangle medial to (7), area of direct herniation
24) Area of indirect herniation

III. ABDOMINAL CAVITY

PERITONEUM

The peritoneum lines the abdominal cavity and is the largest serous membrane in the body. It is composed of a simple squamous epithelium generally termed mesothelium. Areolar connective tissue attaches the peritoneum to the fascia of the abdominal walls. Parietal peritoneum lines the wall while visceral peritoneum covers the organs of the abdomen. The potential space between the visceral and parietal peritoneum is the peritoneal cavity. It contains only enough fluid to moisten the surfaces and prevent friction. A mesentery is the double layer of peritoneal reflection which extends from the viscera to the abdominal wall. Omentum refers to the highly modified mesenteries (anterior and posterior) associated with the stomach and liver. In development the posterior mesentery of the stomach becomes exaggerated and folded upon itself to form a sac known as the omental bursa. This omental bursa forms a subdivision of the peritoneal cavity known as the lesser sac; the rest of the cavity is known as the greater sac. The sacs communicate through a small slit, the Epiploic Foramen of Winslow.

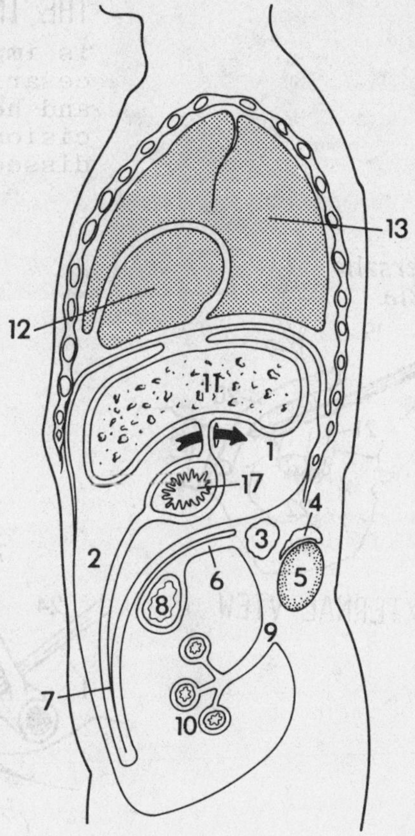

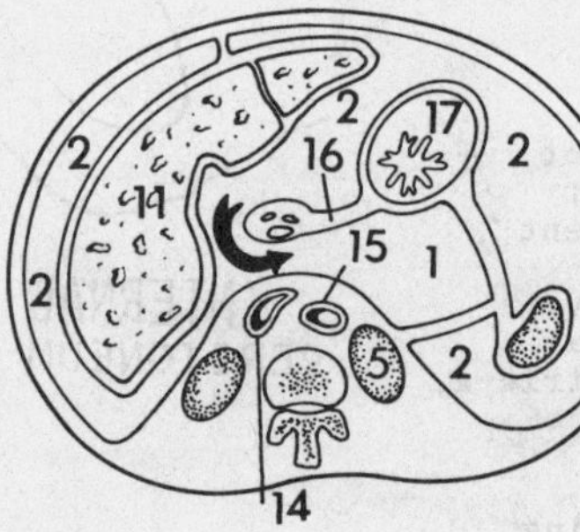

ABDOMINAL SCHEMA:

ARROW IN EPIPLOIC FORAMEN

1) Lesser sac
2) Greater sac
3) Pancreas
4) Adrenal
5) Kidney
6) Transverse mesocolon
7) Greater omentum--the layers may fuse and decrease the size of the lesser sac
8) Colon
9) Mesenteries
10) Small intestine
11) Liver
12) Heart
13) Lungs
14) IVC
15) Aorta
16) Lesser omentum
17) Stomach

IV. ABDOMINAL CONTENTS

VISCERA IN SITU AND PERITONEAL SPECIALIZATIONS

PERITONEAL LIGAMENTS are not true ligaments. They are thickenings of the peritoneum which serve a more specialized supportive role than other portions of the peritoneum and mesenteries.

ANTERIOR VIEWS

1) Liver
2) Falciform ligament
3) Ligamentum teres hepatis
4) Stomach
5) Gallbladder
6) Ascending colon
7) Transverse colon
8) Descending colon
9) Small bowel
10) Greater omentum
11) Gastrophrenic ligament
12) Gastrolienal ligament
13) Lienorenal ligament

LESSER OMENTUM

14) Gastrohepatic ligament
15) Hepatoduodenal ligament
CONTAINS:
Common bile duct, hepatic artery and portal vein

ARROW is in the epiploic foramen

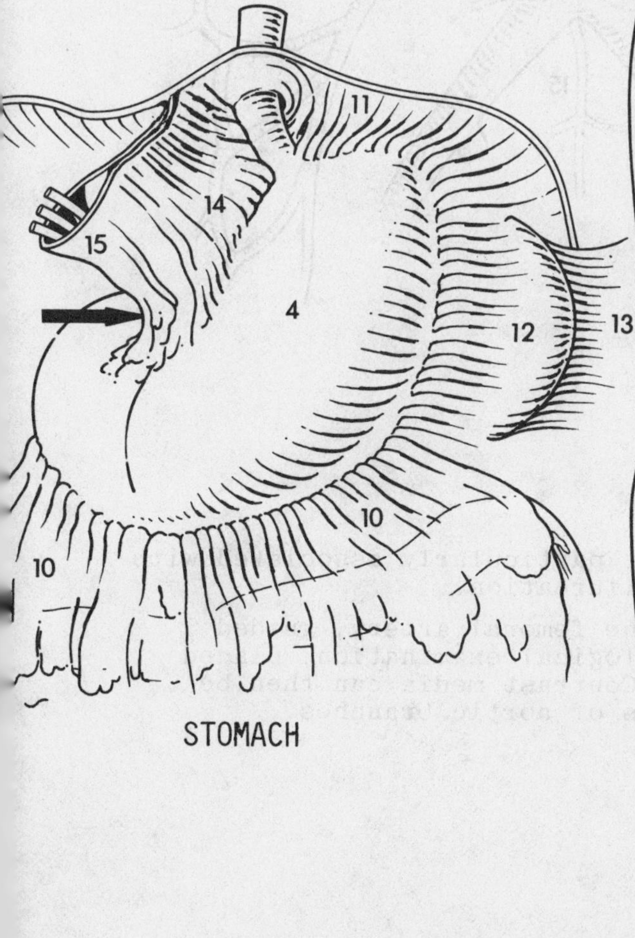

STOMACH

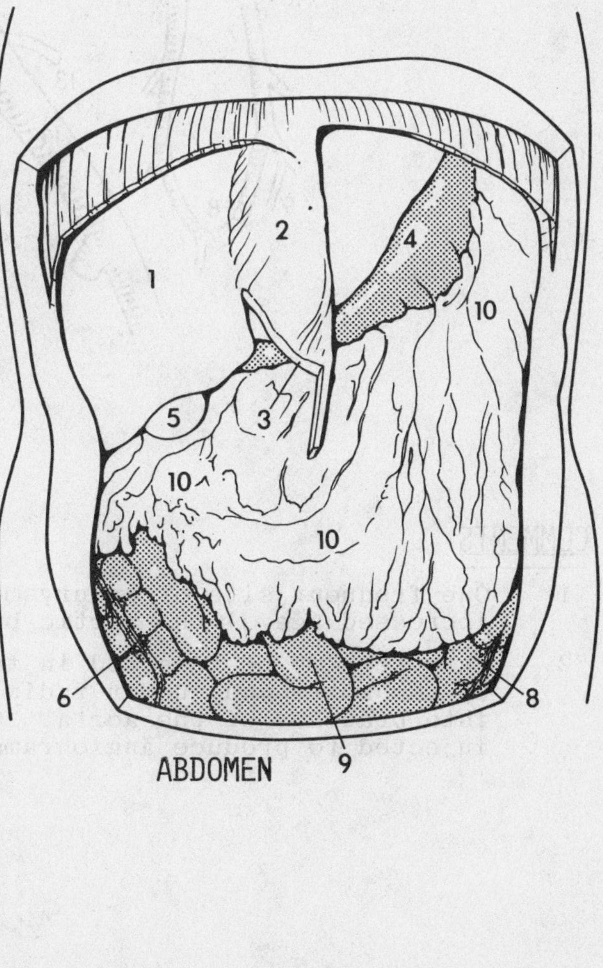

ABDOMEN

ABDOMINAL AORTA

STRUCTURES:

1) Coeliac
2) Inferior phrenic
3) Adrenal
4) Renal
5) Gonadal
6) Superior mesenteric
7) Ileocolic
8) Ileal branch
9) Appendicular
10) Right colic
11) Middle colic
12) Inferior mesenteric
13) Lumbar
14) Common iliac
15) Middle sacral

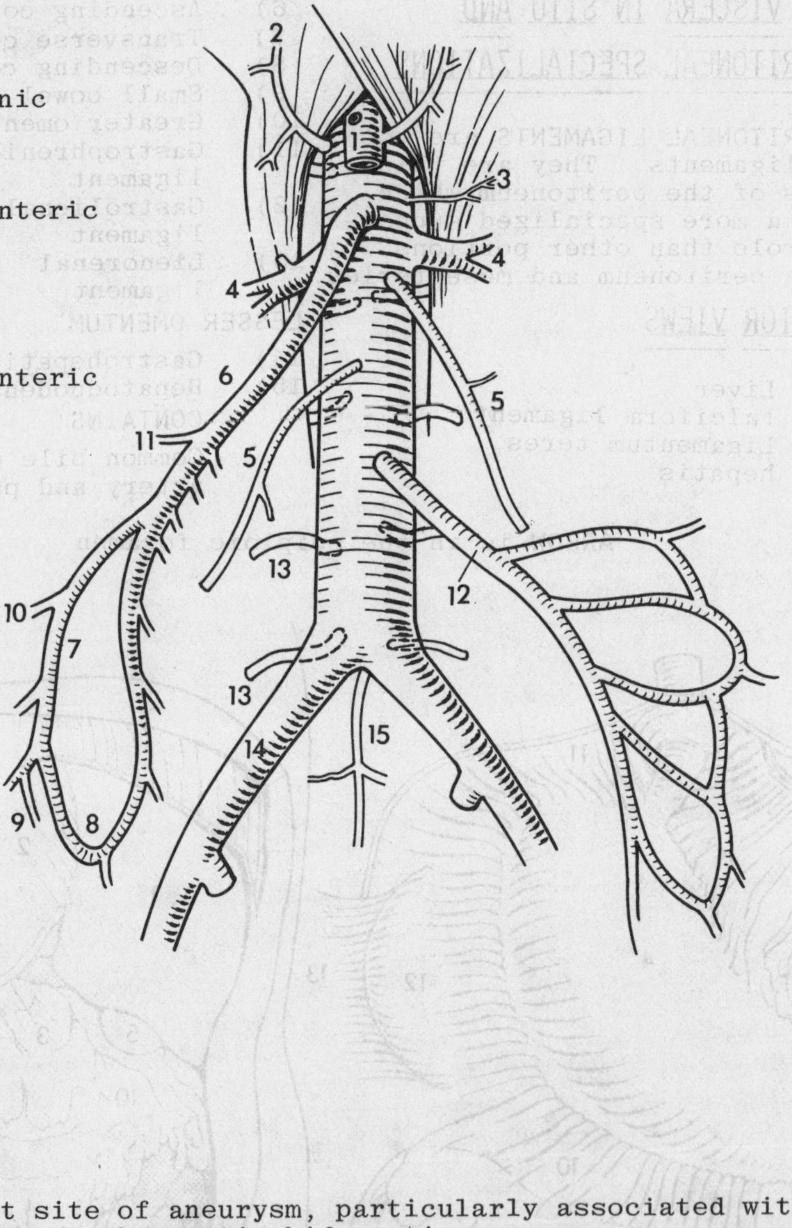

COMMENTS

1 One frequent site of aneurysm, particularly associated with increased age, is the aortic bifurcation.

2 Catheters my be inserted in the femoral artery, guided up the aorta and, under radiological examination, placed into branches of the aorta. Contrast media can then be injected to produce angiograms of aortic branches.

INFERIOR VENA CAVA

STRUCTURES:

1) Hepatic
2) Inferior phrenic
3) Suprarenal
4) Renal
5) Gonadal
6) Lumbar
7) Median sacral
8) Iliolumbar
9) Ascending lumbar
10) Common iliac
11) Hemiazygos
12) Azygos

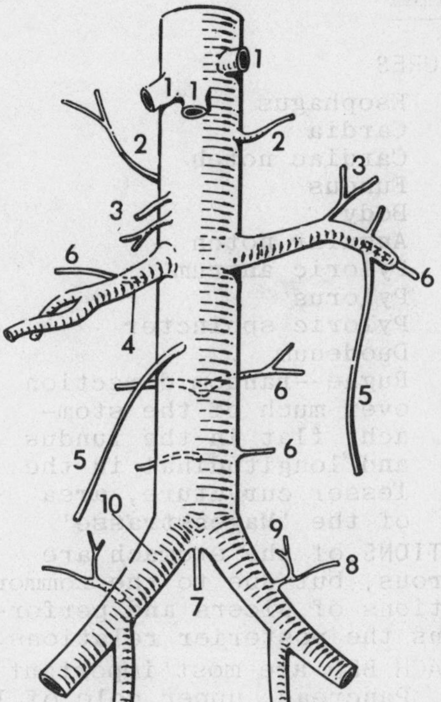

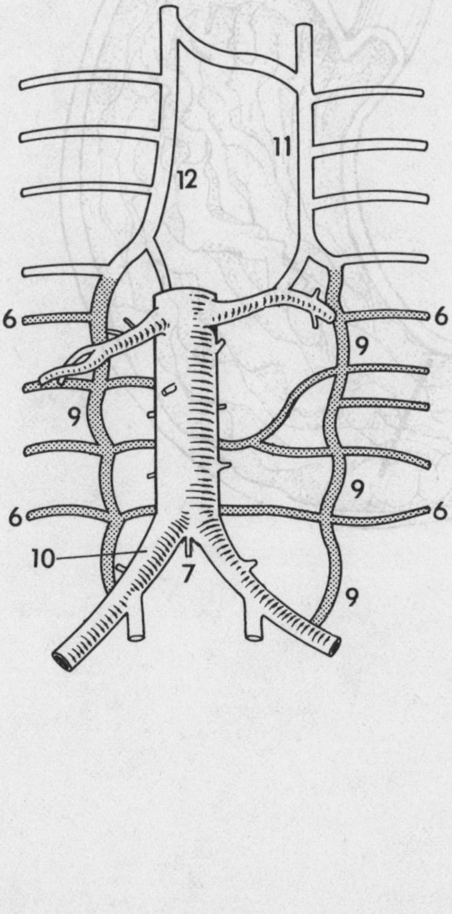

STOMACH

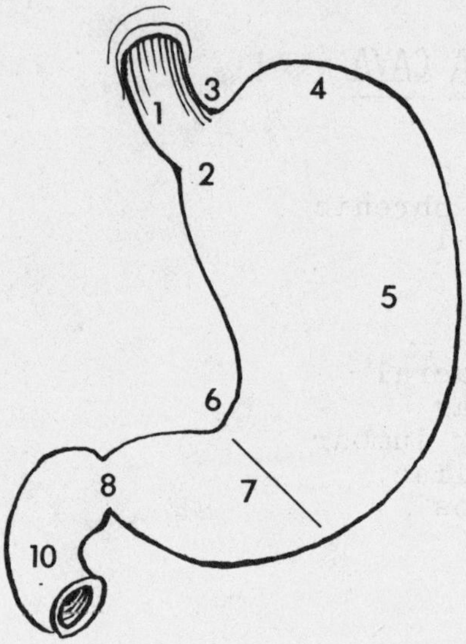

FEATURES

1) Esophagus
2) Cardia
3) Cardiac notch
4) Fundus
5) Body
6) Angular notch
7) Pyloric antrum
8) Pylorus
9) Pyloric sphincter
10) Duodenum
11) Rugae--random direction over much of the stomach, flat in the fundus and longitudinal in the lesser curvature, area of the 'Magenstrasse'

RELATIONS of the stomach are numerous, but due to the common positions of ulcers and perforations the posterior relations, or STOMACH BED, are most important:

Pancreas, upper pole of left kidney and left suprarenal gland, left crus of the diaphragm, left celiac ganglion, gastric surface of the spleen, and posterior wall of the omental bursa.

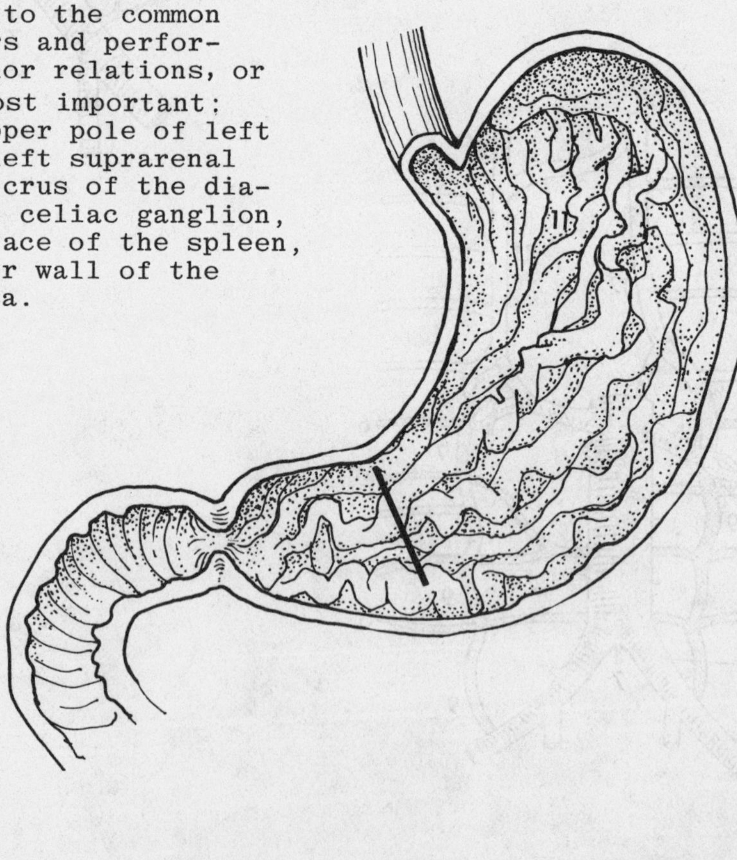

ARTERIES

(12) Celiac trunk
(13) Splenic
(14) Left gastric
(15) Esophageal branch
(16) Common hepatic
(17) Proper hepatic
(18) Right and left hepatic arteries
(19) Right gastric
(20) Gastroduodenal
(21) Anterior superior pancreaticoduodenal
(22) Right gastroepiploic
(23) Left gastroepiploic
(24) Short gastric arteries
(25) Splenic branches
(26) Spleen

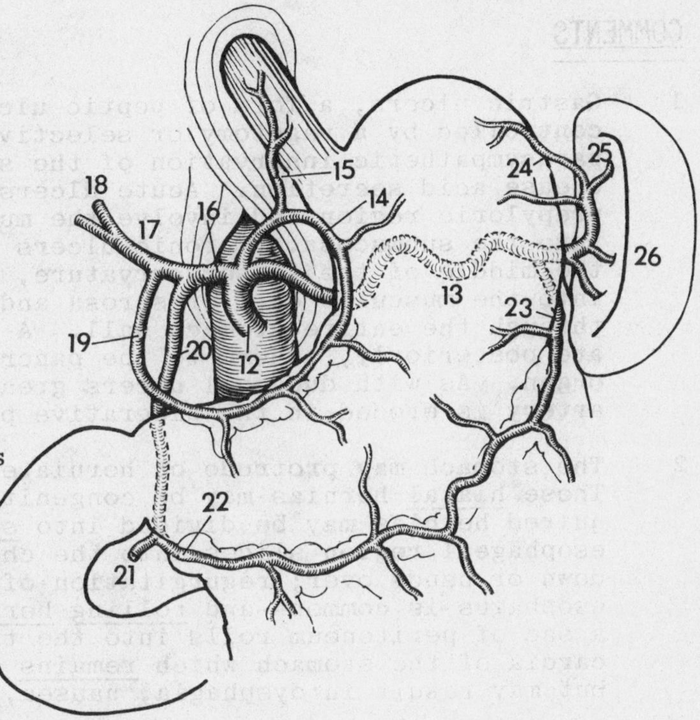

LYMPHATIC DRAINAGE OF THE STOMACH

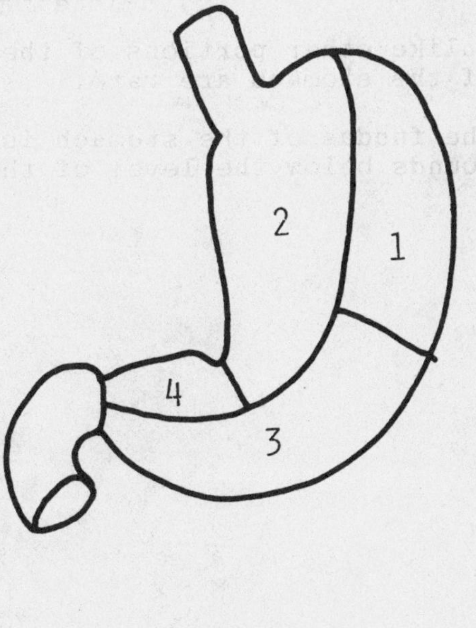

REGION	LYMPHSHED
1	Lower left gastric and paracardial nodes
2	Pancreaticolienal and gastroepiploic nodes
3	Right gastroepiploic nodes
4	Right suprapancreatic nodes

ALL ABOVE DRAIN to coeliac nodes, cisterna chyli and thoracic duct

COMMENTS

1 Gastric ulcers, a form of peptic ulcers or sores, may be controlled by a vagotomy or selective vagotomy to remove parasympathetic innervation of the stomach and thereby decrease acid secretion. Acute ulcers are most common in the prepyloric region and involve the mucosa, possibly extending into the submucosa. Chronic ulcers are generally found in the middle of the lesser curvature, posteriorly. They extend into the muscular coat or serosa and may perforate or erode through the entire stomach wall. A chronic ulcer may perforate posteriorly, adhere to the pancreas and involve that organ. As with duodenal ulcers great danger occurs when an artery is eroded in the ulcerative process.

2 The stomach may protrude or herniate through the diaphragm. These hiatal hernias may be congenital or acquired. Acquired hernias may be divided into sliding hernias--the gastroesophageal region slides into the chest when the patient lies down or bends over; regurgitation of stomach acid into the esophagus is common--and rolling hernias where the fundus and a sac of peritoneum rolls into the thorax adjacent to the cardia of the stomach which remains in place; no regurgitation but may result in dysphagia, nausea, vomiting and pain.

3 Congenital pyloric stenosis or narrowing of the pyloric canal due to hypertrophy of the pylorus is not an uncommon problem. Treatment is usually surgical.

4 Unlike other portions of the gastrointestinal tract, diverticula of the stomach are rare.

5 The fundus of the stomach is liable to injury in penetrating wounds below the level of the left nipple.

SMALL INTESTINE

DUODENUM

First 12", encircles L2, head of pancreas rests in curve

1) Superior part or 'cap', peritonealized by hepatoduodenal ligament
2) Descending part, receives hepatopancreatic ampulla with bile and pancreatic products
3) Horizontal part
4) Ascending part
5) Duodenal-jejunal flexure
6) Suspensory ligament of Treitz

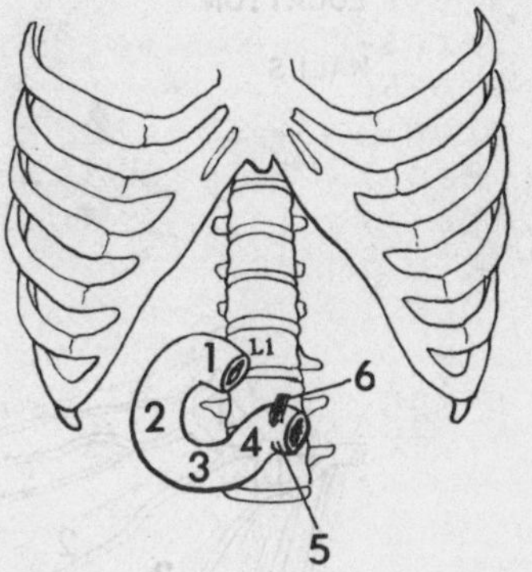

JEJUNUM AND ILEUM

1) Jejunum
2) Ileum
3) Superior mesenteric artery
4) Intestinal branches
5) Iliocolic branch
6) Appendicular artery

Blood to duodenum is from superior mesenteric and coeliac artery--see Pancreas

COMPARISON OF JEJUNUM AND ILEUM

	Jejunum	Ileum
ARTERIES	Few arcades (1) Long vasa recti (2)	Several arcades Short vasa recti
LOCATION	ULQ	LRQ
WALLS	Thicker, more plicae circulares (3), mesenteric fat does not reach gut wall (4)	Thinner, fewer plicae, fat does reach gut wall, Peyer's Patches in gut wall (5) are lymphatic tissue

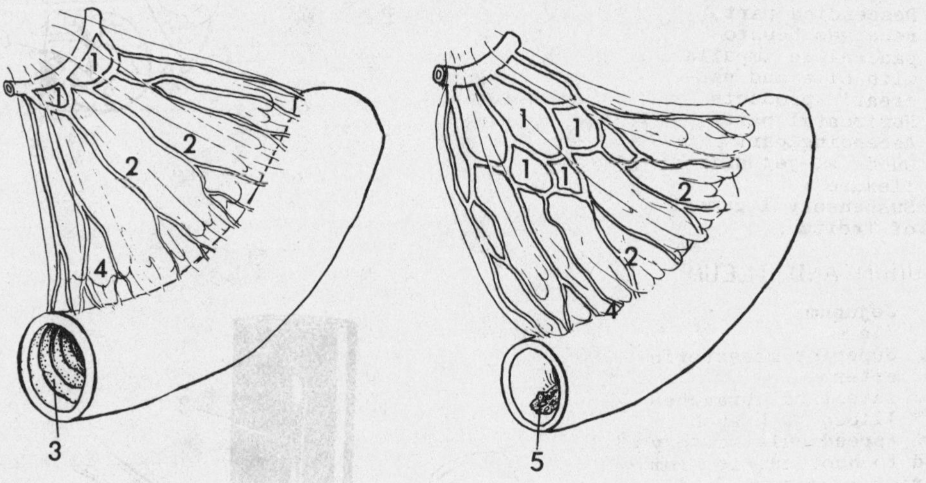

JEJUNUM ILEUM

LYMPHATICS

DUODENUM → SUPRAPANCREATIC NODES → COELIAC NODES → CISTERNA CHYLI

JEJUNUM-ILEUM → MESENTERIC NODES → SUPERIOR MESENTERIC NODES → CISTERNA CHYLI

DUODENUM COMMENTS

1 Duodenal ulcers--another form of peptic ulcer--occur most frequently in the duodenal cap, within 5 cm of the pylorus. If the ulcer perforates through the duodenum peritonitis can result. An ulcer perforating posteriorly may adhere to the pancreas and involve that organ. Severe hemorrhage can occur when pancreatic arteries are eroded.

2 Due to the proximity of the duodenum, pancreas, gallbladder and liver, inflammations and erosions may spread from one to the other with resultant adhesions.

3 Diverticula of the duodenum are common.

JEJUNUM AND ILEUM COMMENTS

1 Diverticula, including Meckel's Diverticulum--a persistent vitelline duct--are common in the small bowel. Inflammation of the diverticula or diverticulitis may eventually result in rupture and peritonitis, or a walling off of the sac.

2 Trauma, carcinoma, enteritis or other problems may necessitate resection of portions of the bowel; other gastrointestinal problems, i.e., persistent gastric ulcers, may be treated by anastomoses with portions of the small bowel; sometimes the ileum may be surgically anastomosed to the rectum. Requirements for all of these manuvers and the various ramifications of these procedures are interesting, but beyond the scope of this study guide. The procedures do, however, require a thorough working knowledge of the normal anatomy of the gastrointestinal tract.

3 Development and rotation of the gut are fraught with potential errors. Results of such errors include stenotic portions of the gut, omphalocoele or external gut, duplication of portions of the gut, volvulus--a twisting of the small bowel around the superior mesenteric artery-- and internal 'hernias' to be discussed later.

4 Diaphragmatic hernias of the small bowel and some proximal colon into the thorax are not rare. A frequent site of this herniation is through the Foramen of Bochdalek in the left posterolateral portion of the diaphragm. The opening is a result of failure of proper development associated with the pleuroperitoneal membrane on that side.

LARGE INTESTINE (COLON)

1) Ascending colon
2) Transverse colon
3) Descending colon
4) Sigmoid colon
5) Rectum
6) Greater omentum last fat reserve
7) Transverse mesocolon
8) Pancreas
9) Hepatocolic ligament
10) Phrenicocolic ligament
11) Mesentery of the sigmoid colon--left ureter lies under the ⌈ shaped attachment
13) Caecum
14) Ileocecal valve
15) Frenulae of valve--in life the sphincter is papilliform rather than labial as in the cadaver
16) Appendix
17) Ostium of appendix
18) Superior mesenteric a.

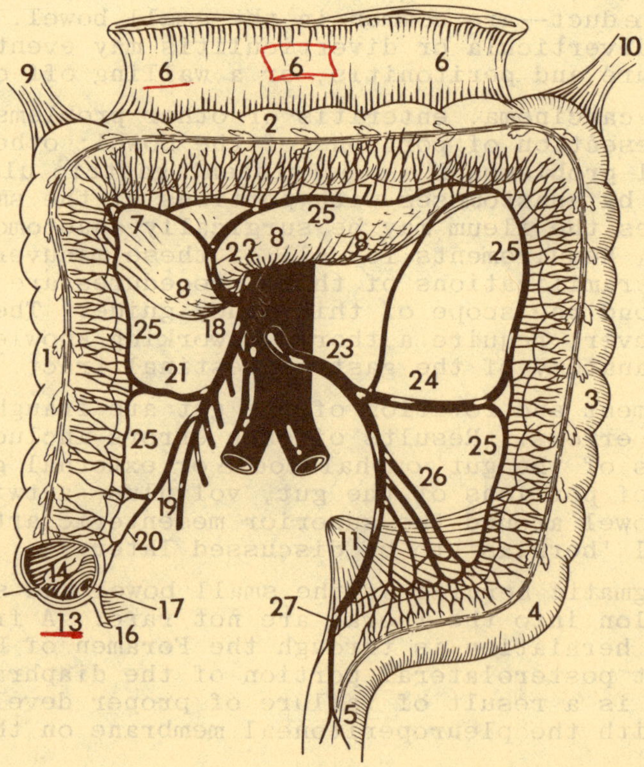

19) Ileocolic a.
20) Appendicular a
21) Right colic a.
22) Middle colic a.
23) Inferior mesenteric a.
24) Left colic a.
25) Marginal a.
26) Sigmoid a.
27) Superior rectal a.

LYMPHATICS

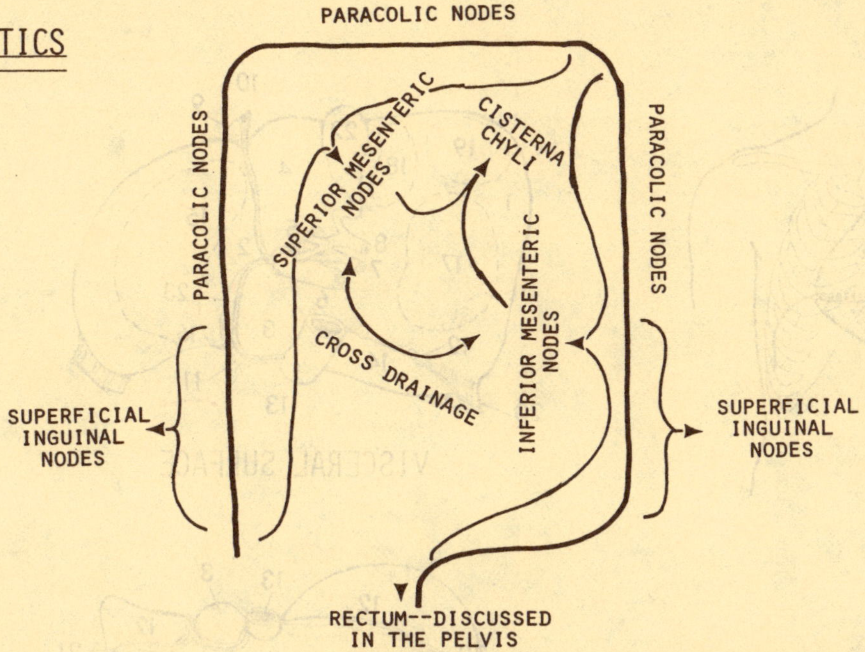

COMMENTS

1 Terminal arteries to the colon pierce the circular muscles between the taenia coli to reach the mucous membrane; a weak site is produced that encourages diverticula--a common occurrence in the colon.

2 Appendicitis is one of the most frequent causes of acute abdominal problems around the world. Often, up to 60% by some sources, the appendix is in the retrocaecal recess. The location of the appendix is beneath McBurney's Point, a spot 1/3 of the way from the ASIS to the umbilicus. A muscle-aponeurosis splitting incision is most frequently used; care must be taken not to jeopardize the iliohypogastric nerve which can underlie the point. Pain from appendicitis is usually vague at onset and becomes localized as the inflammation develops. The RLQ is the usual location for the pain as the peritoneum underlying that quadrant is innervated by somatic nerves and the appendicitis irritates the peritoneum. The phenomenon of <u>rebound pain</u> is particularly evident in appendicitis--there is pain on pressing the area, but even greater pain on release as the peritoneum rebounds from the inflamed organ.

3 Colostomies--a transection of the colon with exteriorization--are routine for carcinomas, ulcerative colitis, trauma and other reasons. They may be temporary, await anastomosis to the rectum or permanently made into an artificial anus.

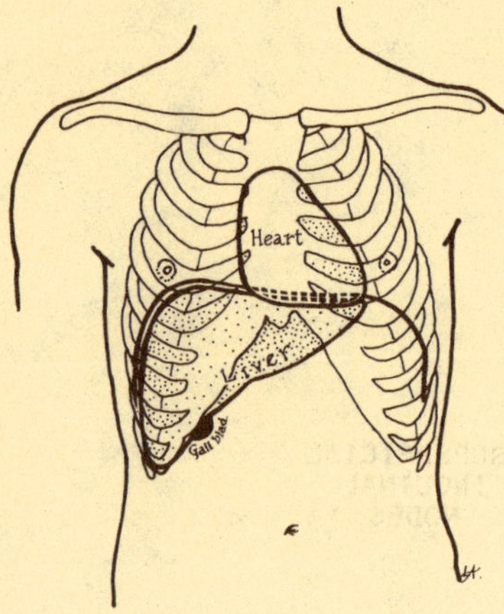

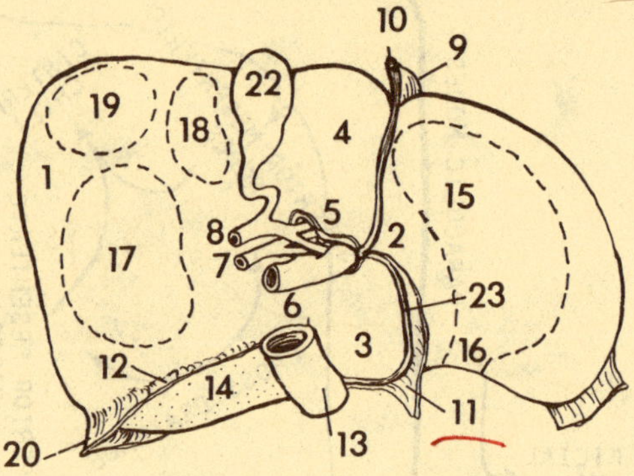

VISCERAL SURFACE

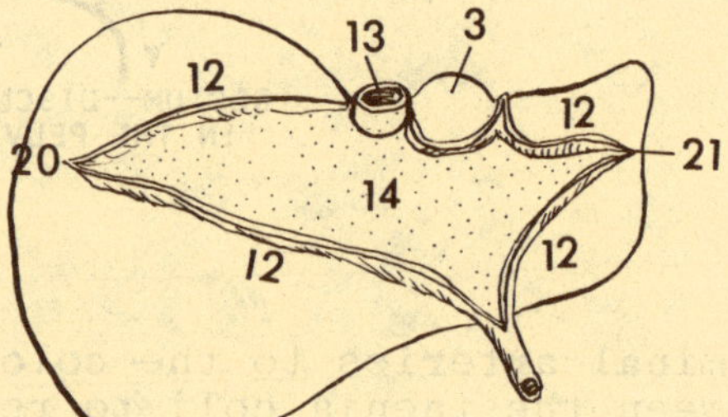

SUPERIOR SURFACE

LIVER

The largest gland in the body;
lies in URQ under subcostal margin

1) Right lobe
2) Left lobe
3) Caudate lobe
4) Quadrate lobe
5) PORTA HEPATIS
   (6) Portal vein
   (7) Hepatic artery
   (8) Common hepatic duct
9) Falciform ligament
10) Ligamentum teres hepatis
11) Lesser omentum
12) Coronary ligament
13) IVC
14) Bare area
15) Stomach impression
16) Esophagus impression
17) Kidney impression
18) Duodenum impression
19) Hepatic flexure-colon impression
20) Right triangular ligament
21) Left triangular ligament
22) Gall bladder
23) Ligamentum venosum

DOTTED LINE INDICATES FUNCTIONAL DIVISION OF RIGHT AND LEFT LIVER BY DIVISIONS OF THE HEPATIC ARTERY INTO RIGHT AND LEFT HEPATIC BRANCHES--NOT ANATOMICAL BY THE FALCIFORM LIGAMENT.

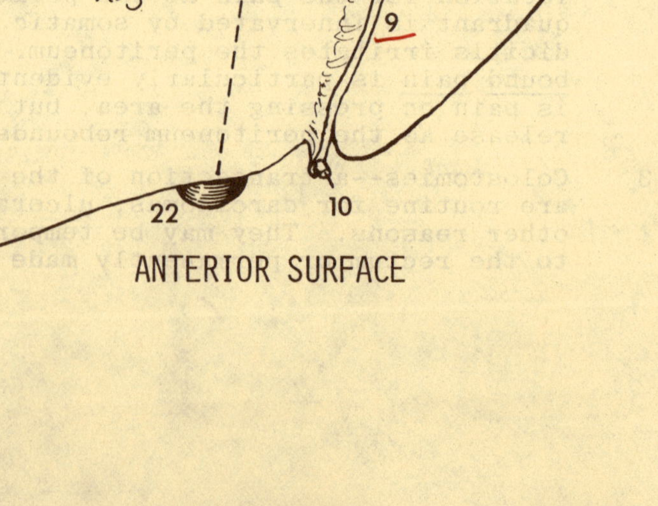

ANTERIOR SURFACE

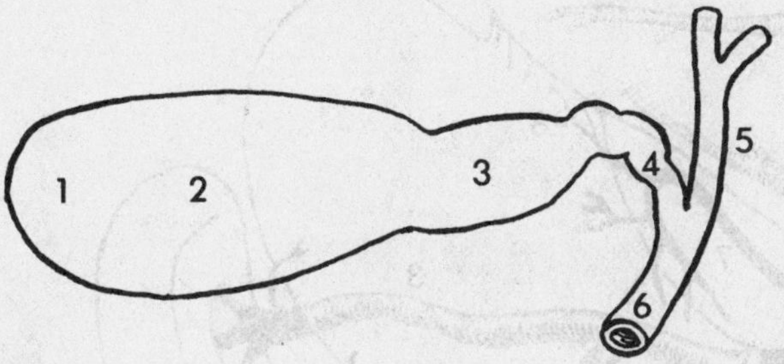

SCHEMATIC VIEW

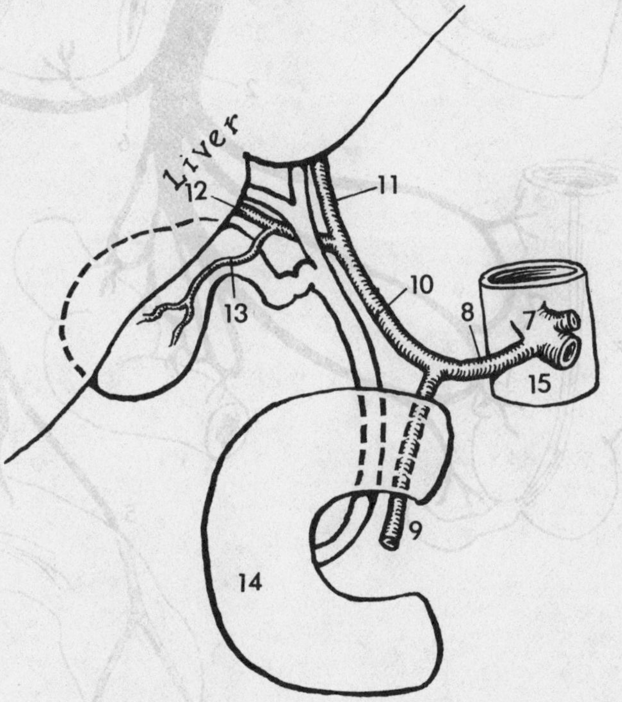

GALLBLADDER

1) Fundus
2) Body
3) Neck
4) Cystic duct
5) Hepatic duct
6) Common bile duct
7) Coeliac trunk
8) Common hepatic artery
9) Gastroduodenal artery
10) Proper hepatic artery
11) Left hepatic artery
12) Right hepatic artery
13) Cystic artery
14) 2nd part of duodenum
15) Aorta

ANTERIOR VIEW

THE PORTAL VEIN in addition to common bile duct and hepatic artery form the structures in the hepatoduodenal ligament.

THE PORTAL SYSTEM

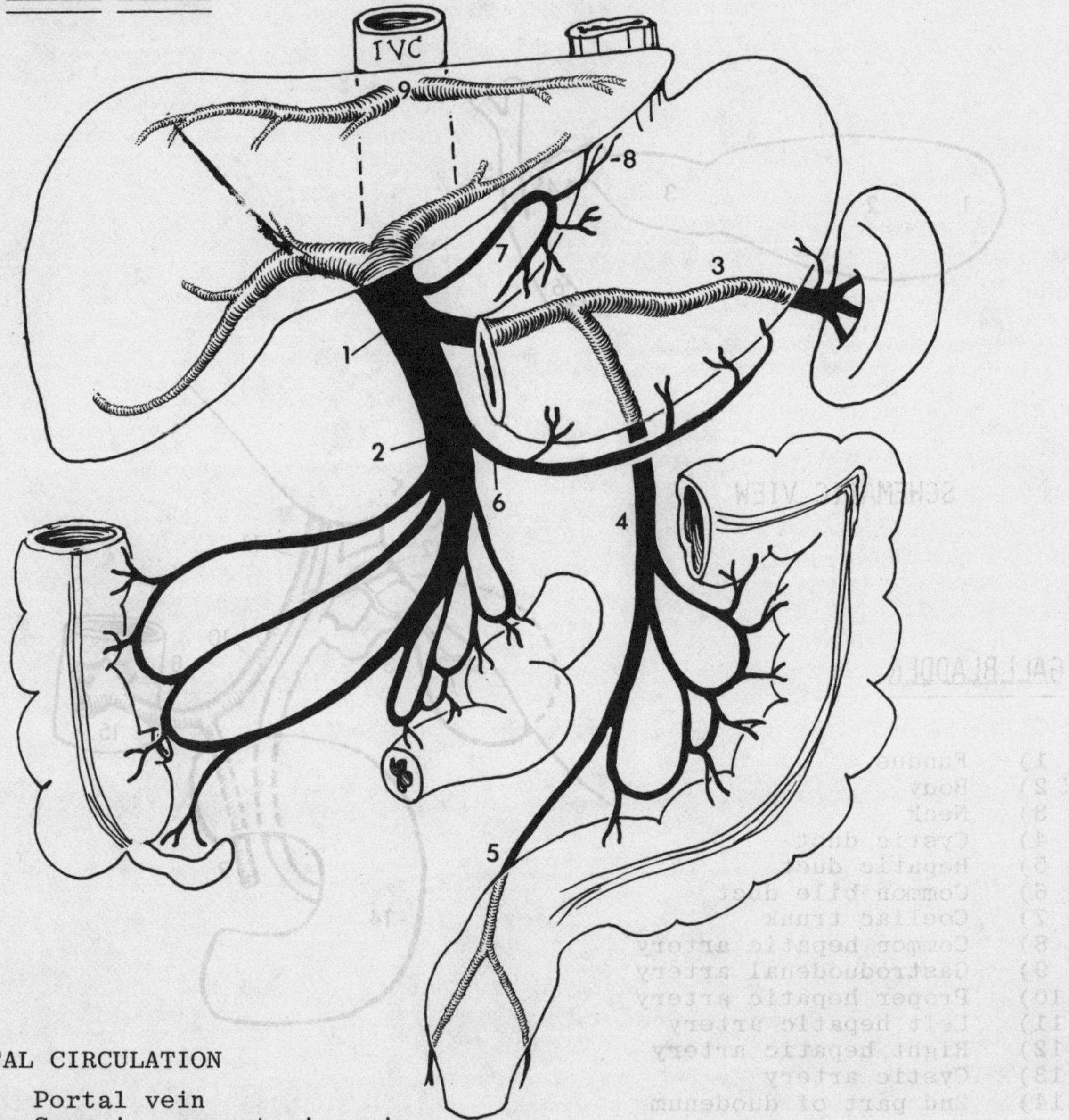

PORTAL CIRCULATION

1) Portal vein
2) Superior mesenteric vein
3) Splenic vein
4) Inferior mesenteric vein
5) Superior rectal vein
6) Right gastroepiploic vein
7) Left gastric vein
8) Esophageal vein
9) Hepatic veins

PORTACAVAL SHUNTS

PORTACAVAL VENOUS ANASTOMOSES are alternate routes for venous blood to return to the IVC in cases of portal vein obstruction. Portal Hypertension is a result. The alternate routes may become dilated, varicose and rupture. Locations of the anastomoses are given below, with clinical signs in BOLD TYPE.

1) Esophageal branches of left gastric vein with esophageal branches of azygos vein
   ESOPHAGEAL VARICES
2) Branches of colic veins and splenic veins with renal veins
3) Superior rectal vein with inferior and middle rectal veins and pelvic venous plexuses
   HEMORRHOIDS
4) Dilatation and opening of old left umbilical vein in ligamentum teres with paraumbilical and epigastric veins
   CAPUT MEDUSAE

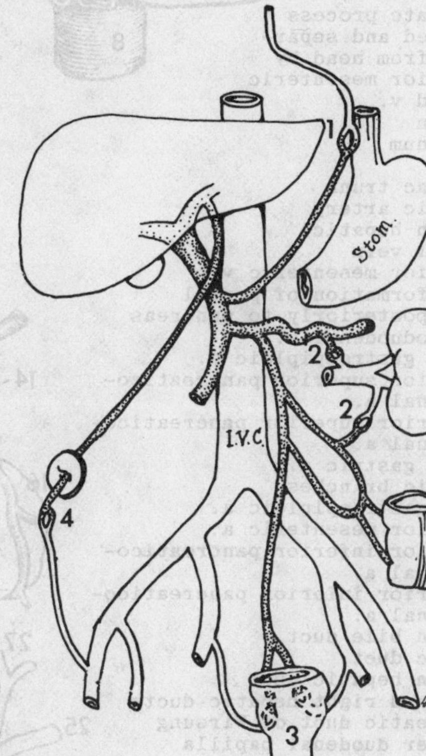

COMMENTS

1 The portal vein transports approximately 70% of the blood that reaches the liver. The remaining 30% is oxygenated blood that reaches the liver via the hepatic artery.

2 The right lobe of the liver receives blood mainly from the intestine. The left lobe, quadrate lobe and caudate lobe receive blood mainly from the stomach and spleen. This distribution may become important in the spread of cancer via the portal system.

3 Portal hypertension commonly results from alcoholism due to cirrhosis of the liver, but may result from conditions such as enlargement of the pancreatic head and compression of the portal vein.

PANCREAS

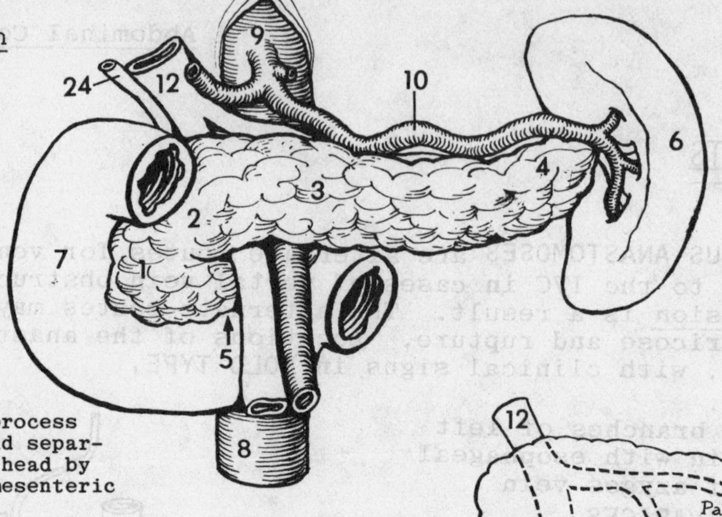

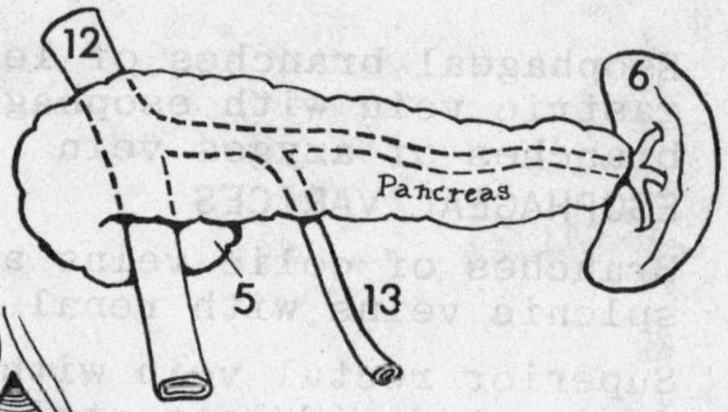

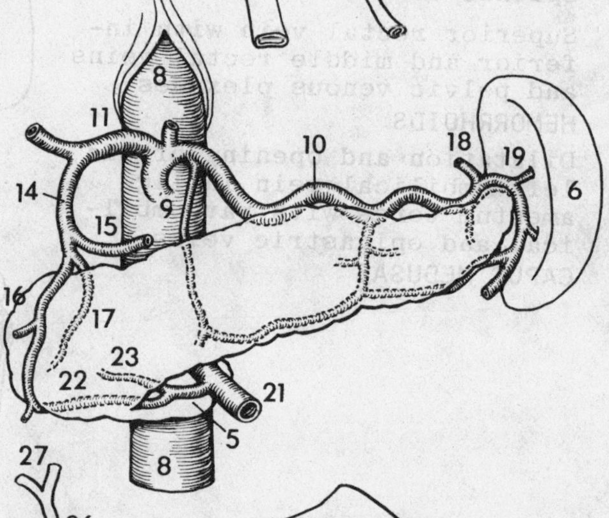

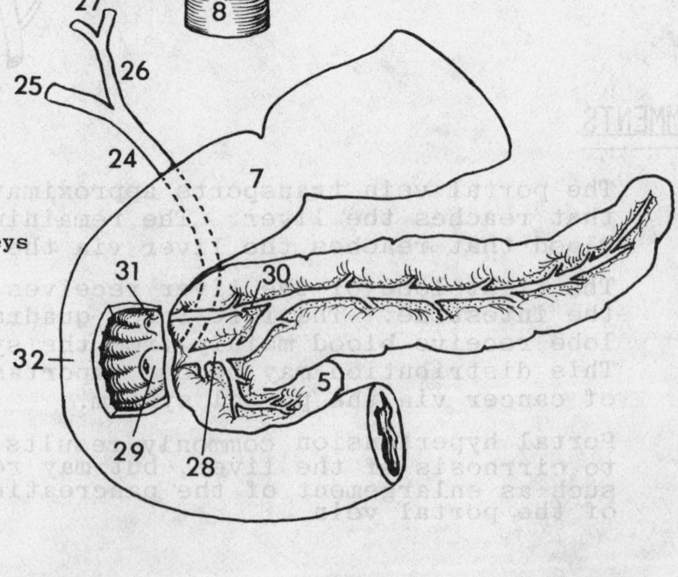

1) Head
2) Neck
3) Body
4) Tail
5) Uncinate process grooved and separated from head by superior mesenteric a. and v.
6) Spleen
7) Duodenum
8) Aorta
9) Coeliac trunk
10) Splenic artery
11) Common hepatic
12) Portal vein
13) Inferior mesenteric v. NOTE formation of portal vein posteriorly to pancreas
14) Gastroduodenal a.
15) Right gastroepiploic a.
16) Anterior superior pancreaticoduodenal a.
17) Posterior superior pancreaticoduodenal a.
18) Short gastric a.
19) Splenic branches
20) Left gastroepiploic a.
21) Superior mesenteric a.
22) Anterior inferior pancreaticoduodenal a.
23) Posterior inferior pancreaticoduodenal a.
24) Common bile duct
25) Cystic duct
26) Common hepatic duct
27) Left and right hepatic ducts
28) Pancreatic duct of Wirsung
29) Greater duodenal papilla
30) Accessory duct of Santorini
31) Lesser duodenal papilla
32) 2nd part of duodenum

RELATIONS

Posterior:	IVC, aorta, hila of kidneys
Anterior:	Transverse mesocolon
Head:	Duodenum
Tail:	Spleen

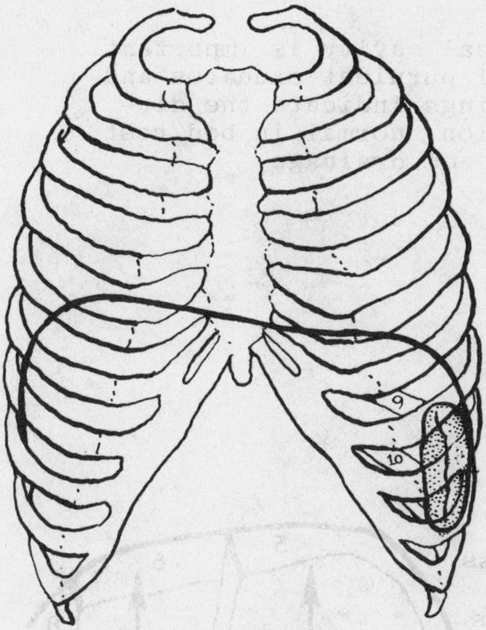

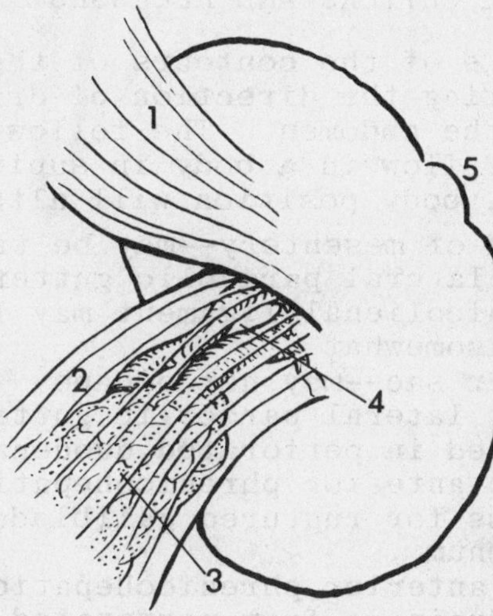

MEDIAL VIEWS

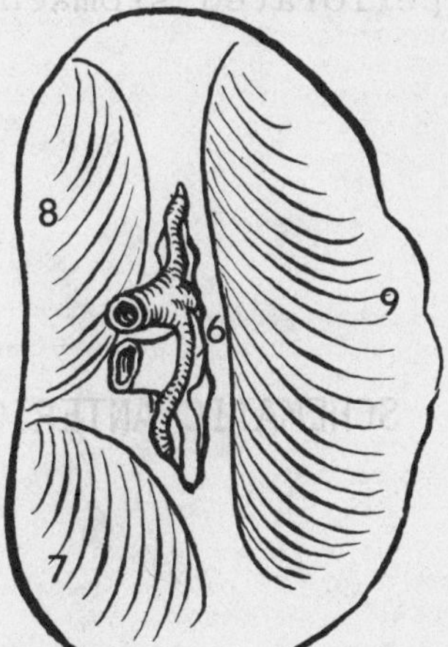

SPLEEN

A lymphatic organ; lies against the diaphragm under protection of the ribs.

1) Gastrosplenic ligament
2) Lienorenal ligament
3) Tail of pancreas
4) Splenic artery and vein
5) Notched superior border
6) Hilum
7) Colonic impression
8) Renal impression
9) Gastric impression

COMMENTS ON THE PERITONEAL CAVITY IN GENERAL

PERITONEAL GUTTERS AND RECESSES

A knowledge of the contours of the abdominal cavity is important in predicting the direction of drainage of purulent exudates and blood in the abdomen. The following drawings indicate the directions of flow in a body in supine position, normal in bed rest. Obviously, body position will alter routes of drainage.

1) Radix of mesentery--may be trapped
2) Left lateral paracolic gutter--phrenicolienal ligament may limit flow somewhat
3) Lesser sac--may not escape
4) Right lateral paracolic gutter favored in perforated duodenal ulcer
5) Right anterior phrenicohepatic recess for ruptured gallbladder, duodenum
6) Left anterior phrenicohepatic recess for drainage from perforated stomach and spleen
7) Right posterior phrenicohepatic recess from right paracolic gutter
8) Left posterior phrenicohepatic recess from perforated stomach

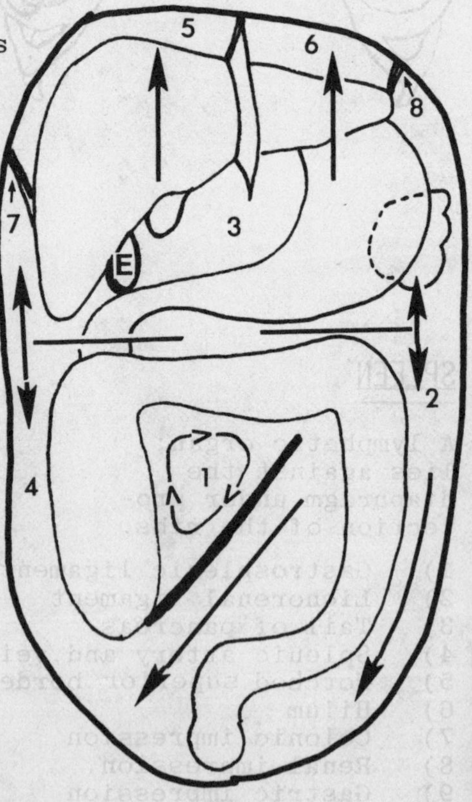

SCHEMATIC ANTERIOR VIEW

The posterior phrenicohepatic recesses (subphrenic spaces) and recto-vesical or recto-uterine pouch of Douglas are the lowest points in a supine individual. A semirecumbent position is often used to insure drainage to the pelvis.

RECESSES

At certain points, generally the transition of portions of the gut from peritoneal to retroperitoneal, the peritoneum forms small pockets or recesses. These occur around the duodenal-jejunal junction superiorly, inferiorly and left laterally; superiorly and inferiorly to the ileocolic junction and posteriorly to the caecum. Knowledge of these points in important because a portion of the small bowel may enter the pouches, enlarge them and be followed by much of the small bowel. These are known as INTERNAL HERNIAS. Strangulation of the gut may occur. At times the small bowel may find its way into the lesser sac through the epiploic foramen. Occasionally the small bowel may pass through perforations in the broad ligament and become strangulated in the female.

HERNIAS

Inguinal hernias have already been discussed as well as internal hernias. However the abdominal contents may herniate at other places:

Femoral	Exits the femoral canal, the hernia sac will be seen at the semilunar hiatus in the upper medial thigh
Epigastric	Through the linea alba
Umbilical	Through the umbilicus
Spigelian	At the semilunar line, below the arcuate line
Post-surgical	Through scars of old incisions
Lumbar	Through Lumbar triangle of Petit--bounded by latissimus, external oblique and iliac crest or through Space of Grynfelt--bounded by internal oblique, 12th rib, serratus posterior inferior and the erector spinae
Obturator	Through the obturator foramen
Sciatic	Through the sciatic foramen
Perineal	Through gaps in the pelvic diaphragm and urogenital diaphragm

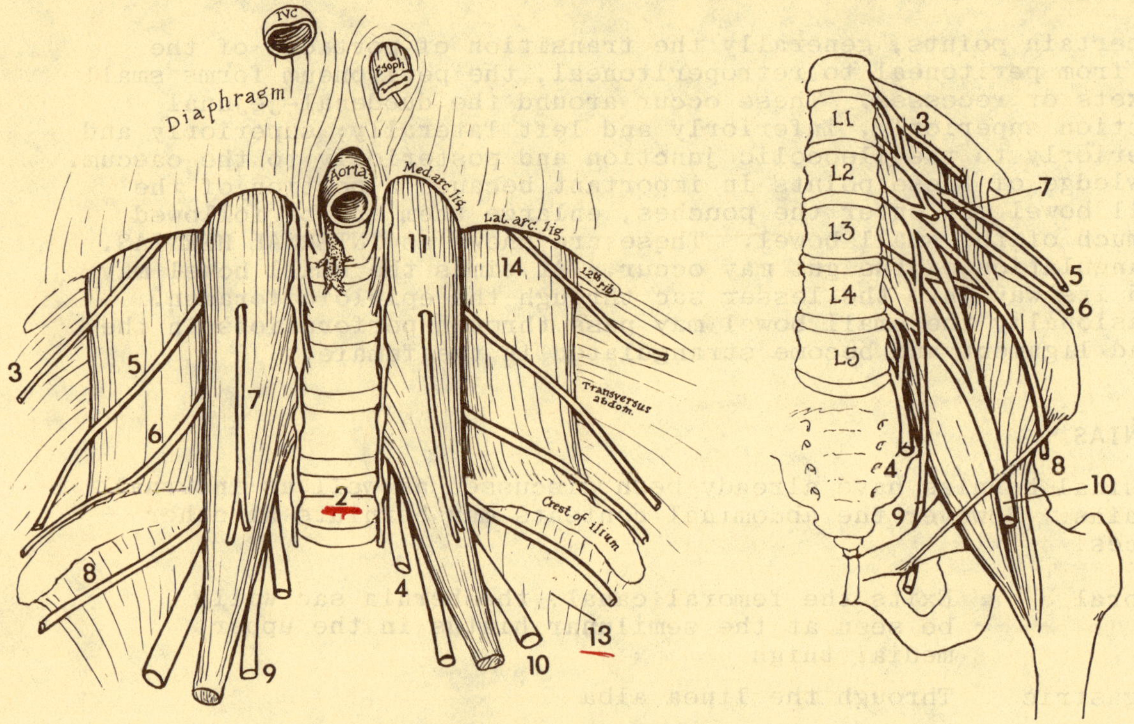

POSTERIOR ABDOMINAL WALL

INTERNAL ASPECTS

1) Cisterna chyli
2) Sympathetic ganglia
3) Subcostal n., T12
4) Lumbosacral trunk, L4,5

LUMBAR PLEXUS--anterior rami L1-4

5) Iliohypogastric L1, T12?
6) Ilioinguinal L1
7) Genitofemoral L1,2
8) Lateral femoral cutaneous L2,3
9) Obturator L2,3,4
10) Femoral L2,3,4

MUSCLES

11) Psoas major
    - O - Bodies, discs, transverse processes T12, L1-5
    - I - Iliopsoas tendon to lesser trochanter of femur
    - A - Flex thigh, vertebral column
    - N - Lumbar plexus
    - B - Lumbar arteries

Psoas minor--60% occurrence
- O - T12-L1
- I - Iliopectineal eminence

13) Iliacus
    - O - Iliac crest and fossa
    - I - Iliopsoas tendon to lesser trochanter of femur
    - A - Flex thigh, flex pelvis
    - N - Femoral
    - B - Iliolumbar artery

14) Quadratus lumborum
    - O - Iliac crest
    - I - 12th rib, transverse processes L1-4
    - A - Holds ribs against traction of diaphragm in respiration, bends trunk
    - N - T12, L1-4
    - B - Lumbar arteries

DIAPHRAGM

ORIGINS
1) Sternal
2) Costal
3) Right crus
   L1-3
4) Left crus
   L1-2
5) Medial arcuate ligament
   --over psoas muscle
6) Lateral arcuate ligament
   --over quadratus lumborum

I - Central tendon of the diaphragm (7)
A - Increases thoracic volume in inspiration
N - Motor--phrenic nerve "C3,4,5 keep the diaphragm alive"; sensory--phrenic and intercostals
B - Inferior and superior musculophrenic, pericardiacophrenic, intercostals

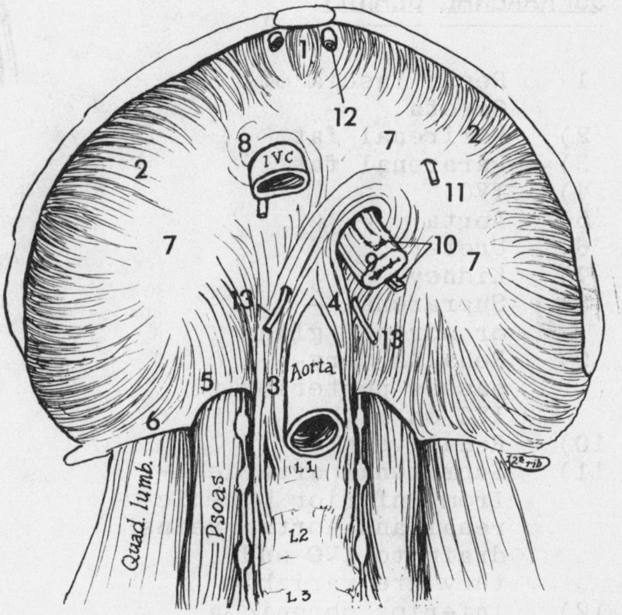

OPENINGS:

Caval Hiatus T8
   Transmits IVC and right phrenic nerve (8)

Esophageal Hiatus T10--
   formed by a sling from the right crus. Transmits esophagus (9), vagus nerves (10), esophageal branch of left gastric artery

Aortic Hiatus T12
   Transmits aorta, thoracic duct, azygos vein

OTHER STRUCTURES

11) Left phrenic nerve
12) Superior epigastric aa., pass between sternal and costal origins of the diaphragm
13) Splanchnic nerves-- pass under medial arcuate ligament

COMMENTS

1. Sole motor innervation of the diaphragm is from the phrenic nerve; since this nerve has a long course through the neck and thorax it is liable to injury, causing paralysis of the diaphragm on the injured side.

2. Development of the diaphragm is complex and herniations of viscera are discussed elsewhere.

KIDNEYS AND SUPRARENAL GLANDS

1) Renal fascia of Gerota
2) Perirenal fat
3) Pararenal fat
4) IVC
5) Aorta
6) Ureter
7) Kidney
8) Suprarenal-- or adrenal gland
9) Renal artery-- passes posterior to IVC on right
10) Renal vein
11) Suprarenal arteries-- from inferior phrenic, renal and aorta; veins drain to IVC and they are variable
12) Inferior phrenic a.
13) Coeliac trunk
14) Superior mesenteric a.
15) Gonadal a.--crosses ureter
16) Gonadal vein--left enters renal vein
17) Inferior mesenteric a.

THE LEFT KIDNEY IS LONGER AND MORE CRANIAL THAN THE RIGHT.

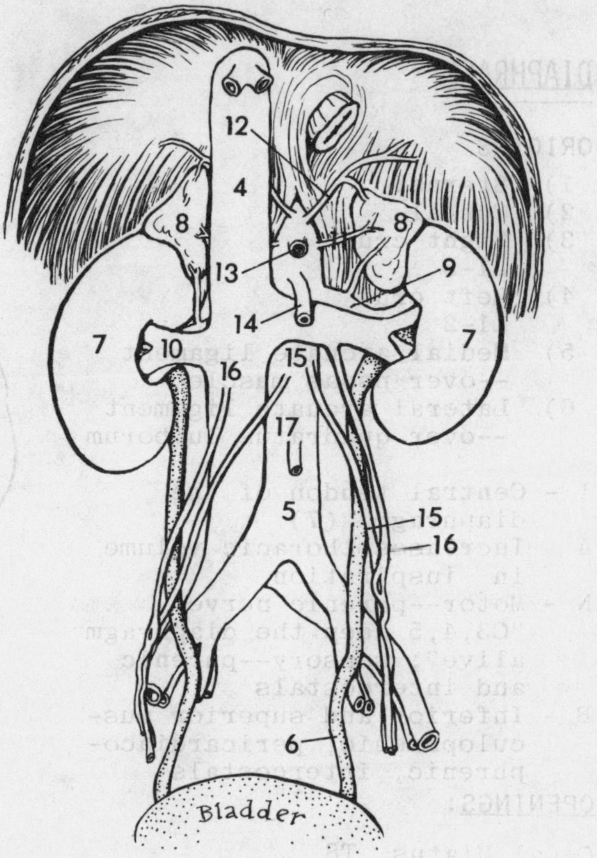

ANTERIOR VIEWS

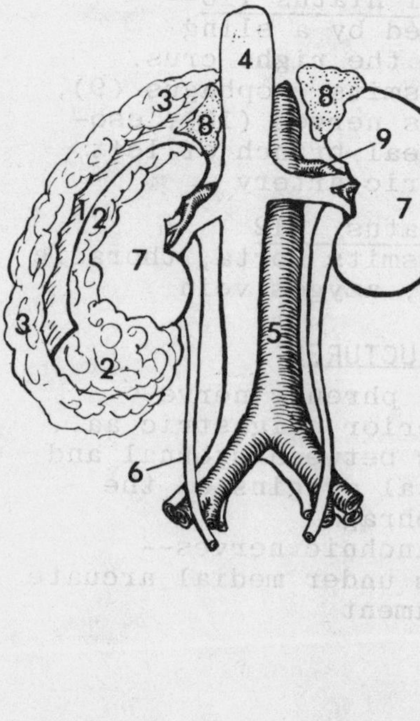

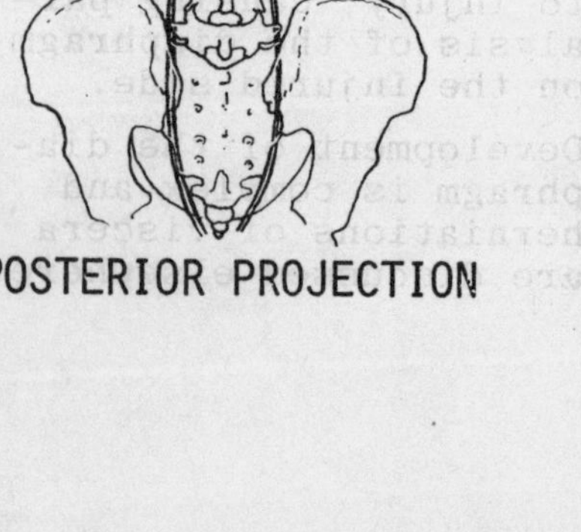

POSTERIOR PROJECTION

CORONAL SECTION OF KIDNEY

1) Cortex
2) Medulla
3) Renal column
4) Renal pyramid
5) Renal papilla
6) Major calyx
7) Minor calyx
8) Renal pelvis
9) Ureter
10) Interlobar artery
11) Arcuate artery
12) Interlobular artery

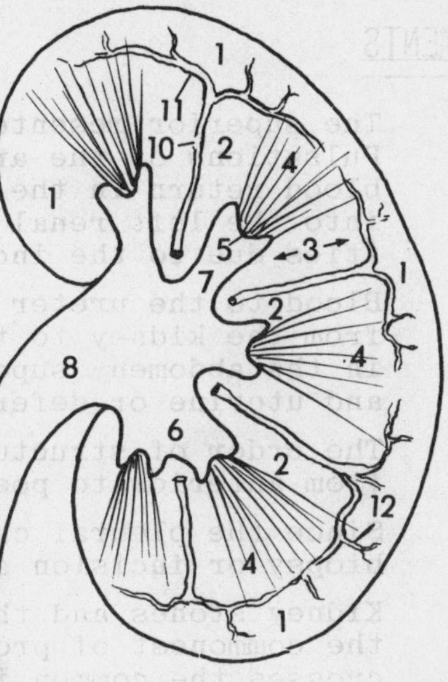

THE RENAL SINUS is the fat filled space surrounding the HILUM, the depression in which the vessels and ureter enter the kidney.

KIDNEY MAP

White areas are covered with peritoneum; stippled areas indicate relationship to abdominal organs

1) Liver
2) Duodenum
3) Colon
4) Pancreas
5) Lienorenal ligament

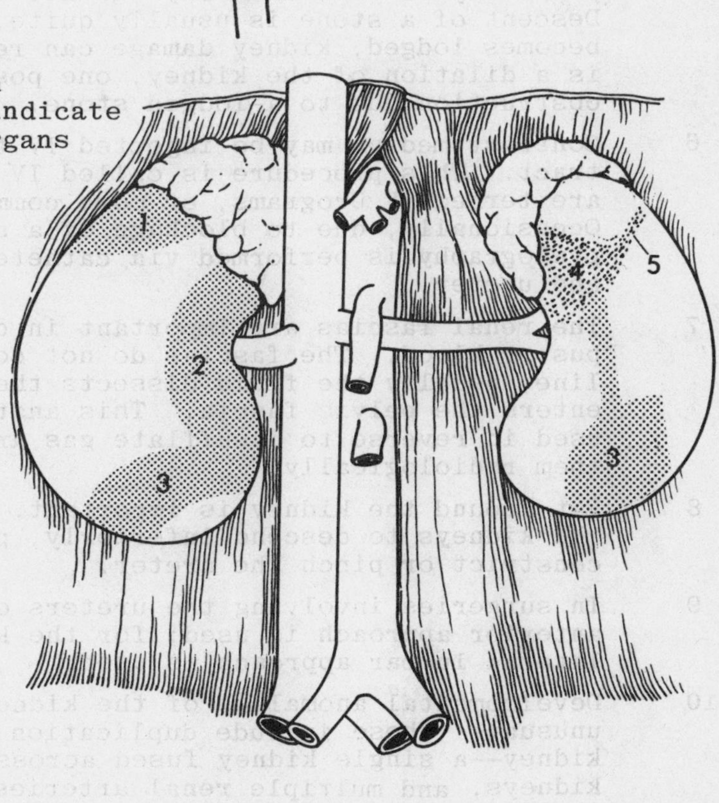

COMMENTS

1 The superior mesenteric artery crosses the left renal vein. Pulsations of the artery may compress the vein and detain blood return in the left renal vein. The gonadal vein drains into the left renal vein and may dilate and form varicosities due to the increased pressure.

2 Blood to the ureter comes from several sources in its course from the kidney to the bladder: renal and gonadal arteries in the abdomen, superior vesicular and iliac in the pelvis and uterine or deferential (depending on the gender).

3 The order of structures entering the hilum of the kidney from anterior to posterior is renal vein, renal artery and pelvis.

4 Since the pleural cavity may extend below T12, a kidney biopsy or incision at that level may enter the pleural space.

5 Kidney stones and their passage down the ureters are among the commonest of problems with the urinary tract. The ureter crosses the common iliac arteries at their bifurcation. This is a site where they may become lodged; this location coincides with the sacroiliac joint, and a stone at this point may be seen on X-ray against the radiolucent joint. Descent of a stone is usually quite painful. If the stone becomes lodged, kidney damage can result. Hydronephrosis is a dilation of the kidney, one possible result of an obstruction due to a lodged stone.

6 Contrast medium may be injected IV to visualize the urinary tract. This procedure is called IV Urography. The X-rays are termed IV Urograms, or more commonly IV Pyelograms. Occasionally, due to blockage of a stone, a retrograde pyelography is performed via catheterization of the bladder and ureter.

7 The renal fascias are important in determining spread of pus or blood. The fascias do not connect across the midline; usually the fluid dissects the fascia inferiorly and enters the pelvic fascia. This anatomical phenomenon is used in reverse to insufflate gas around the kidneys to view them radiologically.

8 Fat around the kidney is important. Its absence can cause the kidneys to descend inferiorly, possibly mobilize and constrict or pinch the ureter.

9 In surgeries involving the ureters or renal vessels, an anterior approach is used; for the kidneys a posterior-lateral lumbar approach is used.

10 Developmental anomalies of the kidneys and ureters are not unusual. These include duplication of ureters, horseshoe kidney--a single kidney fused across the midline-- pelvic kidneys, and multiple renal arteries.

AUTONOMIC INNERVATION

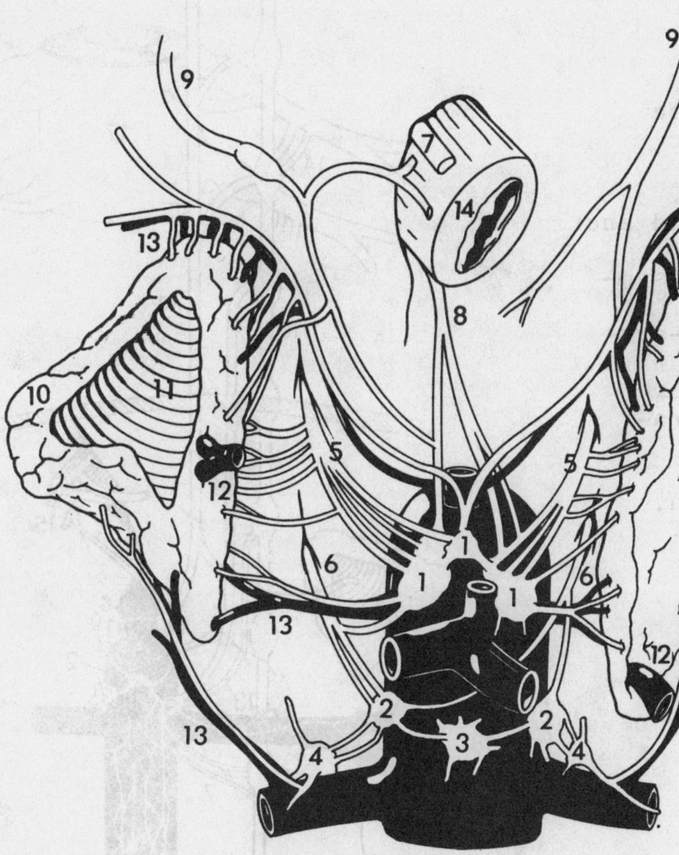

STRUCTURES:

1) Coeliac ganglion
2) Aorticorenal ganglion
3) Superior mesenteric ganglion
4) Renal ganglion
5) Greater splanchnic n.
6) Lesser splanchnic n.
7) Anterior vagal trunk
8) Posterior vagal trunk
9) Phrenic n.
10) Adrenal cortex
11) Adrenal medulla
12) Adrenal vein
13) Adrenal artery
14) Esophagus

AUTONOMIC INNERVATION

STRUCTURES:

1) Coeliac ganglion
2) Superior mesenteric ganglion
3) Aorticorenal ganglion
4) Superior hypogastric plexus or nerve
5) Right hypogastric n.
6) Pelvic plexus--around and on viscera
7) Lumbar splanchnics--sympathetic
8) Sacral splanchnics--sympathetic
9) Pelvic splanchnics or nervi erigentes--parasympathetic
10) Sympathetic chain
11) Greater splanchnic n.
12) Lesser splanchnic n.
13) Least splanchnic n.
14) Rami communicantes
15) Vagal trunk
16) Sciatic n.
17) Spinal cord
18) Kidney
19) Suprarenal gland

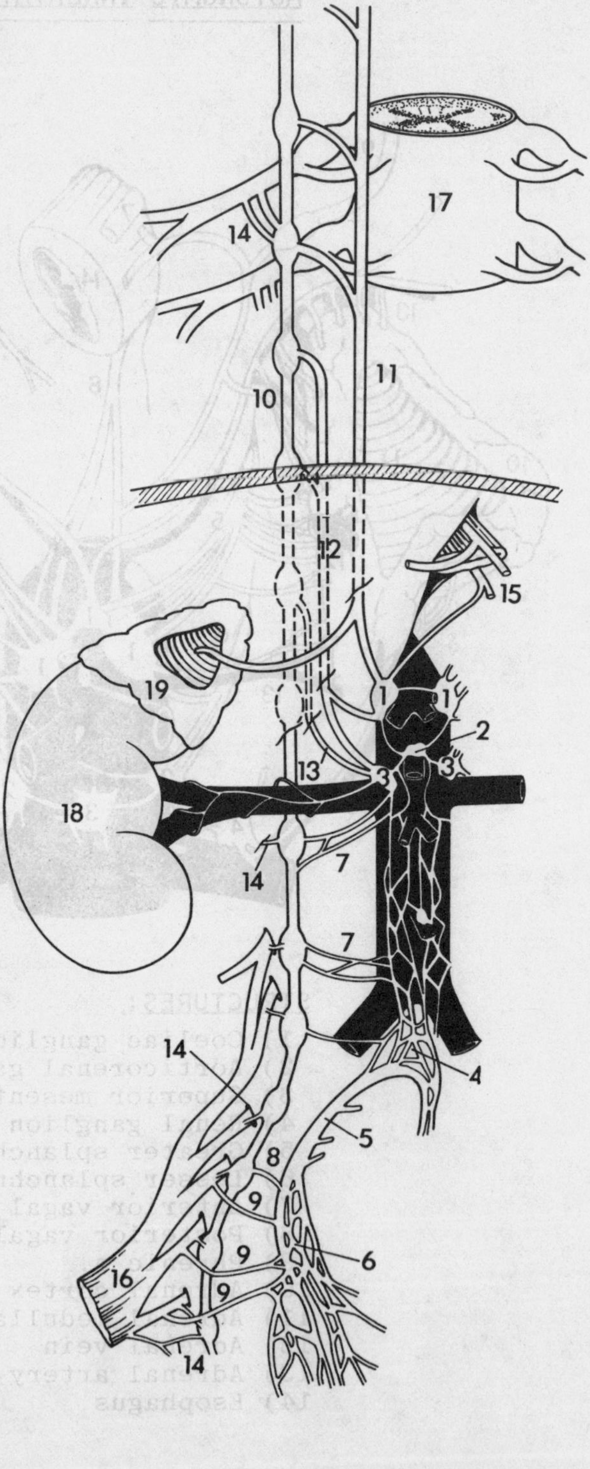

STUDY QUESTIONS: ABDOMEN

1. The peritoneum is divided into how many parts?
   1. two, the parietal and visceral peritoneum

2. Define ascites.
   2. abnormal accumulation of peritoneal fluid. Normally the peritoneal cavity contains only enough fluid to moisten its wall.

3. How large is the peritoneum?
   3. approximately 2 square meters

4. What is found in the mesentery?
   4. vessels and nerves leading to the gut

5. The left umbilical vein forms what structure in the adult?
   5. the round ligament of the liver

6. The lesser omentum is divided into what parts?
   6. the hepatogastric ligament and hepatoduodenal ligament

7. Where are the above structures located?
   7. The hepatogastric ligament is found between the lesser curvature of the stomach and the fissure for the ligamentum venosum of the liver. The hepatoduodenal ligament connects the first part of the duodenum with the porta hepatis

8. What structures are transmitted inside the hepato-duodenal ligament?
   8. portal vein, hepatic artery and common bile duct

9. The dorsal mesentery of the stomach is composed of what structures?
   9. the gastrophrenic, lienorenal, gastrosplenic and phrenicolienal ligament and the greater omentum

10. Where is the omental bursa located?
    10. dorsal to the stomach and lesser omentum. It opens into the peritoneal cavity via the epiploic foramen.

11. What structures surround the epiploic foramen?
    11. Superiorly, the caudate lobe of the liver; inferiorly, the first segment of the duodenum; anteriorly, the hepatoduodenal ligament; posteriorly, the inferior vena cava.

12. The three unpaired branches of the abdominal aorta are:
    12. the celiac trunk, the superior mesenteric artery and inferior mesenteric artery

13. What is the importance of the axis of the superior mesenteric artery in development of the gut?
    13. It is around this axis that the intestine rotates <u>in utero</u>.

14. Name mesenteries which fuse to the dorsal wall after the intestinal rotation is complete.
    14. mesenteries of the ascending and descending colon and duodenum

15. What is the relationship of the superior mesenteric artery to the duodenum before and after intestinal rotation occurs?
    15. Prior to rotation, the superior mesenteric artery is inferior to the duodenum. After rotation is complete, the superior mesenteric artery passes anterior to the 3rd portion of the duodenum.

16. What are the subdivisions of the small intestine?
    16. duodenum, jejunum and ileum

17. What are the gross morphologic differences between the jejunum and ileum?
    17. The jejunum has more numerous plicae circulares, 1-2 arterial arcades in its mesentery and has "windows" in the mesenteric fat. In the ileum, however, the mesenteric fat reaches the margin of the bowel, and it has 3 or 4 arterial arcades in the mesentery.

18. What gross structures distinguish the large from small intestine?

18. taenia coli, haustra coli, appendices epiploicae, plicae semilunares

19. What is the location of the cecum?

19. The cecum is part of the proximal large intestine caudal to the level of the entrance of the ileum.

20. What is the relationship of the taenia coli to the appendix?

20. The three taenia meet at the appendix base.

21. What are the two major aortic branches which supply the large intestine?

21. the superior mesenteric and the inferior mesenteric artery

22. The continuous channel of anastomoses formed between branches of the superior and inferior mesenteric arteries to the large intestine is called ____.

22. the marginal artery

23. Does blood returning from the gut go directly to the inferior vena cava?

23. No, it must pass via the portal vein through the liver, then via the hepatic vein to the IVC.

24-28 Matching

A. Dorsal root ganglia of L-1
B. " " " " T-10
C. " " " " S-2
D. " " " " T-6
E. " " " " T-8

Where are the cell bodies for the following afferent fibers?

24. the stomach _____
25. the appendix _____
26. the transverse colon _____
27. the descending colon _____
28. the rectum _____

24. D
25. E
26. B
27. A
28. C

29. Where is the suspensory ligament of Treitz found?

29. at the duodenal-jejunal juncture

30. What are rugae?

30. Folds of the stomach mucosa. These flatten out in the distended stomach.

31. What vessel is common to the blood supply of the spleen, pancreas and stomach?

31. the splenic artery

32. From the gallbladder, bile is released into the 2nd part (descending) of the duodenum upon certain hormonal stimuli. Trace its path.

32. gallbladder→cystic duct→common bile duct→hepatopancreatic ampulla→duodenum

33. Where is the spiral fold located and what is its importance?

33. The spiral fold is a mucosal fold found in the cystic duct. It functions to keep the duct constantly patent so that bile flows unobstructed in either direction.

34. What are the important relations to the 2nd part of the duodenum?

34. the hilum of the right kidney posteriorly, the head of the pancreas and common bile duct medially, and the transverse mesocolon crossing it in the middle.

35. What is the superior portion of the duodenum called?

35. the duodenal cap or bulb

36. What fetal structure connects the umbilical vein to the IVC?

36. the ductus venosus

37. The left and right hepatic arteries may be aberrant. What is the most common source of these aberrent vessels?

37. Respectively, the left gastric artery and the superior mesenteric artery.

38. What vessels join to form the portal vein?

38. the superior mesenteric vein and splenic vein

39. Why is the renal fascia important?

39. It can retain fluid or injected air around the kidney. It is continuous with the retroperitoneal area of the pelvis.

40. What is the fat inside the renal fascia called? Outside the renal fascia?

40. perirenal fat; pararenal fat

41. The kidney is composed of many subunits. How many major and minor calices and renal papillae are there?

41. 0-3 major; 7-13 minor; 8-18 renal papillae

42. The ureter has a multiple blood supply. What is it?

42. renal artery, gonadal and common iliac arteries, inf. vesicle and middle rectal artery

43. Why is pain from the central portion of the diaphragm referred to the shoulder and base of the neck?

43. The phrenic nerve, sensory and motor to the diaphragm,originates at the same level as sensory nerves to the shoulder - C 3,4,5.

44. At what level do the following pass through the diaphragm - IVC, esophagus, aorta?

44. T8; T10; T12

45. Of the femoral artery, nerve and vein, which is the most medial as they enter the thigh?

45. femoral vein

46. What other structures are transmitted through the diaphragm with the aorta?

46. the thoracic duct and the azygos vein

47. Which layer of superficial abdominal fascia is Colles' fascia (Tunica Dartos) continuous with?

47. Scarpa's fascia (membranous)

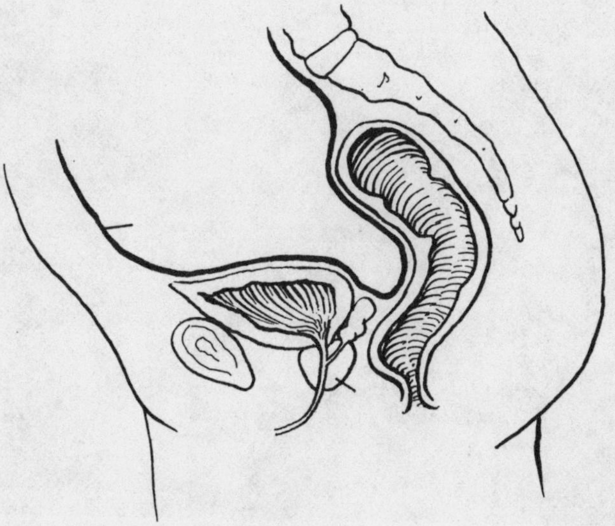

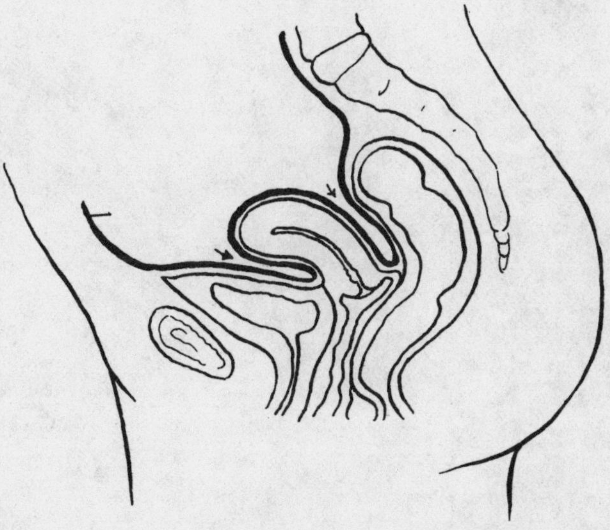

PELVIS & PERINEUM

PELVIS

I. THE BONY PELVIS

GENERAL

The bony pelvis is composed of the sacrum, coccyx and 2 coxal or hip bones (os coxae). Each os coxae represents fused bones named the ilium, ischium and pubis--all three bony elements are represented in the acetabulum, which receives the head of the femur or thigh bone. The bony pelvis is divided by the pelvis brim (arcuate or ilio-pectineal line) into the false pelvis superiorly and the true pelvis inferiorly. The musculature and fascia of the pelvic diaphragm (levator ani and coccygeus muscles) subdivide the bony pelvis into a true pelvic cavity superiorly and the perineum inferiorly.

1) Ischial spine
2) Pelvic brim
3) Iliac fossa
4) Pubis
5) Sacral promontory
6) ASIS
7) Coccyx
8) Acetabulum
9) Obturator foramen
10) Sacral foramina
11) Ischium
12) Pelvic diaphragm
13) Sacroiliac joint
14) Pubic tubercle

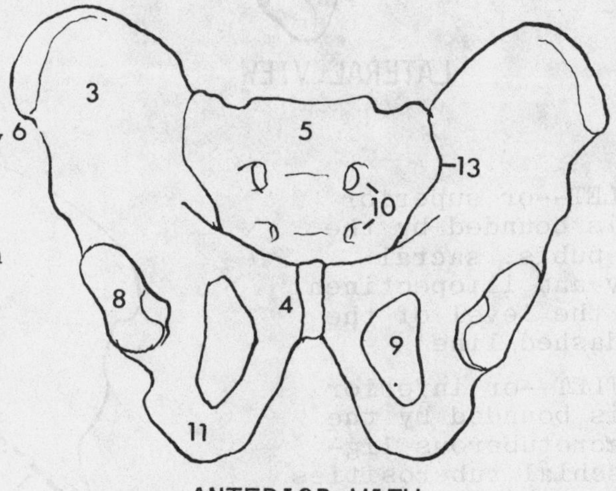

ANTERIOR VIEW

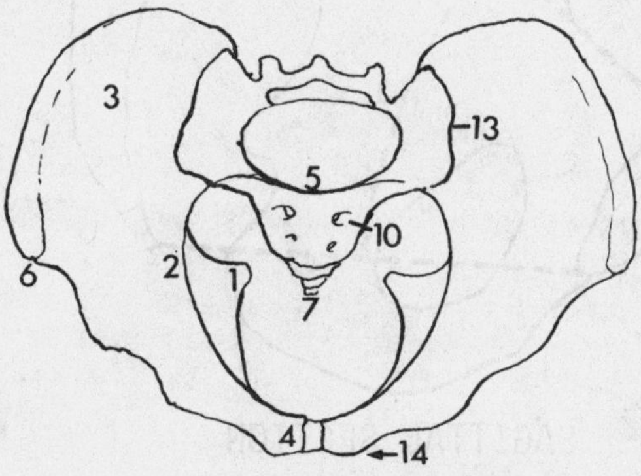

SUPERIOR VIEW

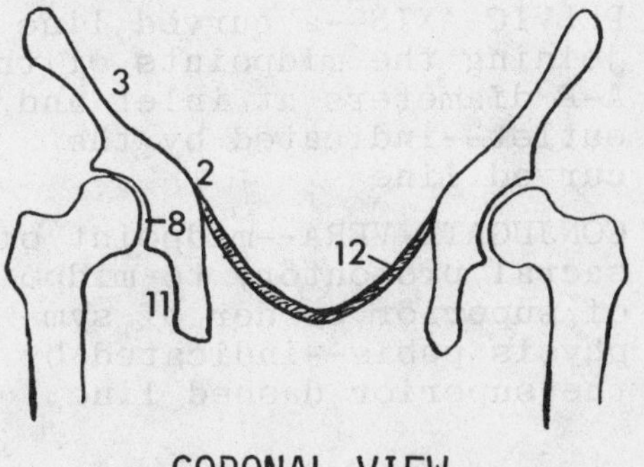

CORONAL VIEW

SCIATIC NOTCHES are converted into foramina by the sacrospinous and sacrotuberous ligaments

1) Greater sciatic foramen
2) Lesser sciatic foramen
3) Sacrospinous ligament
4) Sacrotuberous ligament

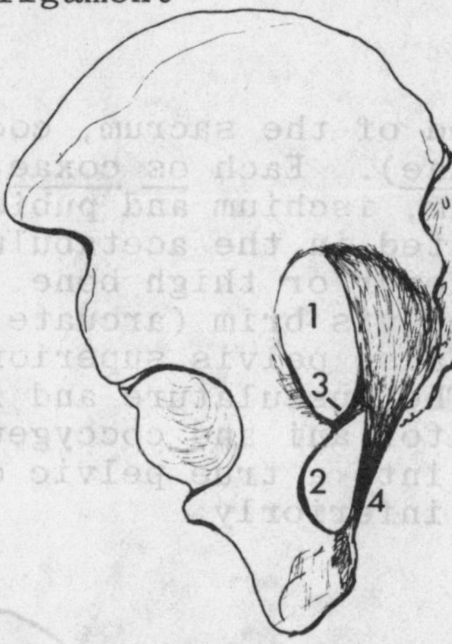

LATERAL VIEW

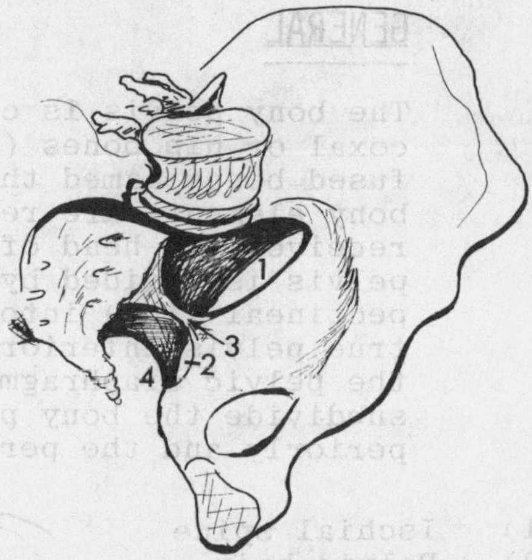

MEDIAL VIEW

DEFINITIONS

1. PELVIC INLET--or superior aperture is bounded by the symphysis pubis, sacral promontory and iliopectineal lines--at the level of the superior dashed line

2. PELVIC OUTLET--or inferior aperture is bounded by the coccyx, sacrotuberous ligaments, ischial tuberosities and pubic arch--at the level of the inferior dashed line

3. PELVIC AXIS--a curved line joining the midpoints of the A-P diameters at inlet and outlet--indicated by the curved line

4. CONJUGATE VERA--midpoint of sacral promontory to midpoint of superior border of symphysis pubis--indicated by the superior dashed line

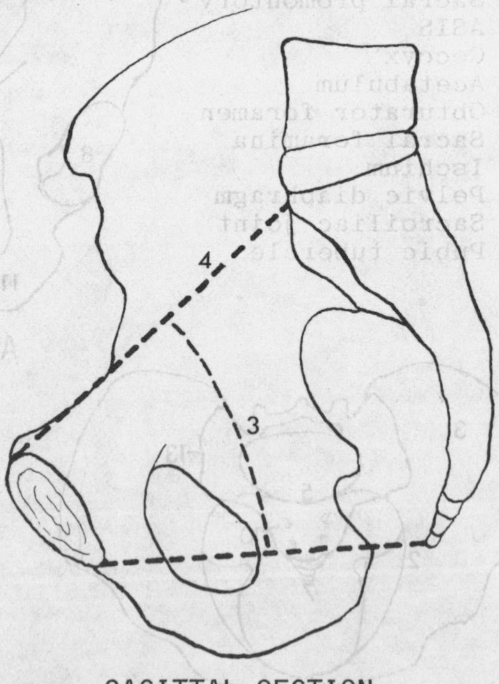

SAGITTAL SECTION

DIFFERENCES IN THE MALE AND FEMALE PELVIS

The male pelvis generally appears heavier than the female, with a narrow and deep true pelvis. In the female, the entire pelvis is tilted anteriorly in comparison to the male. A vertical line from the pubic symphysis intersects the ASIS in the female but not the male; a horizontal line intersects the coccyx in the male but not the female.

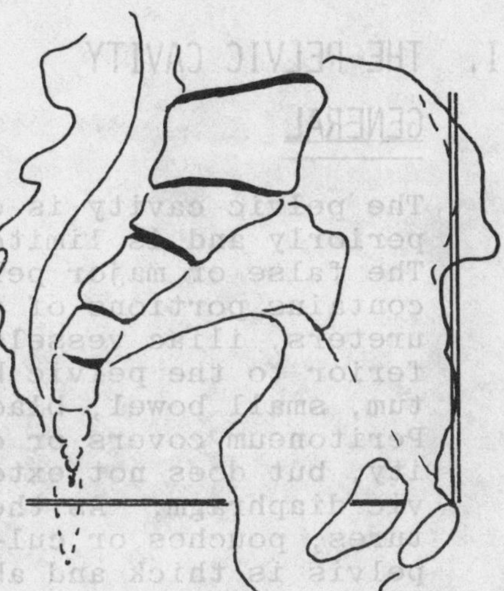

MALE PELVIS is indicated by dashed lines in this composite medial view

SPECIFIC DIFFERENCES

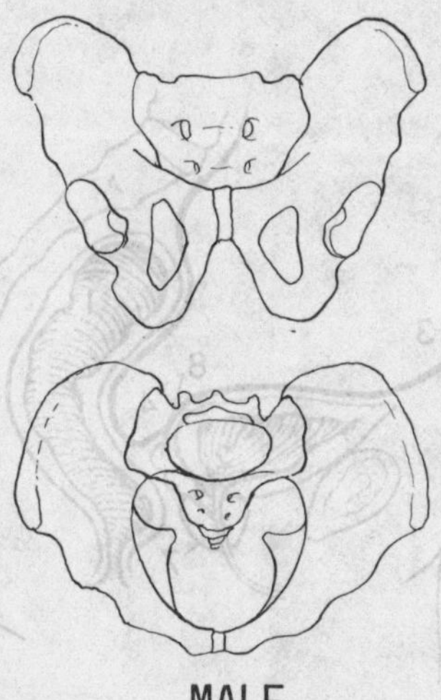

MALE

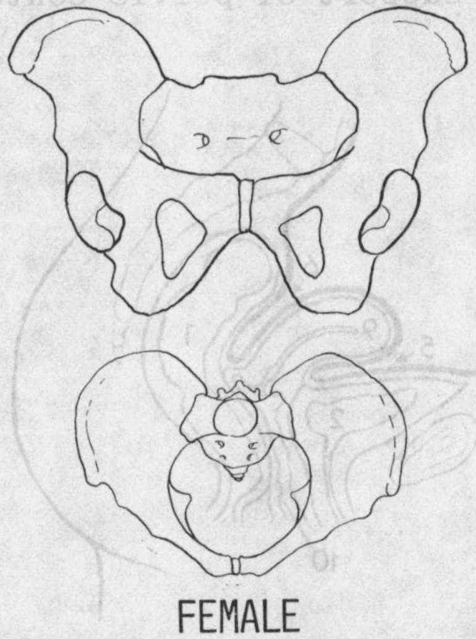

FEMALE

False pelvis:	Narrow	Wide
True pelvis:	Narrow	Wide
Obturator Foramen:	Oval	Triangular
Greater Sciatic Notch:	Narrow	Wide
ASIS:	Inverted	Everted
Superior Aperture:	Heart shaped	Oval
Iliac Fossa:	Deep	Shallow

II. THE PELVIC CAVITY

GENERAL

The pelvic cavity is continuous with the abdominal cavity superiorly and is limited inferiorly by the pelvic diaphragm. The false or major pelvis is superior to the pelvic brim and contains portions of the small bowel, caecum, sigmoid colon, ureters, iliac vessels, while the true or minor pelvis inferior to the pelvic brim contains some sigmoid colon, rectum, small bowel, bladder and internal reproductive organs. Peritoneum covers or caps many structures in the pelvic cavity, but does not extend inferiorly to the level of the pelvic diaphragm. As the peritoneum reflects from various structures, pouches or cul-de-sacs are formed. Fascia in the true pelvis is thick and abundant. This fascia is called the endopelvic fascia and is continuous superiorly with the transversalis fascia. The endopelvic fascia may play a role in the support of pelvic contents.

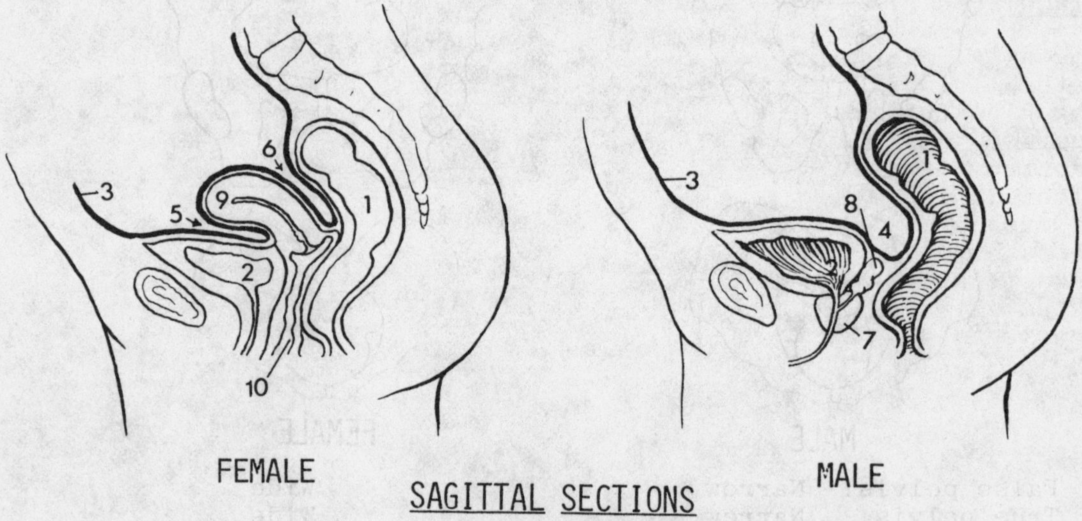

SAGITTAL SECTIONS

1) Rectum
2) Bladder
3) Peritoneum
4) Rectovesical pouch
5) Vesicouterine pouch
6) Rectouterine pouch
7) Prostate
8) Seminal vesical
9) Uterus
10) Vagina

PELVIC WALL AND FLOOR

THE WALL OF THE PELVIS consists of the coxal bones, the obturator internus and piriformis muscles.

1) PIRIFORMIS

O-Anterior sacrum between foramina, only the origin is indicated in the figure
I-Greater trochanter
A-Rotates extended thigh, abducts flexed thigh
N-L5, S1 & 2
B-Lateral sacral, inferior gluteal

2) OBTURATOR INTERNUS

O-Perimeter of obturator foramen, obturator membrane--only the origin is indicated in the figure
I-Medial surface greater trochanter
A-Laterally rotates extended thigh, abducts flexed thigh
N-Nerve to the obturator internus, L5, S1
B-Obturator artery

OTHER STRUCTURES

3) Urogenital diaphragm
4) Urethra
5) Urogenital hiatus--superior to the UG diaphragm, a natural defect in the levator ani muscles, for the prostate, vagina and neck of the bladder
6) Rectum
7) Arcus tendineus

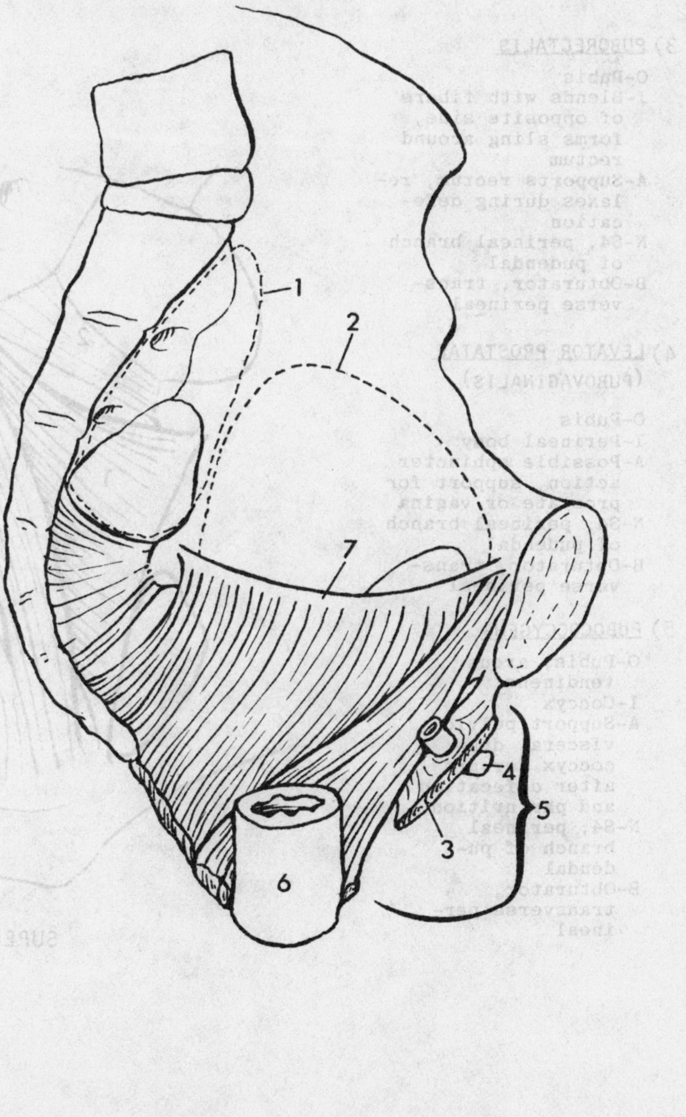

THE FLOOR OF THE PELVIS consists of the pelvic diaphragm, shaped like an inverted pyramid with the rectum at the apex. It is a muscular sling which helps support the viscera; it is composed of the coccygeus and levator ani muscles.

THE LEVATOR ANI can be morphologically divided into the iliococcygeus and pubococcygeus muscles. In general they originate from the ischial spine, arcus tendineus, pubis and insert in the midline and coccyx. THE ARCUS TENDINEUS is a condensation of fascia extending across the obturator internus muscle.

1) ILIOCOCCYGEUS

O-Ischial spine, arcus tendineus
I-Coccyx, median raphe
A-Support pelvic viscera, constrict lower rectum and vagina
N-S4, perineal division of pudendal
B-Obturator, transverse perineal

2) COCCYGEUS

O-Ischial spine
I-Coccyx, lower sacrum
A-May move coccyx after defecation and/or partuition
N-S4,5
B-Lateral sacral, Inferior gluteal

PUBOCOCCYGEUS is described as having 3 portions

3) PUBORECTALIS

O-Pubis
I-Blends with fibers of opposite side, forms sling around rectum
A-Supports rectum, relaxes during defecation
N-S4, perineal branch of pudendal
B-Obturator, transverse perineal

4) LEVATOR PROSTATAE (PUBOVAGINALIS)

O-Pubis
I-Perineal body
A-Possible sphincter action, support for prostate or vagina
N-S4, perineal branch of pudendal
B-Obturator, transverse perineal

5) PUBOCOCCYGEUS

O-Pubis, arcus tendineus
I-Coccyx
A-Support pelvic viscera, draw coccyx forward after defecation and parturition
N-S4, perineal branch of pudendal
B-Obturator, transverse perineal

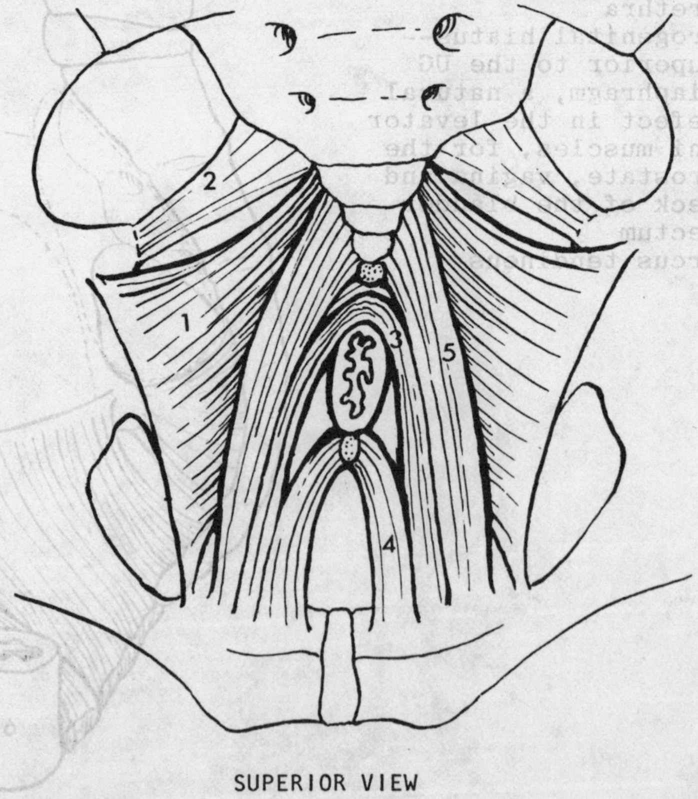

SUPERIOR VIEW

COMMENTS

1. Pelves are classified according to the shape of the pelvic inlet. The most common male shape is termed android, while gynecoid is the common female shape. Anthropoid pelves are shaped as in apes--a long A-P diameter, short lateral; platypelloid pelves are short in A-P diameter, long in lateral diameter.
2. Fractures of the pelvis are common, and may be complicated by injury to pelvic viscera.

COMMENTS

1. The pelvic diaphragm and fascia in women are particularly liable to damage in childbirth. The bladder may herniate into the vagina--cystocele--or cystourethrocele if accompanied by the urethra; these may be accompanied by urinary stress incontinence when intra-abdominal pressure rises during coughing or sneezing.
2. The pelvic diaphragm is the major support of all pelvic viscera. Prolapse--a descent of structures such as the uterus--can result if the diaphragm is injured.

NEURAL ELEMENTS OF THE PELVIS

THE SACRAL PLEXUS is formed by anterior primary rami of L4,5, S1,2,3,4

1) S1
2) S2
3) S3
4) S4
5) Lumbosacral trunk from L4&5
6) Sciatic nerve L4,5, S1,2,3 exits inferior to piriformis
7) Pudendal nerve
8) Greater sciatic foramen
9) Coccygeus muscle and sacrospinous ligament
10) Piriformis muscle
11) Superior gluteal artery
12) Inferior gluteal artery
13) Internal pudendal artery exits greater sciatic foramen and re-enters lesser sciatic foramen

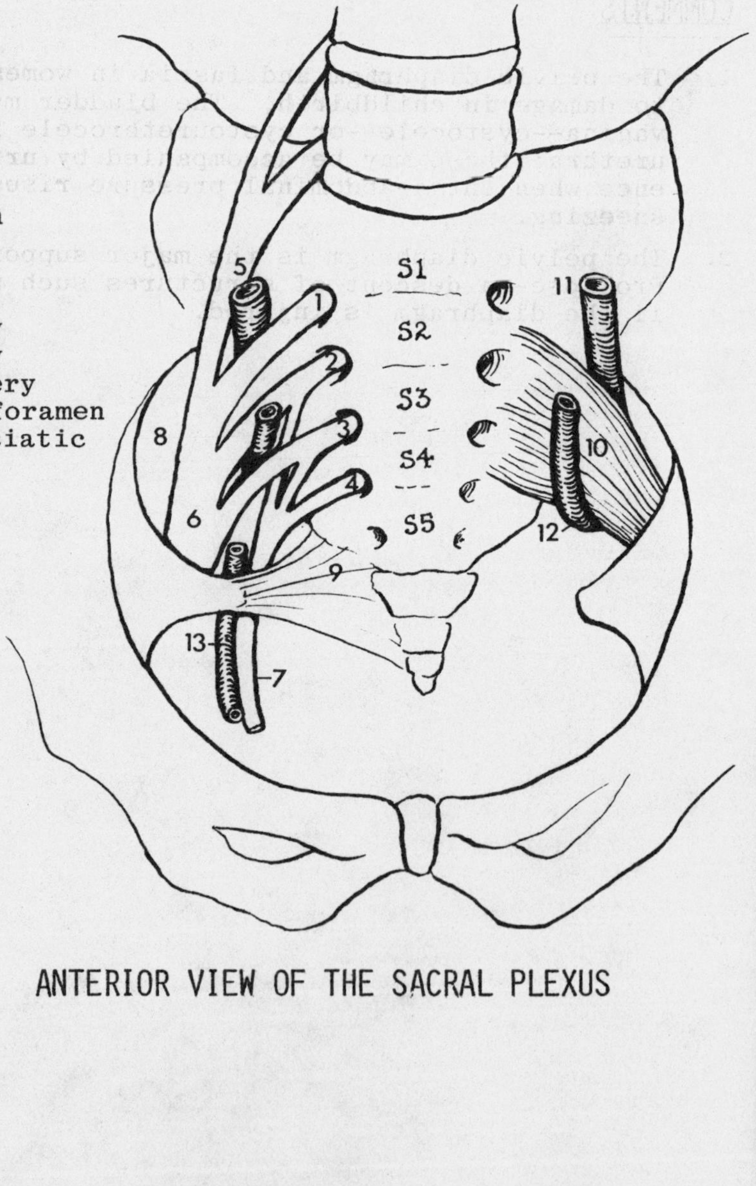

ANTERIOR VIEW OF THE SACRAL PLEXUS

AUTONOMIC INNERVATION

STRUCTURES:

1) Celiac ganglion
2) Superior mesenteric ganglion
3) Aorticorenal ganglion
4) Superior hypogastric plexus or nerve
5) Right hypogastric n.
6) Pelvic plexus--around and on viscera
7) Lumbar splanchnics--sympathetic
8) Sacral splanchnics--sympathetic
9) Pelvic splanchnics or nervi erigentes--parasympathetic
10) Sympathetic chain
11) Greater splanchnic n.
12) Lesser splanchnic n.
13) Least splanchnic n.
14) Rami communicantes
15) Vagal trunk
16) Sciatic n.
17) Spinal cord
18) Kidney
19) Suprarenal gland

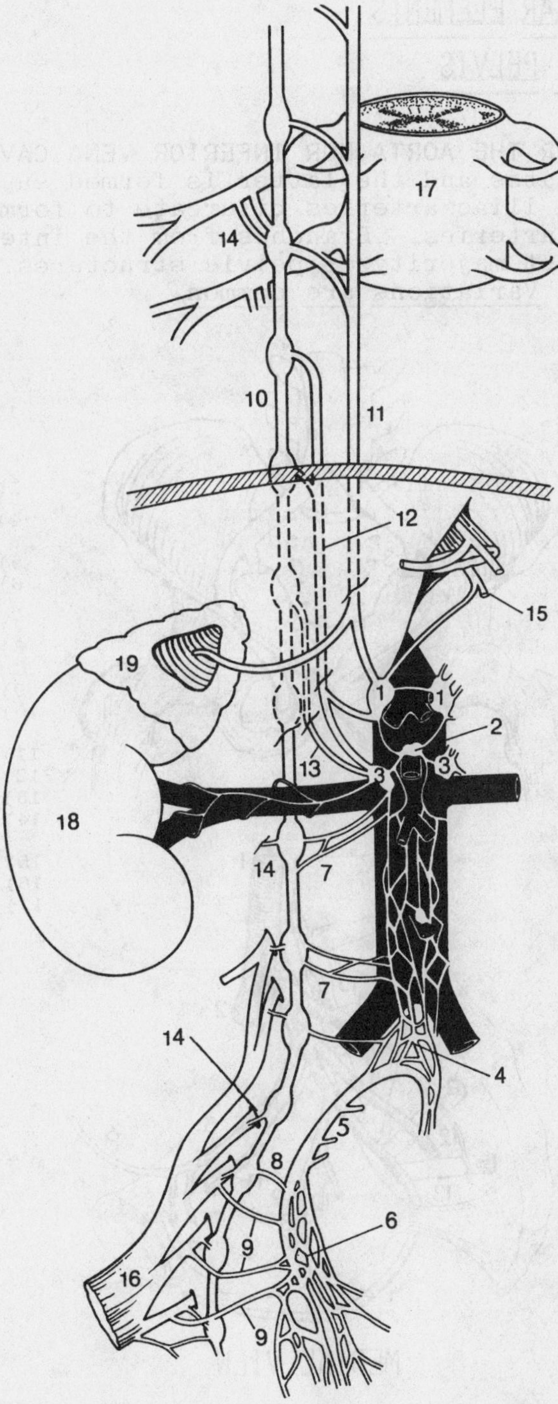

VASCULAR ELEMENTS OF THE PELVIS

NEITHER THE AORTA NOR INFERIOR VENA CAVA is in the pelvis; the former bifurcates and the latter is formed superior to the pelvis. The common iliac arteries bifurcate to form the internal and external iliac arteries. Branches from the internal iliac arteries supply the vast majority of pelvic structures. This is a typical arrangement. Variations are common.

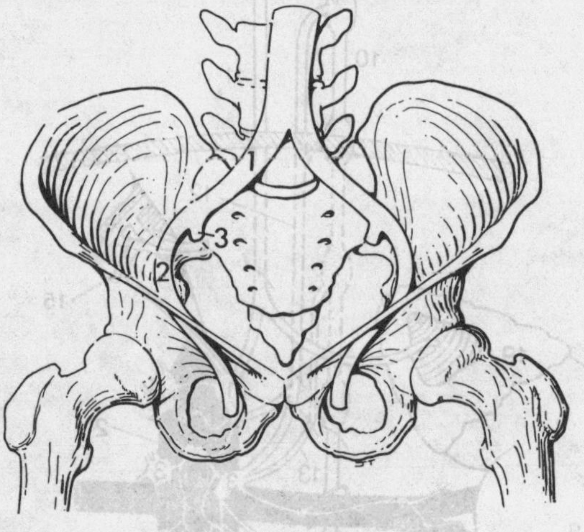

1) Common iliac--divides at the level of the pelvic brim, anterior to the sacroiliac joint
2) External iliac
3) Internal iliac
4) Anterior division of internal iliac
5) Umbilical
6) Obliterated umbilical artery, becomes medial umbilical ligament
7) Superior vesicle
8) Artery to the ductus deferens, may arise from inferior vesicle
9) Inferior vesicle
10) Middle rectal in male, uterine with rectal branches in female
11) Obturator
12) Inferior gluteal
13) Internal pudendal
14) Posterior division of the internal iliac
15) Iliolumbar
16) Superior gluteal
17) Lateral sacral

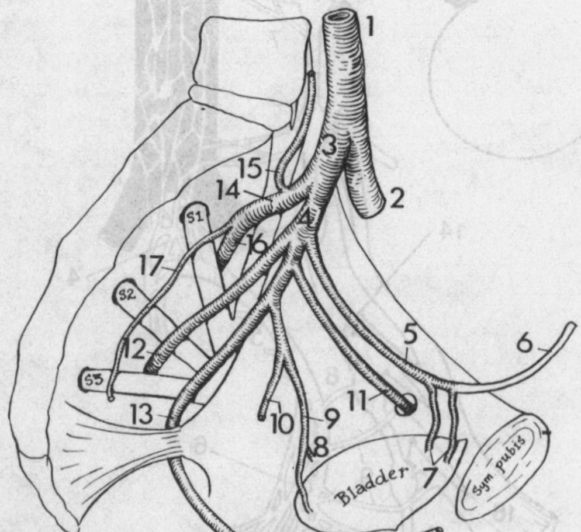

MEDIAL VIEW

PELVIC VASCULATURE

III. PELVIC CONTENTS

RECTUM AND ANAL CANAL

1) Rectum
2) Rectal folds of Houston --two on left, one on right
3) Anal canal--actually in the perineum, below the levator ani and superior portions of the anal columns
4) Anal column
5) Anal fold or valve--connects bases of columns
6) Rectal ampulla--dilatation superior to the levator ani
7) Internal anal sphincter--a thickening in circular muscles of the rectum
8) Longitudinal muscle
9) Levator ani
10) Conjoined muscle of levator ani, longitudinal muscle
11) PUBORECTALIS portion of levator ani forms little of the pelvic diaphragm, contributes a major portion of the ANORECTAL RING superior to the conjoined muscle, relaxes during defecation, coincides with the start of the anal canal

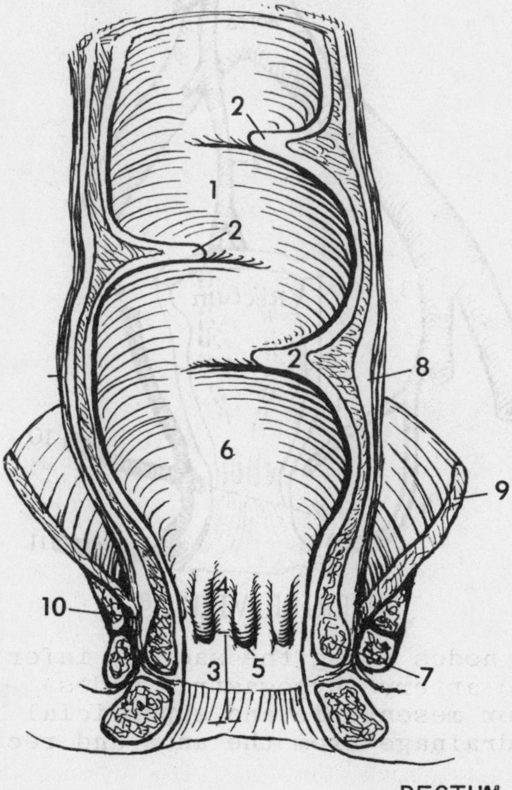

RECTUM
CORONAL VIEW

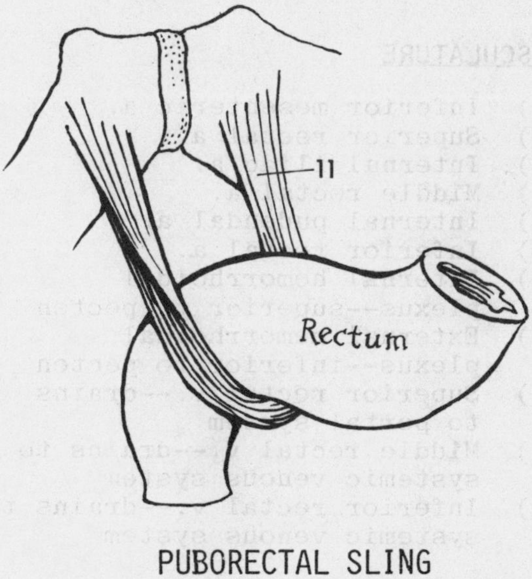

PUBORECTAL SLING

RECTAL RHYMES
--REMEMBER 3

3 parts of external sphincter
3 rectal folds
3 begins opposite S3
3 parts to rectum
- 1st--covered on 3 sides by peritoneum
- 2nd--covered on 1 side by peritoneum
- 3rd--not coverd by peritoneum

SAGITTAL VIEW

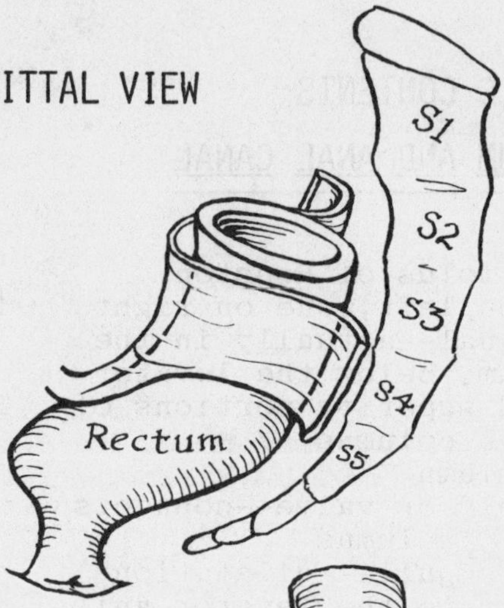

<u>VASCULATURE</u>

1) Inferior mesenteric a.
2) Superior rectal a.
3) Internal iliac a.
4) Middle rectal a.
5) Internal pudendal a.
6) Inferior rectal a.
7) Internal hemorrhoidal plexus--superior to pecten
8) External hemorrhoidal plexus--inferior to pecten
9) Superior rectal v.--drains to portal system
10) Middle rectal v.--drains to systemic venous system
11) Inferior rectal v.--drains to systemic venous system

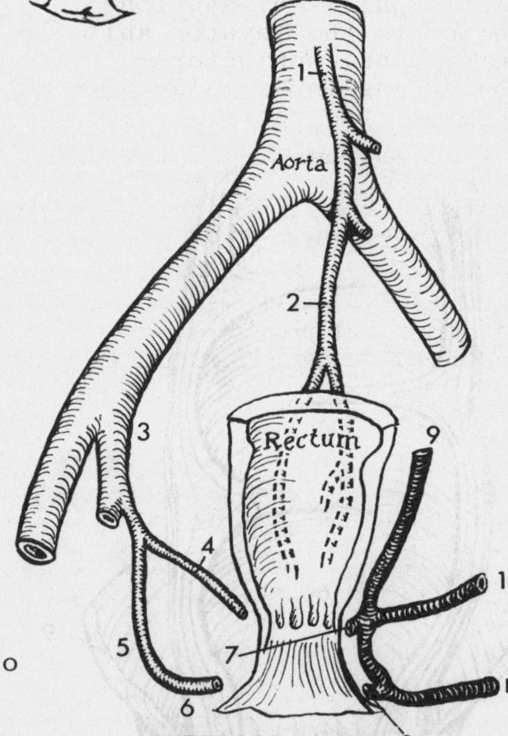

ANTERIOR VIEW

<u>LYMPHATICS</u>

Lymphatics from the rectum drain to nodes along the sacrum, inferior mesenteric artery and internal iliac artery (hypogastric nodes). Lymph from the anus drains to inferior mesenteric and superficial inguinal nodes. The common nodes to drainage from the anus and rectum are the <u>inferior mesenteric nodes</u>.

COMMENTS

1. Hemorrhoid is a term used to refer to varicosities of the hemorrhoidal plexuses. Internal hemorrhoids are dilatations of the internal or superior hemorrhoidal plexus while external hemorrhoids are dilatations of the external or inferior plexus. Various factors may cause the hemorrhoids: upright posture, childbirth, straining during defecation, tumors and portal hypertension.
2. Occasionally during development, the anus does not form properly and the infant may have an imperforate anus or there may be more complex defects involving the region.
3. Carcinomas may necessitate removal of the rectum or anastomoses of other portions of the intestines to the superior portion of the rectum to maintain a normal bowel movement. Colostomies may be performed to create an artificial anus.
4. Anorectal abscesses occur when bacteria invade and infect the anal glands or crypts. The resultant abscess perforates into perianal and perirectal tissues and eventually to the surface of the skin. The abscess then drains and leaves a fistula connecting the skin to the rectum. Various forms and positions of the abscesses may occur.
5. Prolapse is a protrusion of the rectum or its mucosa through the anus. If all layers of the rectal wall 'turn inside out,' the term procidentia is used. The causes are unknown; the condition is most frequent in the elderly and children.

BLADDER

1) Trigone--smooth area bounded by ureteric orifices and urethral orifice
2) Ureteric orifice
3) Interureteric crest
4) Internal urethral orifice
5) Uvula--small in females, larger in males, often due to enlargement of middle lobe of prostate.
6) Apex
7) Ejaculatory duct
8) Prostate
9) Retropubic space of Retzius
10) Aorta
11) Internal iliac a.
12) Umbilical a.
13) Superior vesical a. --in the adult the proximal portion of the umbilical artery remains as the superior vesical a., while the distal portion becomes the medial umbilical ligament
14) Urachus--becomes median umbilical ligament
15) Umbilical vein--becomes ligamentum teres hepatis

FUNDUS of the bladder is the dome-shaped portion postero-superior to a line between the apex and ureteric crest, is highly expansile.

BASE of the bladder is the portion which rests on the prostate of the male and urogenital diaphragm-levator ani in the female.

NECK of the bladder is the area around the internal urethral orifice.

BODY of the bladder is the region surrounding the ureteric orifices.

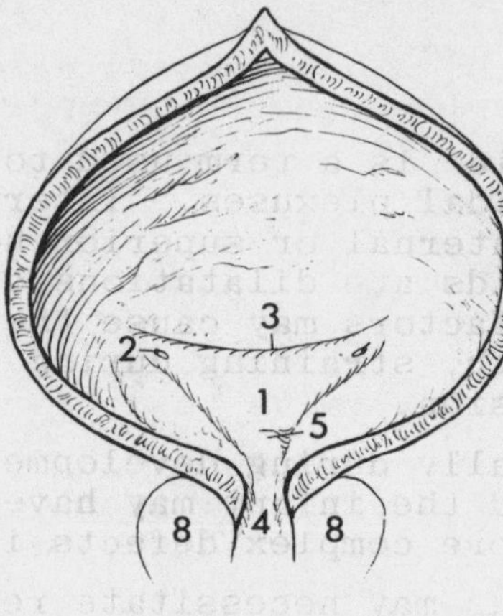

FRONTAL VIEW

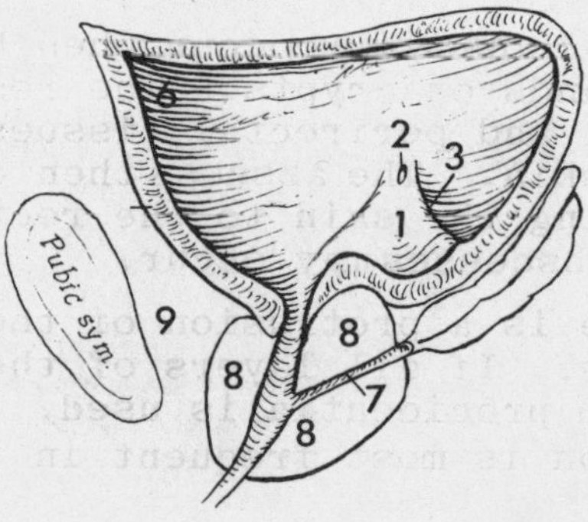

SAGITTAL VIEW

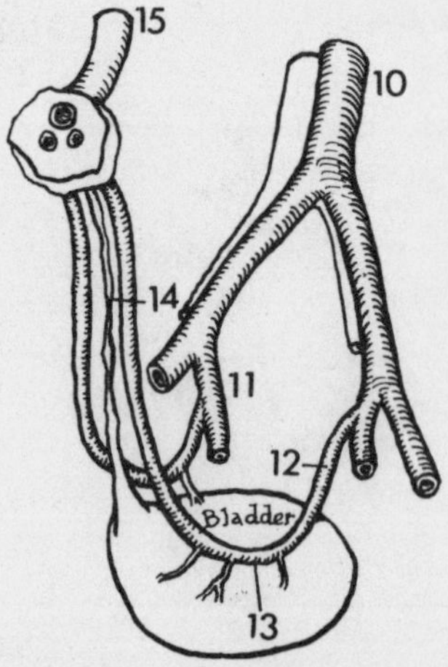

FETAL STRUCTURES

LYMPHATICS of the bladder drain anteriorly to the external iliac nodes and posteriorly to the hypogastric (internal iliac) nodes. They converge in the common iliac nodes.

COMMENTS

1. The course of the ureters through the bladder wall is oblique, forming a flap valve to discourage regurgitation of urine.

2. Catheterization of the bladder may lead to bladder, ureteric and kidney infections; sterile procedures are a must. In insertion of the catheter, one must keep in mind the different directions of turns in the urethra--trauma to the urethra and bladder is possible.

3. Cystoscopy of the bladder is the introduction of a tubular viewing apparatus through the urethra into the bladder. Cystourethrography is the introduction of radio-opaque material into the bladder for the visualization of the bladder and urethra. Still or cinematographic methods are used to inspect for filling defects, strictures, etc.

4. Bladder stones or calculi are not common in the U.S. anymore, but that is not true around the world. Many times they are the result of renal calculi which are not expelled, or may form around foreign objects. They may be removed via a suprapubic incision or crushed and removed transurethrally.

5. Diverticula of the bladder are rarely seen. A patent urachus that extends to the umbilicus and resultant 'weeping navel' are possible but not common.

6. Dense condensations of pelvic fascia around the bladder base are termed 'true' ligaments of the bladder. The major ones, or 'pillars' of the bladder, extend laterally to the fascia over, the levator ani and anteriorly to the pubis.

7. In the female, a thin areolar septum, the vesicovaginal septum separates the bladder from the vagina. In the male, however, a heavier fascia--Denonvilliers Fascia or prostatoperitoneal membrane--is connected to the seminal vesicles and ductus deferens and separates the bladder from the rectum.

THE MALE PELVIS

UNLIKE THE FEMALE, the only reproductive structures to be considered are the ductus deferens, seminal vesicles and the prostate.

1) Bladder
2) Ureter
3) Epididymis
4) Inguinal canal
5) Left and right lateral lobes of prostate
6) Isthmus--joins lateral lobes of prostate
7) Middle or median lobe of prostate
8) Posterior lobe
9) Urethral crest
10) Seminal colliculus
11) Prostatic sinuses
12) Prostatic utricle
13) Orifice of ejaculatory duct
14) Sphincter urethrae of urogenital diaphragm
15) Bulbourethral gland
16) Bulb of penis

URETHRA
17) Prostatic
18) Membranous
19) Penile

20) Seminal vesicle
21) Ductus deferens
22) Ampulla of ductus
23) Ejaculatory duct

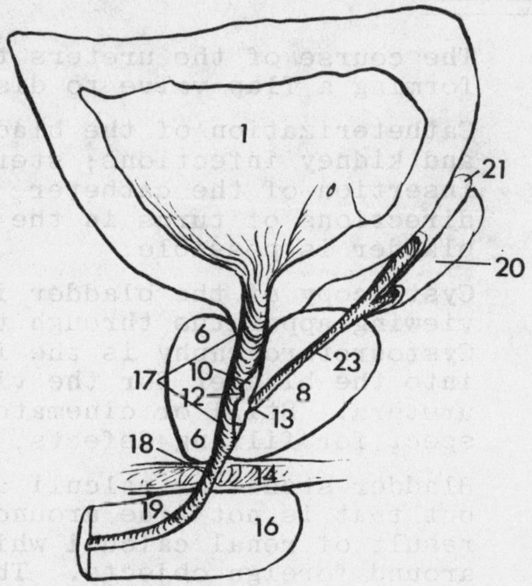

SAGITTAL SECTION

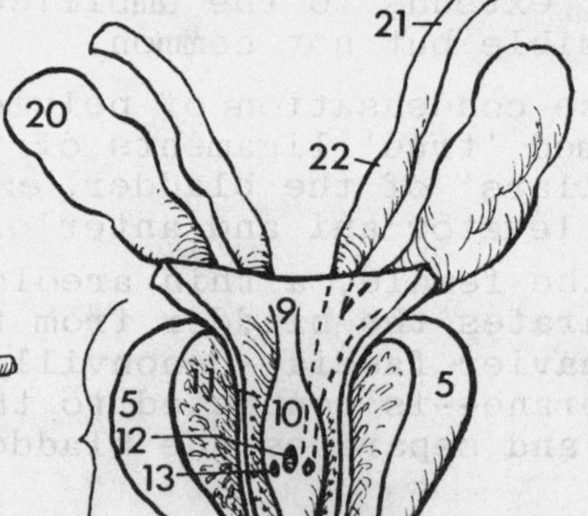

CORONAL SECTION

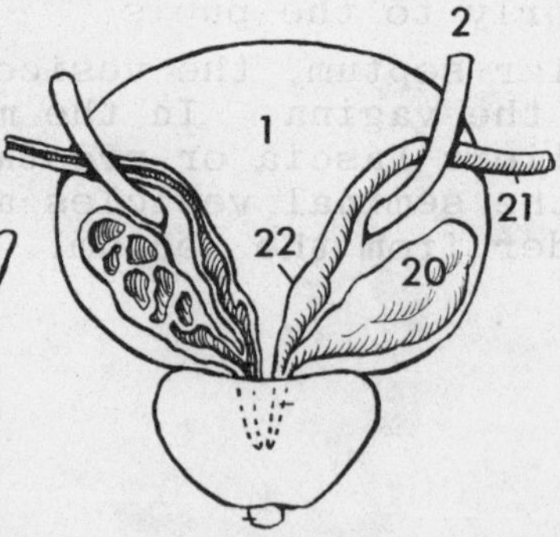

POSTERIOR VIEW

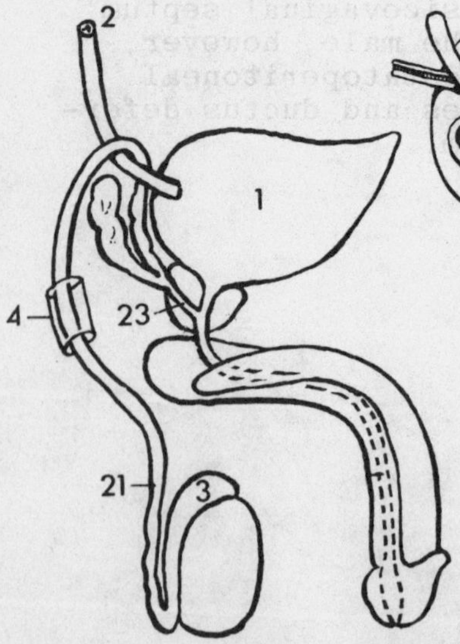

SCHEMATIC VIEW

COMMENTS

1. A plexus of veins surrounds the prostate and lower urinary bladder. This plexus is the Plexus of Santorini, prostatic plexus or vesico-prostatic plexus. It communicates posteriorly with veins that enter the vertebral plexus and constitutes one important route of metastasis of prostatic carcinomas.

2. Benign hypertrophy is most commonly the bilobular type involving both lateral lobes or trilobular involving the middle lobe in addition to the lateral ones. As a result, the individual encounters increasing difficulty in voiding urine, the muscular wall of the bladder can thicken and increased pressure in the bladder can dilate the ureters and cause hydronephrosis of the kidney. Treatment includes removal of the enlargements.

3. Chronic prostatitis is an infection of the prostate that may result from urinary infections, catheterization infections, or infections in other foci. It can cause pyuria, sexual dysfunction, scrotal and back pains. Treatment may include antibiotics, transurethral resection or possibly radical removal of the prostate.

4. Cancer of the prostate occurs in up to 14-46% of men above 50 years of age. It is usually slow growing and is most common in the the posterior lobe. The most frequent sites of metastases are the liver, bony pelvis and sacrum. Even though the primary lesion in the prostate may grow slowly, metastases may occur rapidly.

THE FEMALE PELVIS

THE BROAD LIGAMENT extends like a wall from left to right across the female pelvis; it divides the pelvis into anterior and posterior compartments. The ovaries, uterus and uterine tubes are covered by this broad ligament and therefore exist within the ligament, and not in the anterior or posterior compartments. The female reproductive organs are associated with various ligaments which are fascial in nature, not like the true ligaments which connect bone to bone.

STRUCTURES:
1) Uterus
2) Cervix
3) Vagina
4) Bladder
5) Urethra
6) Rectum
7) Suspensory ligament of the ovary
8) Ovary
9) Uterine tube
10) Ureter
11) Aorta
12) Common iliac a.
13) External iliac a.
14) Internal iliac a.
15) Superior vesical a.
16) Broad ligament

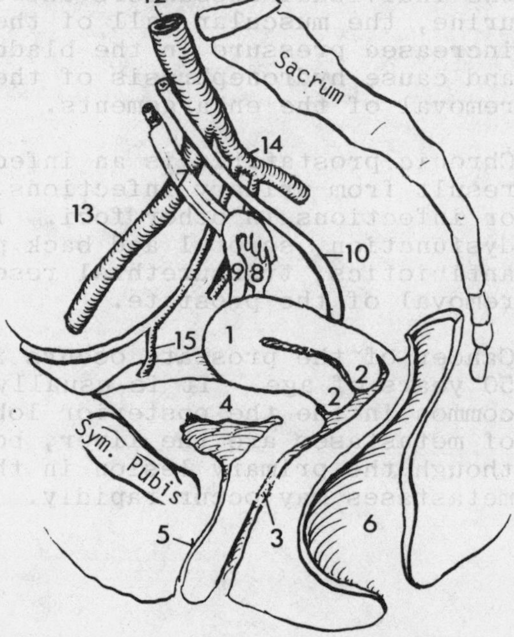

SAGITTAL VIEW

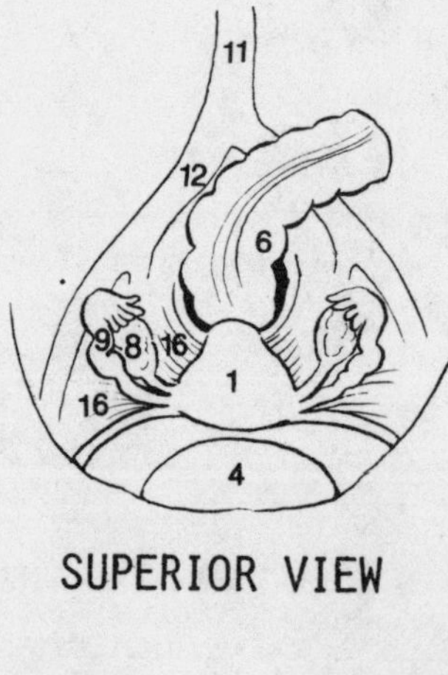

SUPERIOR VIEW

UTERUS, UTERINE TUBE AND OVARIES

UTERUS
1) Fundus
2) Body
3) Isthmus
4) Cervix
5) Lumen
6) Cervical canal
7) Uterine ostium --external os
8) Vagina

UTERINE TUBE AND OVARY
9) Uterine tube
10) Isthmus--narrow medial 1/3 of tube
11) Ampulla--lateral 2/3 of tube
12) Infundibulum--funnel-shaped end of tube
13) Abdominal ostium
14) Fimbriae
15) Ovarian fimbria--an especially long fimbria attached to the ovary
16) Ovary
17) Ovarian ligament
18) Suspensory ligament of the ovary

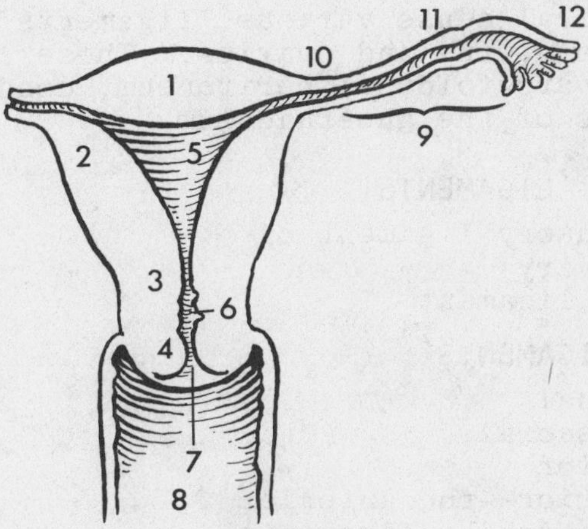

CORONAL SECTION

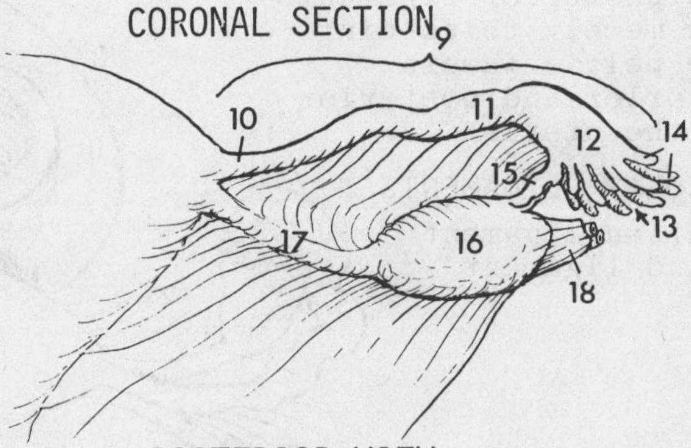

POSTERIOR VIEW

ANGLES OF THE UTERUS

1) THE VERSION ANGLE of the uterus refers to the angle with which the cervix meets the vagina. The normal position is tilted anteriorly or ANTEVERTED.

2) THE FLEXION ANGLE of the uterus refers to the angle with which the body of the uterus meets the cervix. The normal position is tilted anteriorly or ANTEFLEXED.

THESE TWO ANTERIORLY ORIENTED BENDS PLACE THE UTERUS IN A NEARLY HORIZONTAL POSITION SUPERIOR TO THE BLADDER.

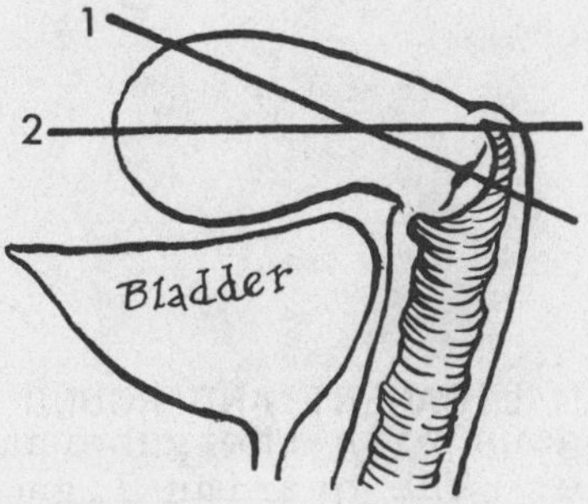

SCHEMATIC LATERAL VIEW

'LIGAMENTS' OF THE UTERUS AND OVARIES

In a classical sense various 'ligaments' have been described which support the uterus and ovaries. These ligaments are not true ligaments but are folds of peritoneum, condensations of pelvic fascia or remnants of the gubernaculum.

PERITONEAL 'LIGAMENTS'

1) Suspensory ligament of the ovary
2) Broad ligament

FASCIAL 'LIGAMENTS'

Cardinal
3) Uterosacral
4) Anterior
5) Posterior--the anterior and posterior ligaments are merely thickenings of the pelvic fascia anterior and posterior to the uterus

GUBERNACULAR 'LIGAMENTS'

6) Ovarian ligament
7) Round ligament

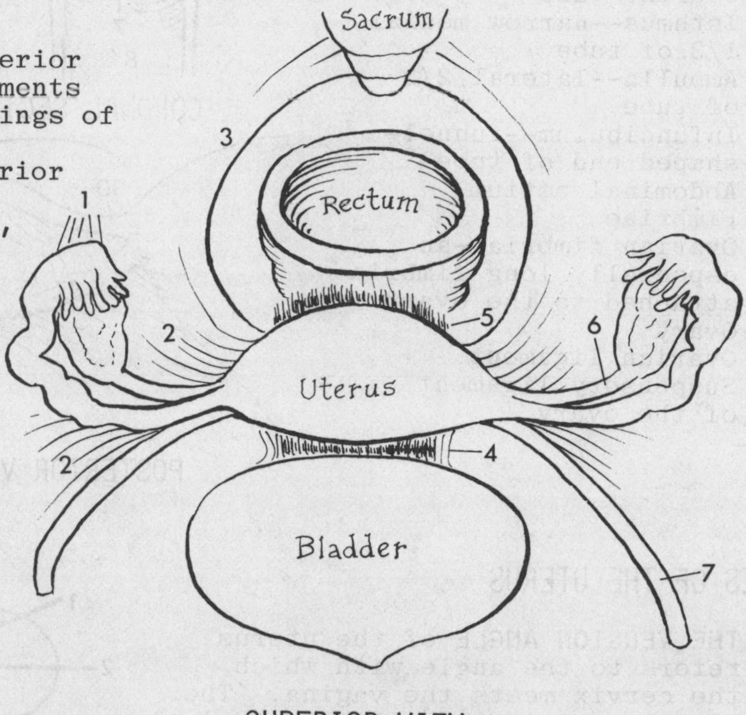

SUPERIOR VIEW

THE OVARIAN LIGAMENT AND ROUND LIGAMENT are homologous with the gubernaculum testis in the male. The ovarian ligament extends from the ovary to the lateral angle of the uterus; the round ligament of the uterus continues from there, passes through the inguinal canal and ends in the labia majora.

THE BROAD LIGAMENT

THE BROAD LIGAMENT is a double layer of peritoneum formed by the draping of the peritoneum over the uterus, uterine tube and ovary. It is composed of three parts--the mesometrium, mesosalpinx and mesovarium.

THE CARDINAL LIGAMENTS are dense condensations of fascia extending laterally from the level of the cervix, surrounding the uterine vessels. The cardinal ligaments are inferior to and between the leaves of peritoneum which form the base of the broad ligament. The fascia of the cardinal ligament is a portion of dense pelvic fascia which ensheaths the upper vagina and uterus. At the level of the uterus it is termed the parametrium.

1) Mesometrium
2) Mesosalpinx
3) Mesovarium
4) Suspensory ligament --a fold of peritoneum over the ovarian vessels
5) Ovarian ligament
6) Round ligament
7) Internal inguinal ring
8) Uterine tube
9) Ureter
10) Uterine artery
11) Peritoneum--cut edge
12) Vesicouterine pouch
13) Rectouterine pouch
14) Ovary
15) Ovarian artery
16) Cardinal ligament

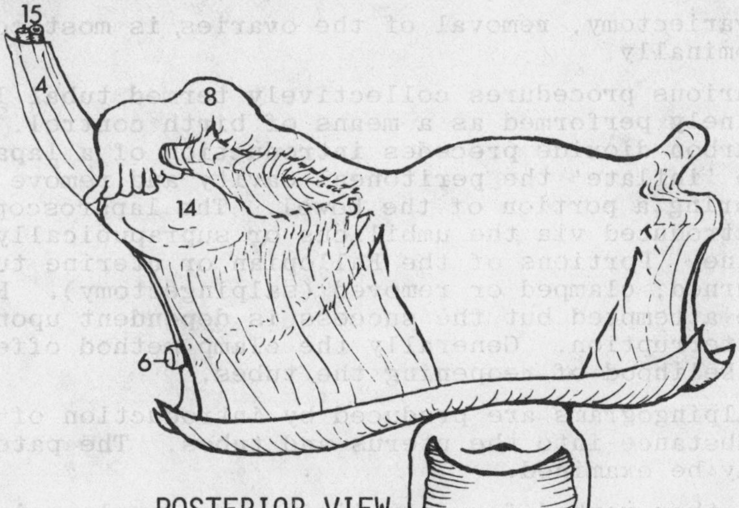

POSTERIOR VIEW

SAGITTAL SECTIONS

THROUGH OVARY NEAR UTERUS

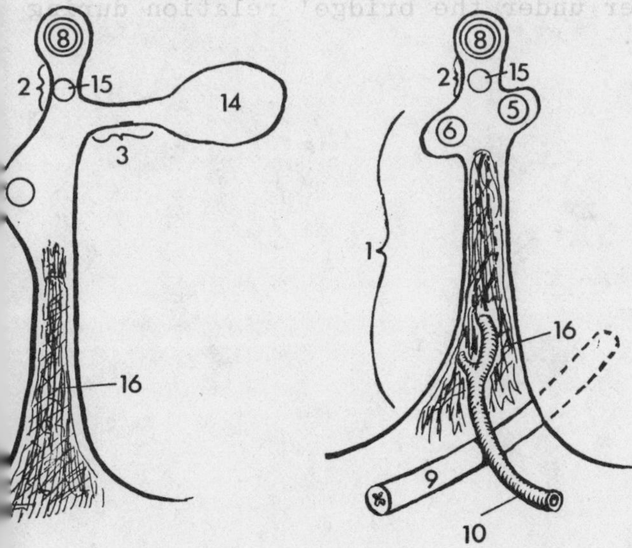

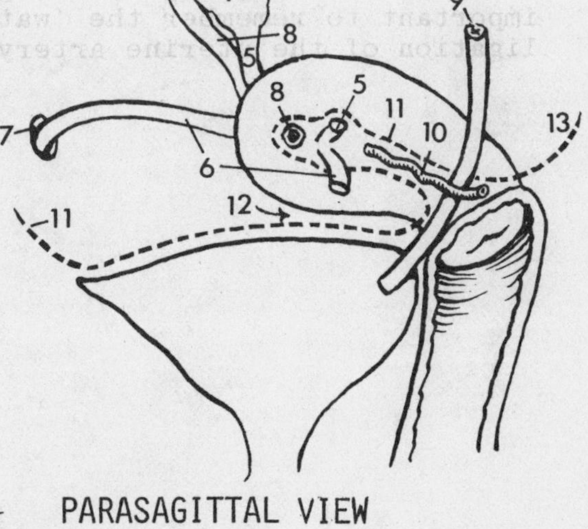

PARASAGITTAL VIEW

COMMENTS

1. Hysterectomy is removal of the uterus, which can be performed by an abdominal or vaginal approach; the technique usually depends on the surgeon. In general the vaginal approach is best; there is no incision, and vaginoplasty to repair cystoceles may be done simultaneously. The presence of complications such as fibroids necessitates an abdominal, supra-pubic approach.
2. Ovariectomy, removal of the ovaries, is most commonly done abdominally.
3. Various procedures collectively termed tubal ligations are routinely performed as a means of birth control. An injection of carbon dioxide precedes introduction of a laparoscope in order to 'inflate' the peritoneal cavity and remove the risk of puncturing a portion of the bowel. The laparoscope is commonly introduced via the umbilicus or suprapubically below the bikini line. Portions of the Fallopian or uterine tube may be cut, burned, clamped or removed (salpingectomy). Reanastomosis may be attempted but the success is dependent upon the method of interruption. Generally the clamp method offers the greatest likelihood of reopening the tubes.
4. Salpingograms are produced by introduction of a radio-opaque substance into the uterus and tubes. The patency of the tubes may be examined.
5. Another method for visualizing the female pelvis is via the introduction of a culdoscope through the posterior fornix of the vagina. The posterior fornix is separated from the pelvic cavity only by the wall of the vagina and peritoneum and offers a readily accessible entry into the pelvis.
6. The ureter is in danger of ligation during hysterectomy. It is important to remember the 'water under the bridge' relation during ligation of the uterine artery.

VAGINA

Superiorly to the urogenital diaphragm the vagina is flattened anteriorly-posteriorly so there is an anterior and posterior wall of the vagina. Inferiorly to the urogenital diaphragm the vagina and vestibule are flattened laterally so there are lateral walls of the vagina. The cervix protrudes into the vagina in such a manner as to produce a circular pocket around the cervix. This 'pocket' is referred to as the fornix. Although the fornix is continuous around the cervix it is usually referred to as having anterior, posterior and lateral fornices.

1) Posterior fornix
2) Anterior fornix
3) Vagina
4) Vestibule of the vagina

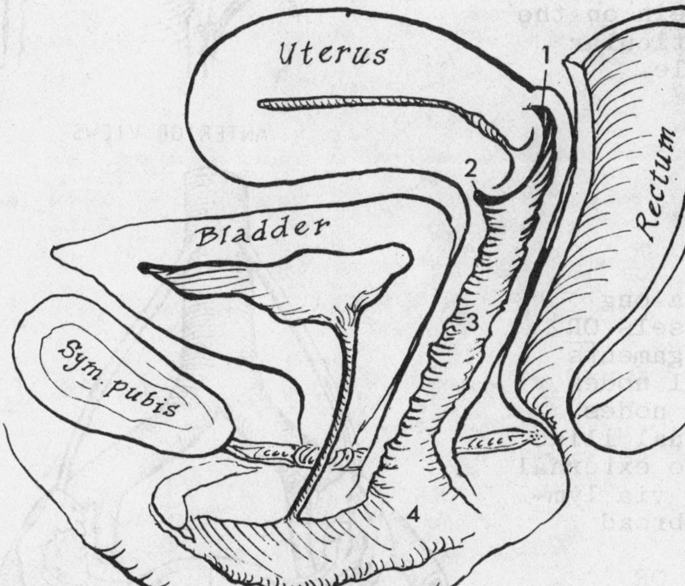

SAGITTAL VIEW

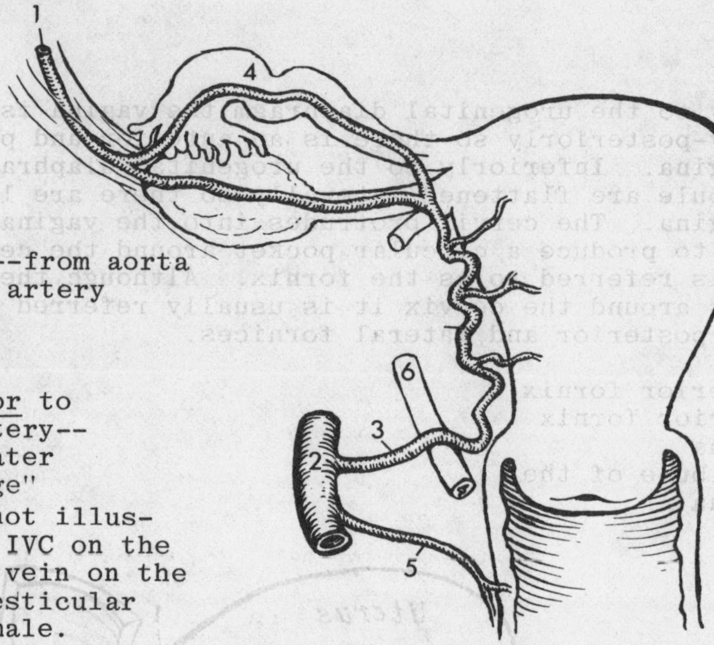

VASCULATURE

1) Ovarian artery--from aorta
2) Internal iliac artery
3) Uterine artery
4) Tubal branches
5) Vaginal artery
6) Ureter--inferior to the uterine artery--remember by "water under the bridge"

THE OVARIAN VEINS, not illustrated, drain to the IVC on the right and the renal vein on the left, just as the testicular veins drain in the male.

ANTERIOR VIEWS

LYMPHATIC DRAINAGE

REGION	ROUTE
1	Lymphatics along ovarian vessels OR to round ligaments and inguinal nodes
2	Hypogastric nodes along internal iliac artery OR to external iliac nodes via lymphatics in broad ligament
3	Hypogastric OR sacral nodes
4	External iliac nodes or femoral nodes

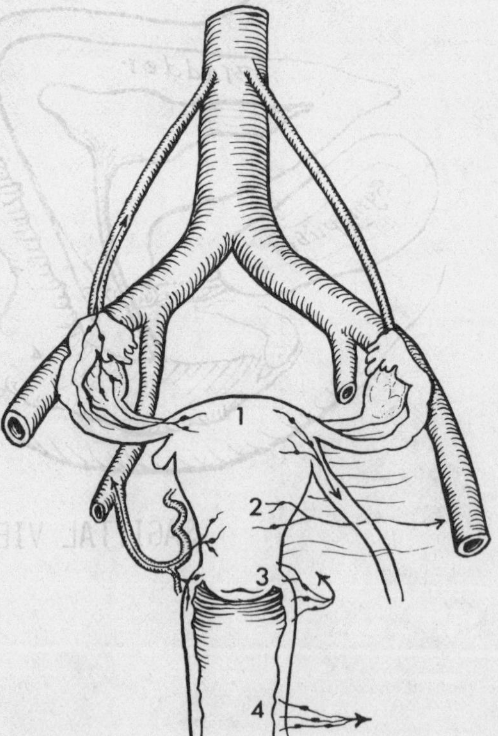

REMEMBER alternate routes become well established as normal routes are blocked by invasive processes.

THE PELVIC EXAMINATION

Palpation of the female pelvic organs, known as the bimanual pelvic examination, is an important diagnostic tool in clinical medicine to reveal diseases and disorders of the lower urinary tract, reproductive organs, and rectum. The examiner places the index and second finger into the vagina until the tips of the fingers are facing anteriorly and touching the cervix. At the same time, the fingers of the other hand are pressed deep into the abdomen allowing the organs to be palpated between the two hands. To obtain additional information the rectopelvic examination may be performed by placing the index finger into the vagina while inserting the second finger into the rectum. Again the other hand presses deeply into the abdomen and the pelvis is palpated. When the examiner judges the pelvic or rectopelvic examination is undesirable for the patient, as in the case of the virgin whose hymen is intact, enough may be learned by a rectal examination in the bimanual method to make or exclude diagnoses.

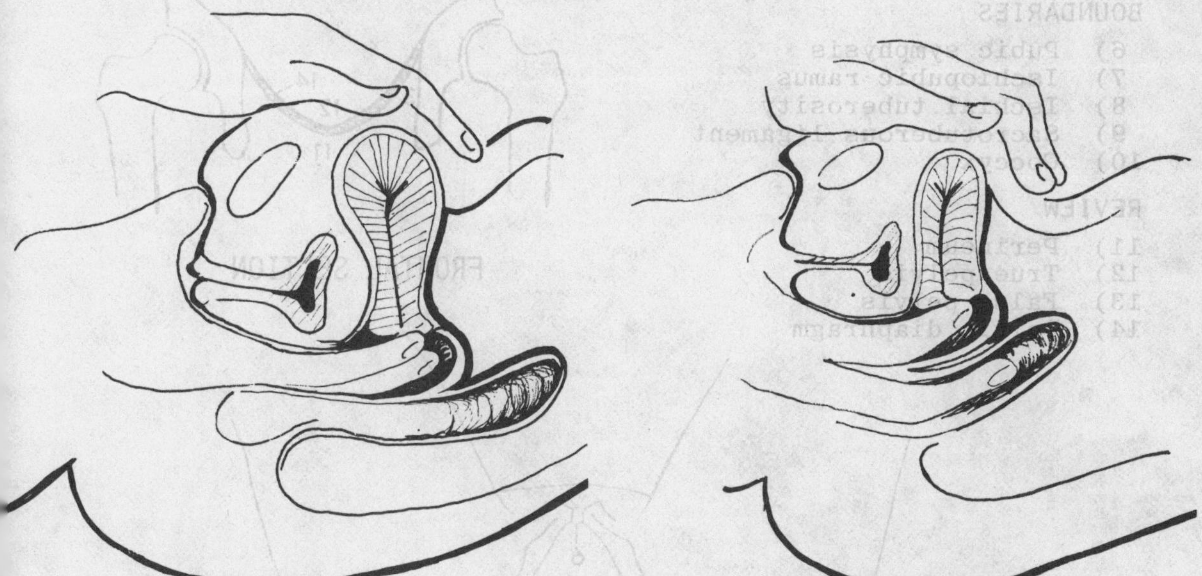

PERINEUM

I. SURFACE CORRELATIONS AND DEFINITIONS

The perineum is the diamond-shaped area enclosing the external genitalia and anus. It is composed of the urogenital triangle containing the external genitalia, and the anal triangle; the two triangles do not lie in the same plane, but make a 35° angle to each other. The boundaries of the perineum are also the boundaries of the pelvic outlet.

1) Urogenital--UG--triangle
2) Anal triangle
3) Ischiorectal fossae
4) Perineal body
5) Coccygeal body--along with the perineal body these are two condensations of fascia important for muscle attachment

BOUNDARIES

6) Pubic symphysis
7) Ischiopubic ramus
8) Ischial tuberosity
9) Sacrotuberous ligament
10) Coccyx

REVIEW

11) Perineum
12) True pelvis
13) False pelvis
14) Pelvic diaphragm

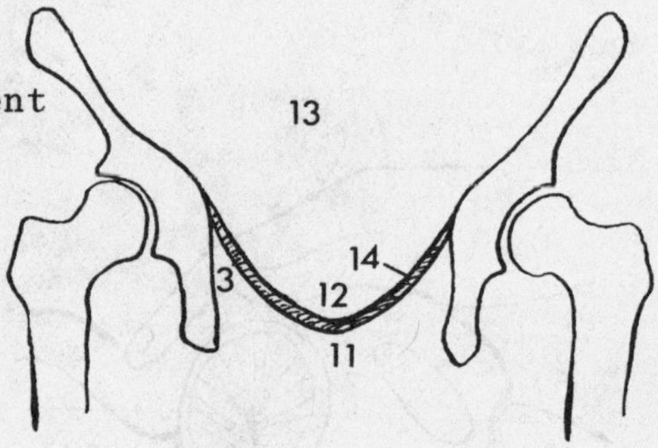

FRONTAL SECTION

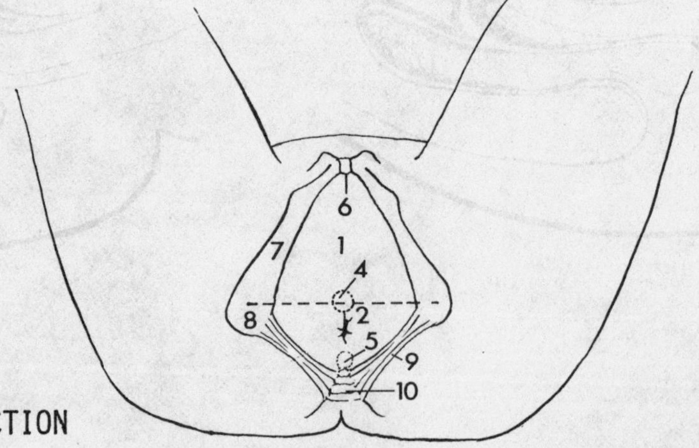

SURFACE PROJECTION

II. ANAL TRIANGLE

ANAL CANAL AND ANUS

The anal canal is inferior to the level of the levator ani and begins at the level of the puborectalis externally or superior end of the anal columns internally. It is divided into 3 regions: superior, transitional or intermediate, and cutaneous.

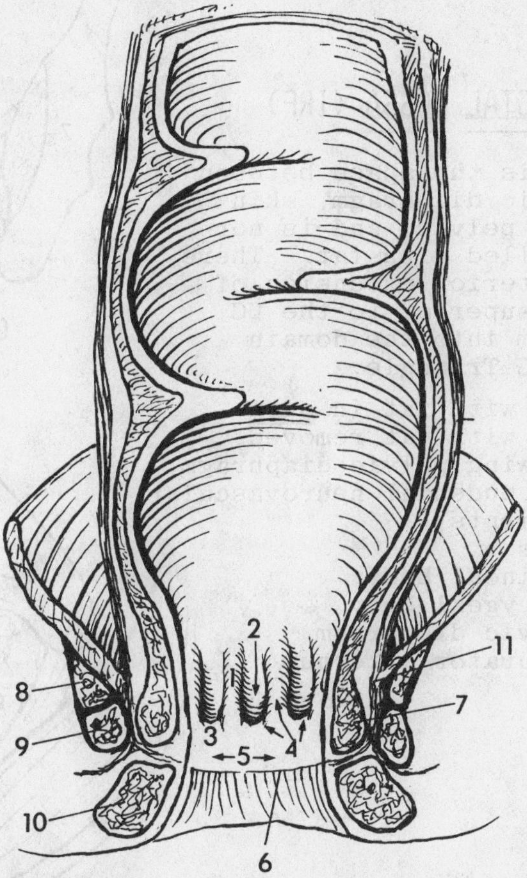

SUPERIOR: Extends from superior end of anal columns to base, covered by true anal mucosa derived from endoderm.

1) Anal column
2) Anal sinus--deep to anal valve

TRANSITIONAL: Characterized by the pecten, a smooth hairless band 2 cm wide; site of the proctodeal membrane, covered by epithelium derived from ectoderm.

3) Anal valve
4) Pectinate line--along superior border of anal columns and superiorly along bases of anal columns; marks division of blood supply, venous and lymphatic drainage, and transition of endoderm to ectoderm.
5) Pecten
6) Anocutaneous verge--or White Line of Hilton--indicates inferior limit transitional zone, also demarcates most inferior border of internal anal sphincter and superior border of subcutaneous external sphincter.

CUTANEOUS: Contains pigmented skin, hairs and glands.

ALSO:

7) Internal sphincter
8) Deep external sphincter
9) Superficial external sphincter
0) Subcutaneous external sphincter
1) Conjoined longitudinal muscle
2) Perineal body
3) Coccygeal body

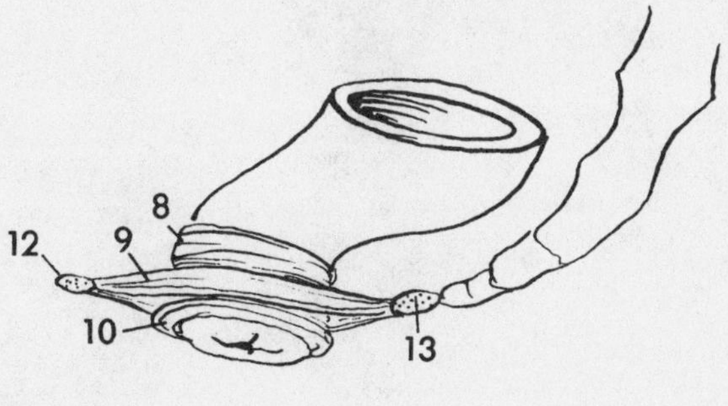

ISCHIORECTAL FOSSA (IRF)

The IRF is the space between the pelvic diaphragm, skin and bony pelvis, and is normally filled with fat. There is an anterior extension of the IRF superior to the UG diaphragm into the domain of the UG Triangle.

1) IRF with fat *in situ*
2) IRF with fat removed, showing pelvic diaphragm and pudendal neurovascular elements
3) Anus
4) Perineal body
5) Coccygeal body
6) Pelvic diaphragm
7) Obturator internus

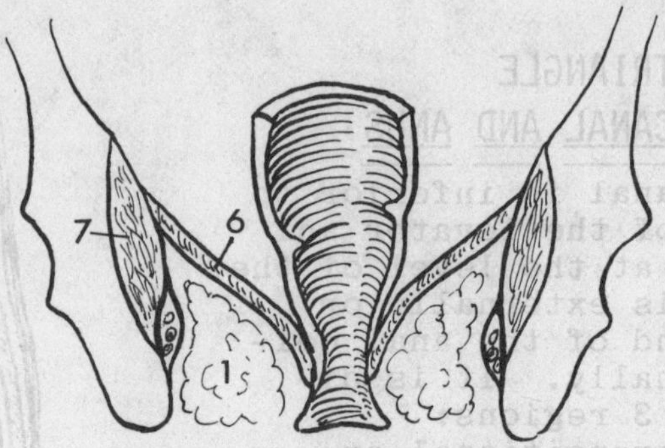

CORONAL VIEW IRF

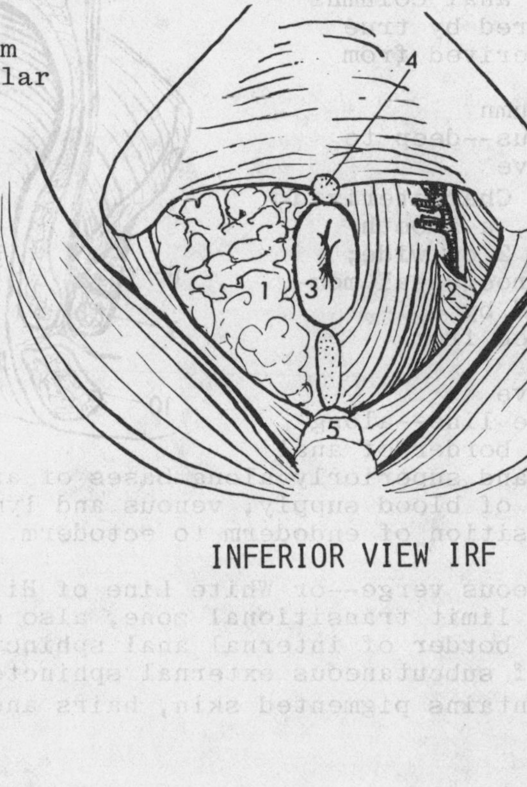

INFERIOR VIEW IRF

COMMENTS

1. The anus offers an important means for palpation of pelvic contents, especially in the male. An enlarged prostate may be detected as well as an inflamed appendix projecting into the pelvis.
2. Proctoscopy or sigmoidoproctoscopy is the introdution of a viewing apparatus--the proctoscope--into the rectum and/or sigmoid colon for diagnostic and surgical procedures.
3. Hemorrhoids, already discussed, may often be seen protruding from or originating near the anus. Skin tabs, remnants of thrombosed hemorrhoids, may also be observed.
4. The fat of the ischiorectal fossa functions to cushion the ischial tuberosities when sitting and offers support and elastic recoil to the rectal ampulla and anal canal in defecation.
5. Anal glands open into the anal sinuses--Crypts of Morgagni--and may function in the lubrication of feces.

I. UROGENITAL TRIANGLE

EXTERNAL GENITALIA

FEMALE

1) Mons pubis--fat pad
2) Labia majora
3) Labia minora
4) Anterior commissure
5) Posterior commissure
6) Prepuce
7) Frenulum of clitoris
8) Clitoris
9) Vaginal orifice
10) Hymenal caruncles
11) External urethral meatus
12) Frenulum pudendi--or fourchette--where the two labia minora meet posteriorly
13) Perineal body
14) Anus
15) Bulb of the vestibule
16) Greater vestibular glands--of Bartholin--open into the vestibule

THE VESTIBULE is the space enclosed by the labia minora.

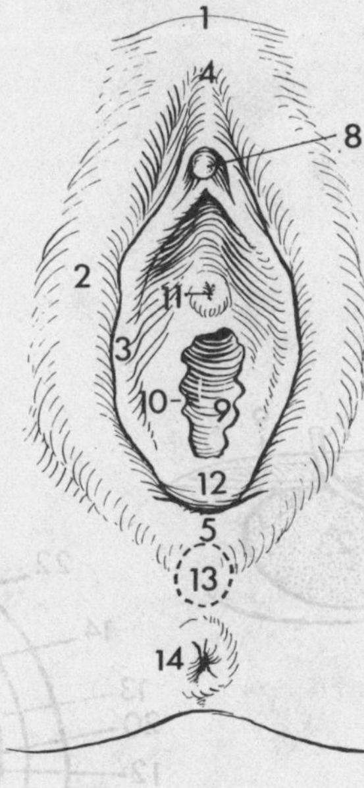

EXTERNAL GENITALIA

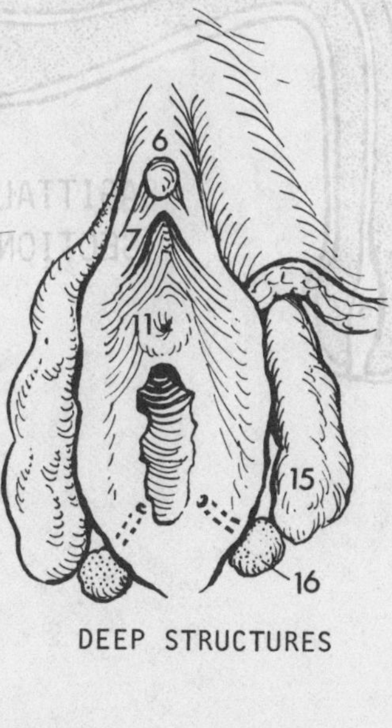

DEEP STRUCTURES

MALE

1) Shaft of penis
2) Corona
3) Glans
4) Prepuce or foreskin
5) Frenulum
6) External urethral meatus
7) Urethra
8) Navicular fossa
9) UG diaphragm
10) Scrotum
11) Corpus spongiosum
12) Corpus cavernosum
13) Buck's fascia
14) Colles' fascia (Dartos)
15) Superficial dorsal vein
16) Deep dorsal vein
17) Dorsal artery
18) Dorsal nerve
19) Deep artery
20) Tunica albuginea
21) Septum
22) Skin
23) Bulb of the penis

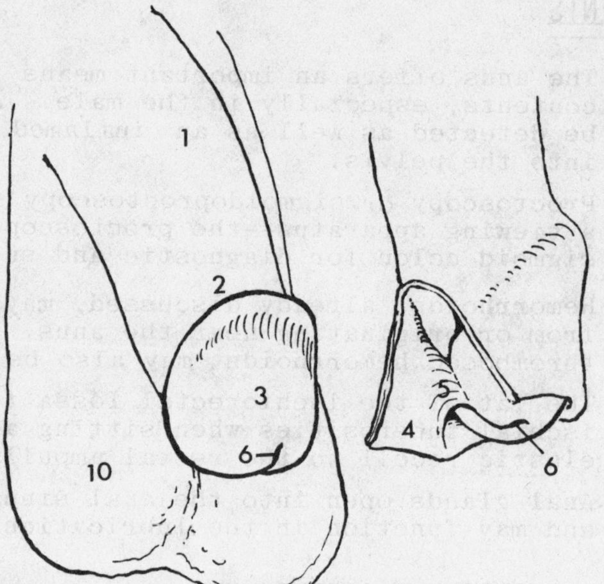

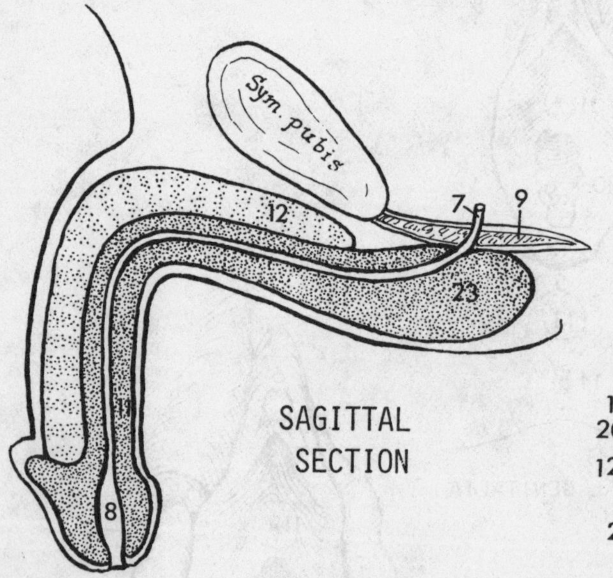

SAGITTAL SECTION

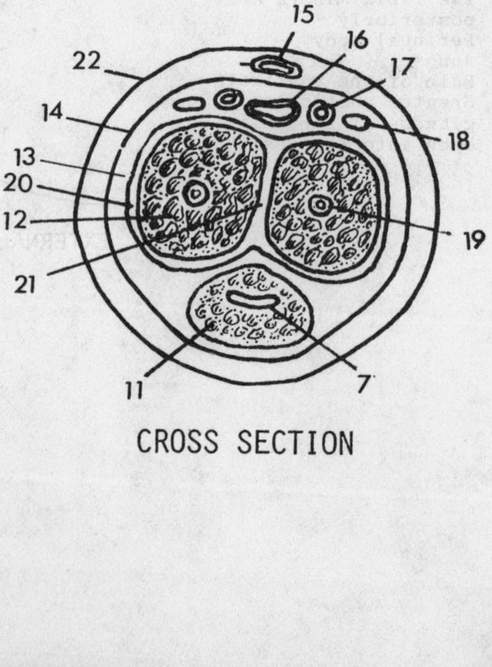

CROSS SECTION

TESTICULAR DETAIL:

1) Testicle
2) Head of epididymis
3) Body of epididymis
4) Tail of epididymis
5) Scrotal sac
6) Parietal layer tunica vaginalis
7) Visceral layer tunica vaginalis
8) Sinus of epididymis
9) Tunica albuginea
10) Testicular septum
11) Testicular lobules
12) Convoluted seminiferous tubules
13) Straight seminiferous tubules
14) Rete testis
15) Efferent ductule
16) Ductus deferens

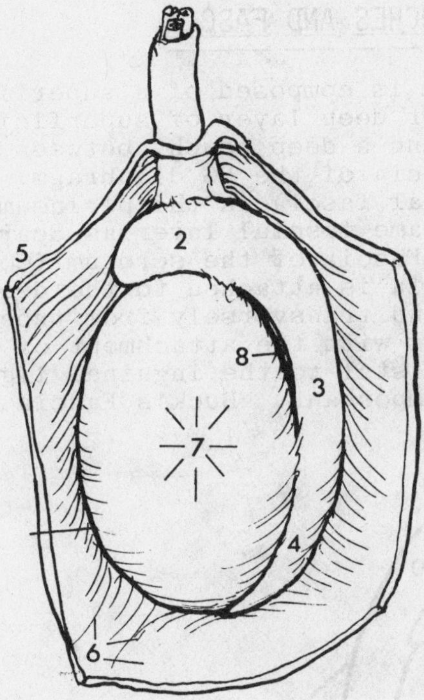

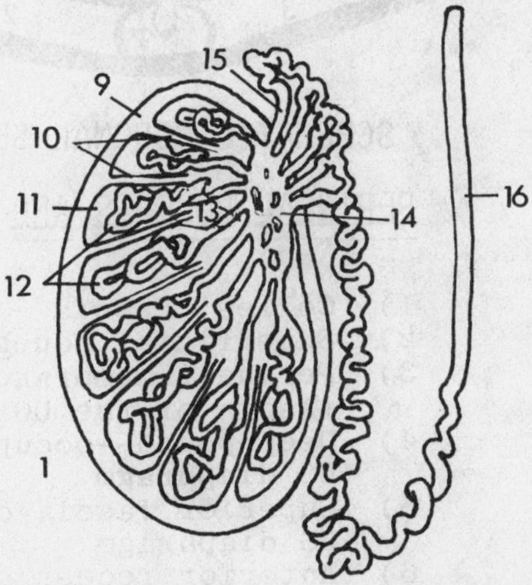

PERINEAL POUCHES AND FASCIA

The perineum is composed of a superficial pouch, between the membranous or deep layer of superficial fascia and the perineal membrane, and a deep pouch, between the perineal membrane and superior fascia of the UG diaphragm. The deep or membranous layer of superficial fascia in the perineum is called Colles' Fascia and is the same fascial layer as Scarpa's Fascia of the abdomen and Dartos Fascia of the scrotum (Colles'=Scarpa's=Dartos). Colles' Fascia is attached to the deep fascia along the ischio-pubic rami and transversely from tuberosity to tuberosity. This is continuous with the attachment of Scarpa's fascia to the fascia lata, just distal to the inguinal ligament. These relations are clinically important. Buck's Fascia is deep to Colles' Fascia.

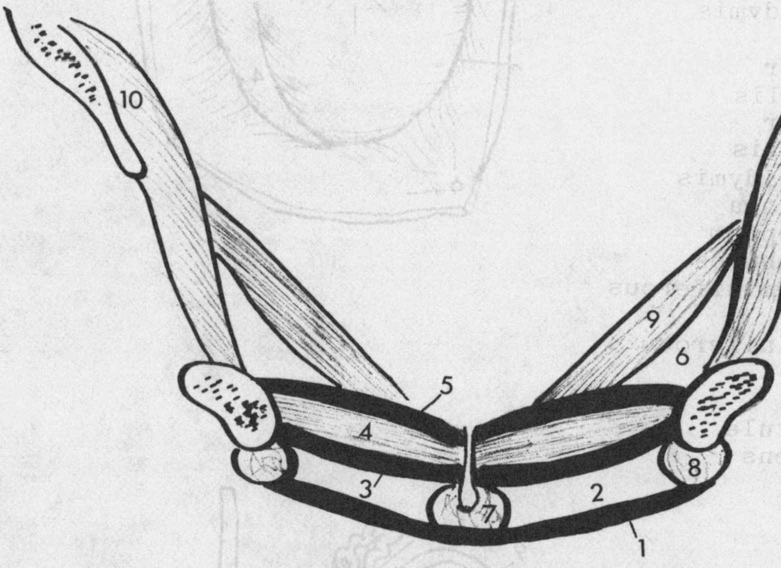

SCHEMATIC CORONAL SECTION

PERINEAL POUCHES AND FASCIA

1) Colles' Fascia
2) Superficial pouch
3) Perineal membrane--inferior fascia of the UG diaphragm
4) Deep pouch--occupied by the UG diaphragm
5) Superior fascia of the UG diaphragm
6) Anterior recess of the IRF
7) Bulb of the penis or vestibule
8) Crus of the penis or clitoris
9) Pelvic Diaphragm
10) Obturator internus

CONTENTS OF THE SUPERFICIAL POUCH

FEMALE

1) Superficial transverse perineus m.
2) Bulbospongiosus m.
3) Ischiocavernosus m.
4) Perineal membrane
5) Perineal body
6) Pudendal nerve
7) Inferior rectal n.
8) Perineal n.
9) Nerve of the clitoris
10) Internal pudendal a.
11) Inferior rectal a.
12) Perineal a.
13) Artery of the clitoris
14) Vestibular bulb
15) Clitoris
16) Crus of the clitoris
17) Coccygeal body

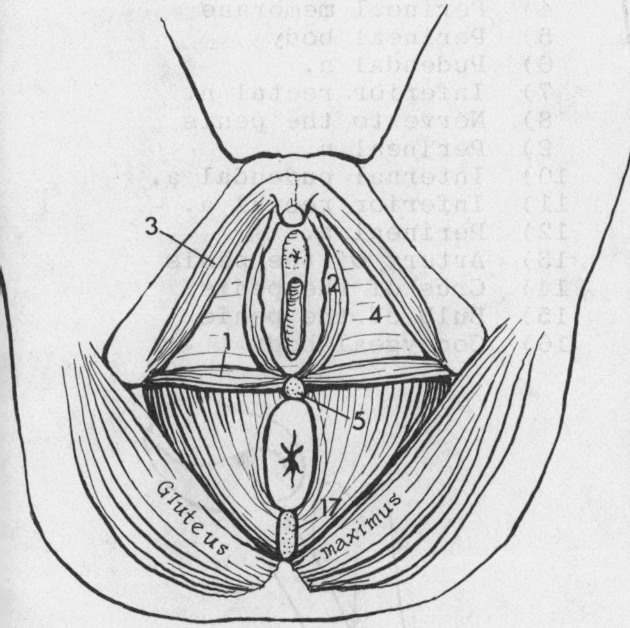

MUSCLES

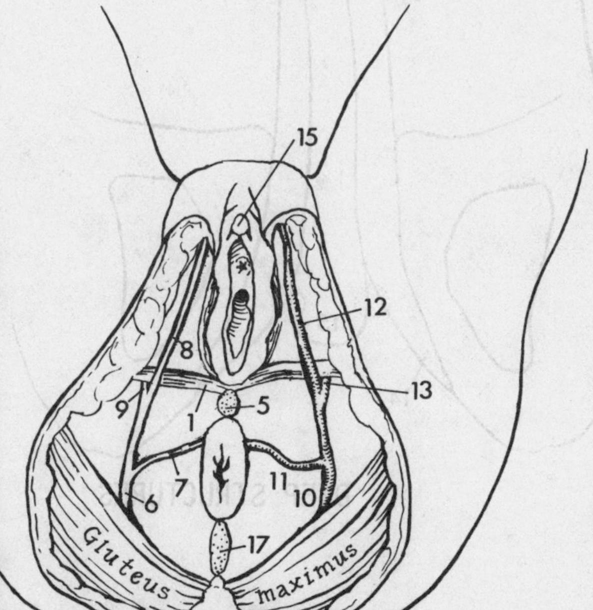

NERVES AND ARTERIES

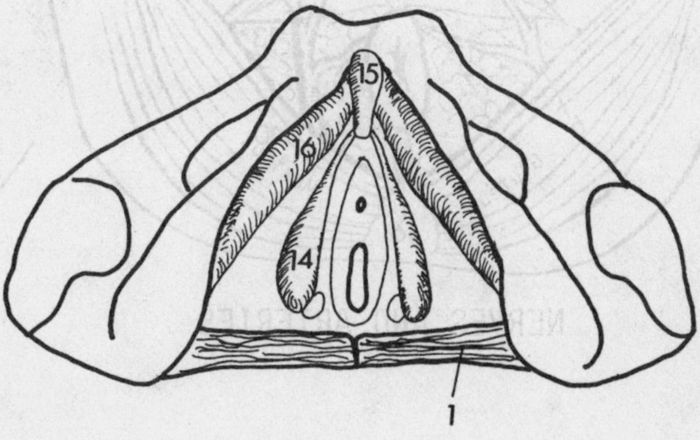

DEEP STRUCTURES

MALE

1) Superficial transverse perineus m.
2) Ischiocavernosus m.
3) Bulbospongiosus m.
4) Perineal membrane
5) Perineal body
6) Pudendal n.
7) Inferior rectal n.
8) Nerve to the penis
9) Perineal n.
10) Internal pudendal a.
11) Inferior rectal a.
12) Perineal a.
13) Artery of the penis
14) Crus of the penis
15) Bulb of the penis
16) Coccygeal body

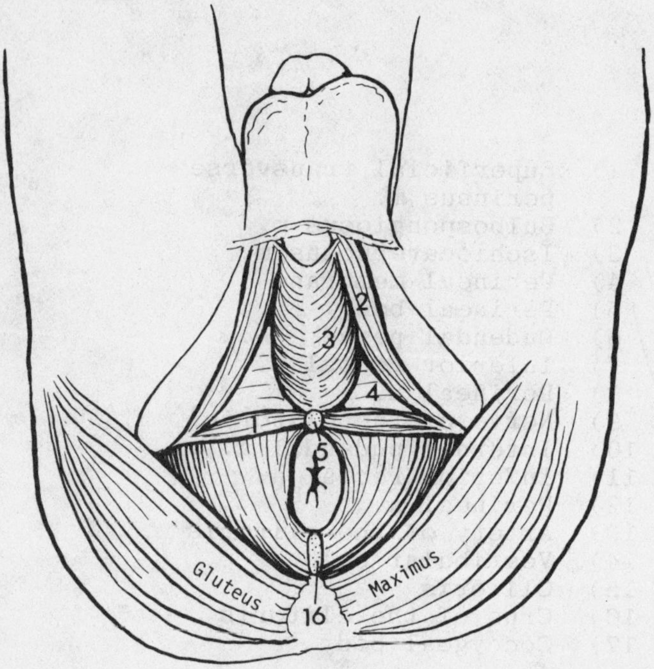

MUSCLES

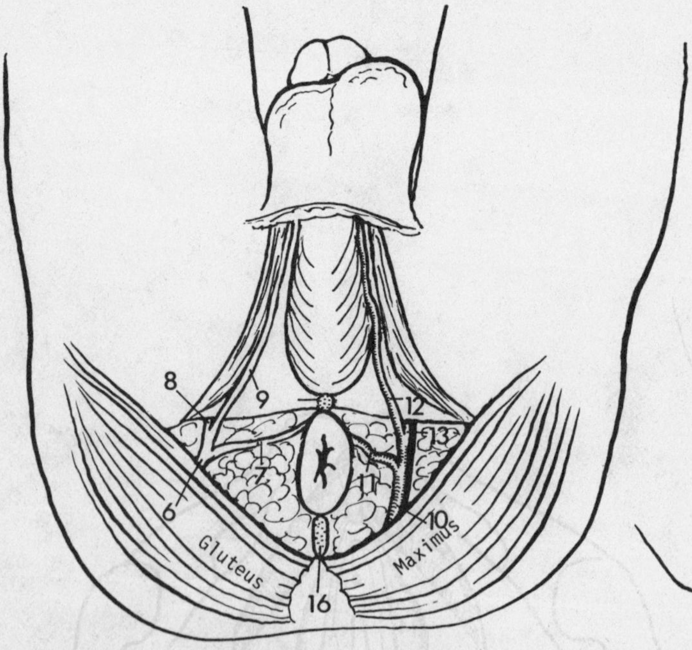

NERVES AND ARTERIES

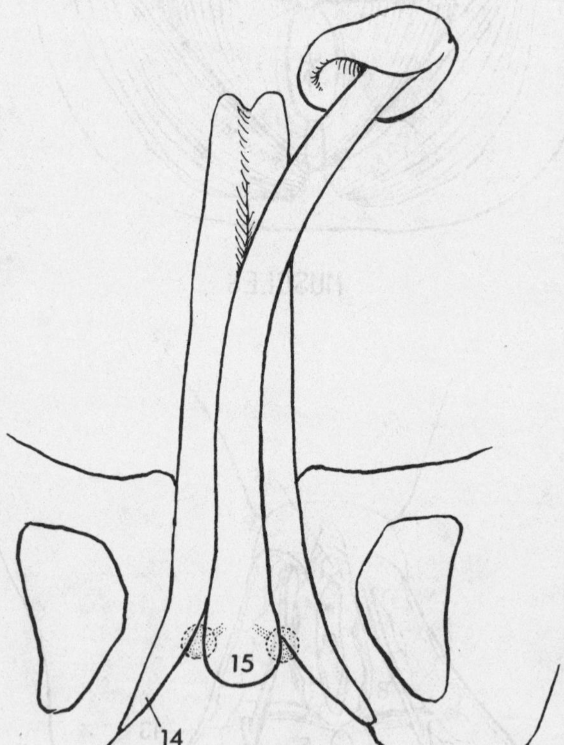

DEEP STRUCTURES

CONTENTS OF THE DEEP POUCH

1) Perineal body
2) Superficial transverse perineus m.
3) Perineal membrane
4) Deep artery of the penis
5) Artery to the bulb
6) Urethra
7) Transverse ligament of the pelvis
8) Arcuate ligament of the pubis
9) Aperture for the deep dorsal vein
10) Sphincter urethrae
11) Deep transverse perineus m.
12) Pudendal n.
13) Dorsal nerve of the penis
14) Deep dorsal vein of the penis
15) Dorsal artery of the penis
16) Internal pudendal a.

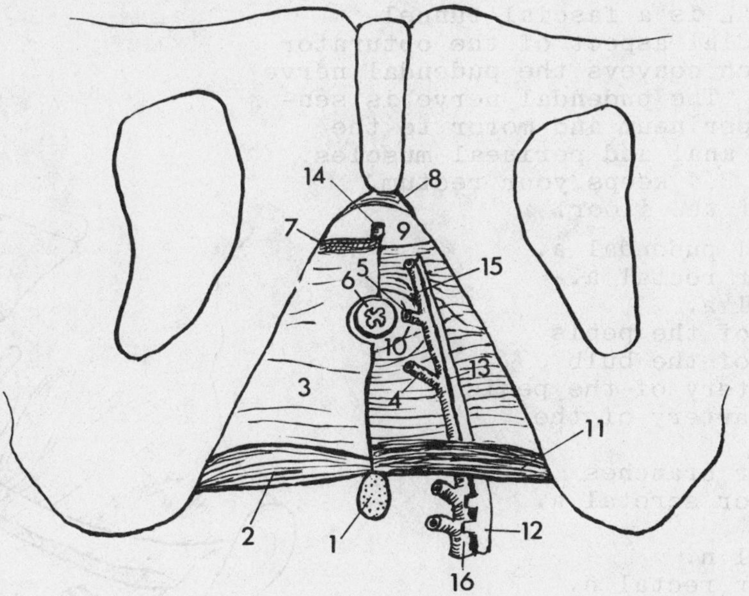

THE PUDENDAL CANAL OF ALCOCK

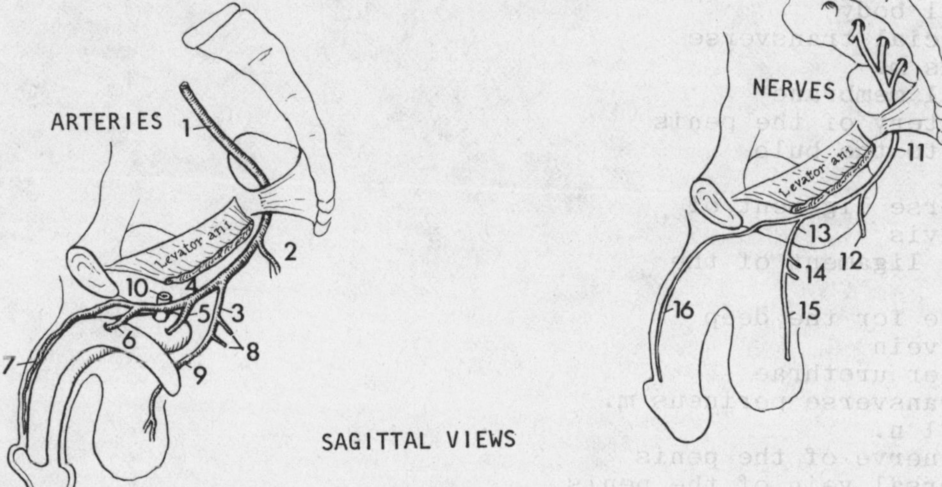

SAGITTAL VIEWS

ALCOCK'S CANAL is a fascial tunnel along the medial aspect of the obturator internus which conveys the pudendal nerve and vessels. The pudendal nerve is sensory to the perineum and motor to the levator ani, anal and perineal muscles.
REMEMBER: S2,3,4 keeps your rectum off the floor.

1) Internal pudendal a.
2) Inferior rectal a.
3) Perineal a.
4) Artery of the penis
5) Artery of the bulb
6) Deep artery of the penis
7) Dorsal artery of the penis
8) Muscular branches
9) Posterior scrotal a.
10) Urethra
11) Pudendal n.
12) Inferior rectal n.
13) Perineal n.
14) Muscular branches
15) Posterior scrotal n.
16) Dorsal nerve of the penis
17) Pudendal canal

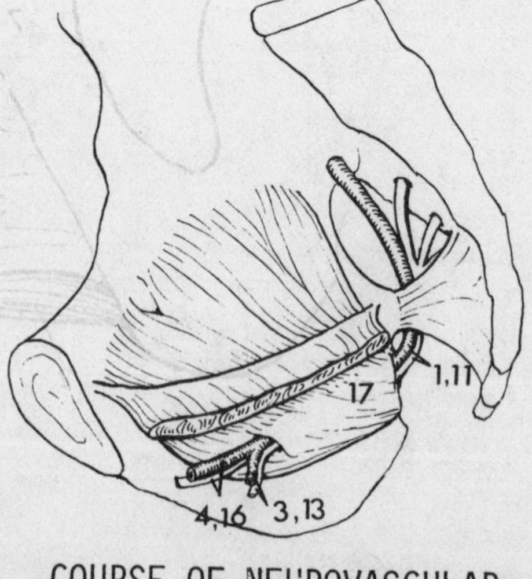

COURSE OF NEUROVASCULAR ELEMENTS IN THE PUDENDAL CANAL

COMMENTS

1. The hymen is a thin tissue plate which separates the vagina from the vestibule. A virginal hymen is usually perforate, after intercourse and childbirth it is represented by small tabs of tissue known as hymeneal caruncles. Occasionally a hymen may be imperforate, which will not allow menses to flow and therefore result in a hematocolpos in which the vagina engorges with blood and may also engorge the uterus and tubes. Hymen may also be septate or present a fenestrated cribiform appearance.
2. The perineal body is important as an attachment point of muscle fibers and helps maintain the integrity of the perineum. Episiotomies--a cut in the posterior or posterior-lateral portion of the vulva--are made to facilitate deliveries and prevent tearing of perineal structures because a cut is controlled and heals better than a tear. Prolapse of the vagina and cervix may occur if the perineal body is damaged.
3. The formation of the normal adult female reproductive system is complex and various developmental errors may occur. These include failure of the uterine horns to fuse with resultant bicornuate uterus (duplex uterus). The duplication may include merely a septum in the uterine cavity or complete duplication of the uterus and cervix.
4. Female to male trans-sexual operations are preceded by testosterone therapy and mastectomy. A complete hysterectomy is performed but the vagina remains. The clitoris is enlarged by the testosterone treatment and some surgeons will 'free up' the clitoris to act as a penis. More commonly a tube of skin derived from the abdomen or thigh is used to construct a penis. A semi-rigid plastic rod may be implanted to simulate an erect state. Construction of penile urethras has been met with moderate success but the urethral meatus usually remains in the vestibule.
5. Circumcision of male infants is a common practice involving the removal of the prepuce or foreskin of the penis. This eliminates conditions such as phimosis--a relatively immobile prepuce--and paraphimosis--a constriction of the glans--and resultant problems.
6. The male urethra is usually divided into a prostatic, membranous and penile urethra while some authorities also define a bulbous urethra in the penile bulb. Trauma commonly tears the urethra at the juncture of the prostatic-membranous and membranous-penile urethra; blood and urine will then extravasate into surrounding tissues. Duplications and diverticula of the urethra occur.
7. Vasectomy, a common means of birth control, is performed on a section of the vas (ductus deferens) between the inguinal

ring and epididymis. A section of the vas about 1" long is removed. Re-anastomosis depends upon the length of the segment removed and other factors such as method of ligation.

8. Male to female trans-sexual operations are preceded by estrogen therapy. The testes are removed, the scrotum and underlying crura of the penis are converted into labia majora and bulb of the vestibule. Erectile tissues in the shaft of the penis are removed except for a small portion at the base of the penis. The penile skin is turned 'inside out' and 'stuffed' into the pelvis to serve as a vagina--the glans serves as a cervix. The reserved portion of penile erectile tissue becomes a clitoris; the urethra is rerouted into the new vestibule.

9. Skin of the perineum is innervated mainly by the pudendal nerve, with some contribution by the posterior femoral cutaneous nerve and minor contribution from the labial branch of the ilioinguinal nerve. A pudendal nerve block is a common form of local anesthesia in gynecologic and obstetric procedures. For this procedure, a needle is guided, with aid of vaginal palpation, inferiorly and posteriorly to the ischial spine and anesthetic is injected around the pudendal nerve before it enters Alcock's Canal. The needle is then directed to the ischial tuberosity and anesthetic is injected around the posterior femoral cutaneous nerve.

10. Trauma to the penis can rupture the urethra and Buck's fascia and allow urine to extravasate into the tissues. The attachment of Colles' and Scarpa's fascia forms a pouch which limits the urine to the penis, scrotum, UG triangle and lower abdomen. If infections erode the wall of the urethra the urine will extravasate but will be limited to the penis by Buck's fascia. In time, however, Buck's fascia will be breached.

STUDY QUESTIONS: PELVIS AND PERINEUM

1. What bones form the os coxae?	1. ilium, ischium, and pubis
2. Describe the divisions of the pelvis.	2. The bony pelvis is divided by the arcuate (iliopectineal) line. The false pelvis is superior to this line and the true pelvis is inferior to it. The pelvic diaphragm separates the bony pelvis into the true pelvic cavity superiorly and the perineum inferiorly.
3. Are the greater and lesser sciatic foramen true bony foramina?	3. No, the greater and lesser sciatic notches are bordered by the sacrospinous and sacrotuberous ligaments which serve to create the foramen.
4. Describe the fascia in the pelvis.	4. Fascia in the true pelvis is called endopelvic fascia. It is continuous superiorly with the transversalis fascia and helps support the pelvic contents.
5. What is the urogenital hiatus?	5. a defect in the levator ani muscle which allows for passage of the neck of the bladder and prostate in the male and bladder and vagina in the female.
6. What are the parts of the levator ani muscle?	6. puborectalis, pubococcygeus, and iliococcygeus
7. Does the pudendal nerve re-enter the pelvis? Does the sciatic nerve?	7. The pudendal nerve re-enters the pelvis via the lesser sciatic foramen. The sciatic nerve does not.
8. What arteries pass through the sacral plexus?	8. superior gluteal artery, inferior gluteal artery and the pudendal artery
9. What is the ganglion impar?	9. the convergence of the sympathetic chains in front of the coccyx
10. What is the origin of the pelvic splanchnic nerves?	10. S2 - S4
11. To what lymph nodes do all the pelvic organs drain?	11. the internal iliac (hypogastric) nodes
12. Where do the common iliac arteries bifurcate?	12. at the level of the pelvic brim anterior to the sacroiliac joint
13. From what structure do the medial umbilical ligaments form?	13 the umbilical arteries
14. What structure is formed by the obliteration of the umbilical vein? the urachus?	14. teres hepatis ligament, median umbilical ligament
15. Prostatic carcinoma is most commonly found in what lobe of the prostate?	15. posterior lobe
16. What structures are contained within the broad ligament?	16. ovaries, uterus, and uterine tubes
17. What is the normal position of the uterus?	17. anteverted and anteflexed
18. What are the three parts of the broad ligament?	18. mesometrium, mesosalpinx, and mesovarium
19. What structure in the male is analogous to the round ligament?	19. the gubernaculum

20. Into what structure do the ovarian veins drain on the right? the left?

20. the IVC, the renal veins (the same is true of the testicular veins in the male).

21. What is the relationship of the ureter to the uterine artery?

21. It is inferior and posterior to the artery "Bridge over water".

22. What is the boundary of the pelvic outlet?

22. the pubic arch, ischial tuberosities, sacrotuberous ligaments and coccyx

23. What is the male analogue of the bulb of the vestibule?

23. the corpus spongiosum

24. What is the relationship of Buck's and Camper's (Dartos') fascia of the penis?

24. Camper's is superficial to Buck's fascia.

25. Describe the passage of sperm from the testis to the urethra.

25. seminiferous tubule→rete testis→ efferent ductules→epididymis→vas deferens→ampulla→ejaculatory duct→ urethra

26. What is the perineal membrane?

26. the inferior fascia of the UG diaphragm

27. Where is the pudendal canal? What are its contents?

27. The pudendal canal is located along the medial aspect of the obturator internus. Enclosed within it is the pudendal nerve and vessels.

28. What is the blood supply to the vagina?

28. the vaginal artery and the vaginal branch of the uterine artery

29. What is the name of the widest portion of the fallopian tube? The narrowest?

20. the ampulla, the isthmus

30. What structures converge to form the ejaculatory ducts?

30. the ducts of the seminal vesicles and the ductus deferens.

31. What is the importance of the pudendal nerve?

31. It is the sole somatic nerve of the perineum supplying all the perineal musculature and providing sensory innervation to most of the associated skin.

32. What is the pampiniform plexus?

32. Testicular veins arising from the testis and epididymis anastomose and form the plexus. The plexus traverses the inguinal canal and ends near the deep inguinal ring.

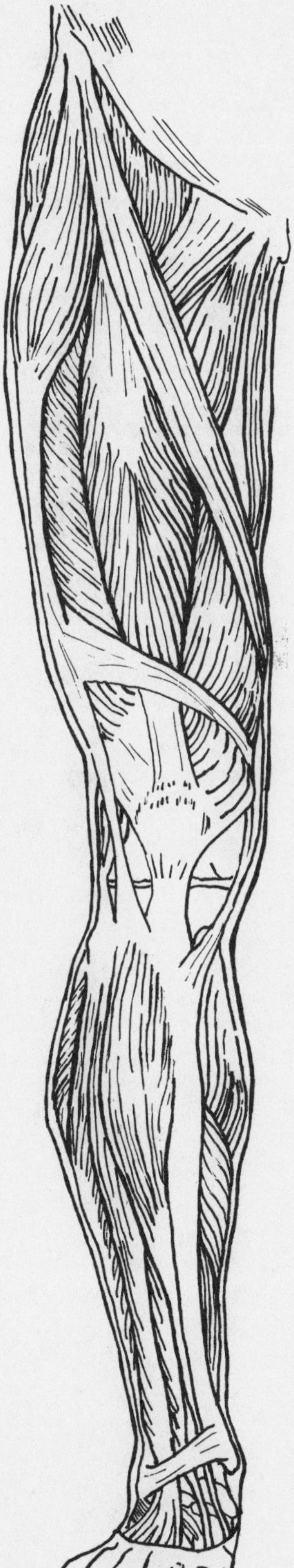

LOWER EXTREMITY

LOWER EXTREMITY

LOWER EXTREMITY

I. SCHEME OF NEUROVASCULAR ELEMENTS

<u>ARTERIES</u> <u>AND</u> <u>NERVES</u>

1) Inguinal ligament
2) External iliac a.
3) Femoral n.
4) Femoral a.
5) Obturator n.
6) Medial femoral circumflex a.
7) Deep femoral a.
8) Lateral femoral circumflex a.
9) Popliteal a.
10) Common peroneal n.
11) Anterior tibial a.
12) Dorsalis pedis a.
13) Arcuate a.
14) Deep plantar a.
15) Superior gluteal a. & n.
16) Inferior gluteal a. & n.
17) Sciatic n.
18) Perforating branches deep femoral a.
19) Tibial n.
20) Peroneal a.
21) Posterior tibial a.
22) Medial plantar a.& n.
23) Lateral plantar a. & n.
24) Plantar arch
25) Deep plantar a.

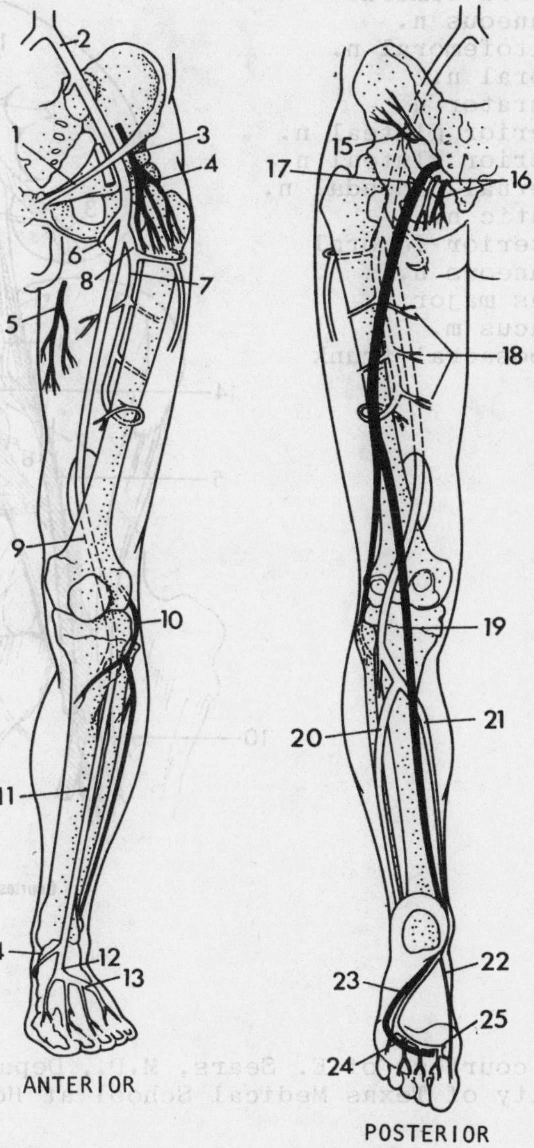

LUMBOSACRAL PLEXUS

1) Iliohypogastric n.
2) Ilioinguinal n.
3) Lateral femoral cutaneous n.
4) Genitofemoral n.
5) Femoral n.
6) Obturator n.
7) Superior gluteal n.
8) Inferior gluteal n.
9) Internal pudendal n.
10) Sciatic n.
11) Posterior femoral cutaneous n.
12) Psoas major m.
13) Iliacus m.
14) Lumbosacral trunk

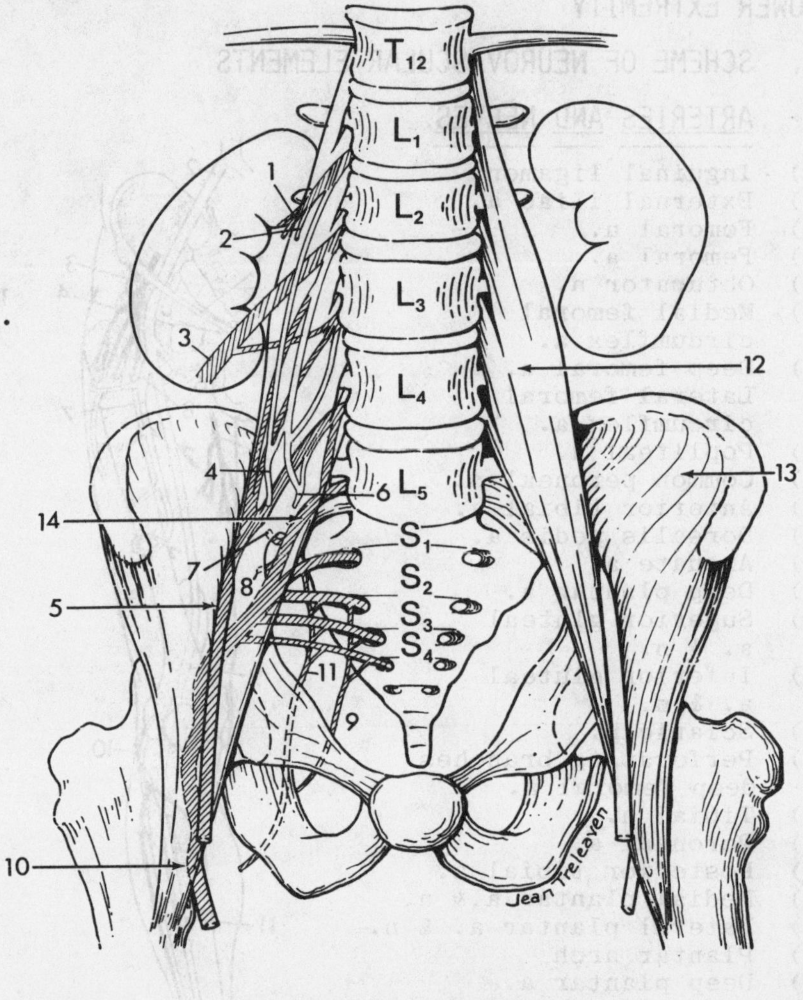

Courtesy of Dr. E. Sears, Department of Neurology

Drawing courtesy of E. Sears, M.D., Department of Neurology, The University of Texas Medical School at Houston

II. GLUTEAL REGION

OSTEOLOGY

HIP BONE, FEMUR PROXIMAL TIBIA AND FIBULA

1) Ilium
2) Ischium
3) Pubis--these first three bones were originally separate but fuse by puberty
4) Superior ramus pubis
5) Inferior ramus pubis
6) Ischial ramus--or ischiopubic ramus
7) Acetabulum
8) Posterior gluteal line
9) Anterior gluteal line
10) Inferior gluteal line
11) Anterior superior iliac spine--ASIS
12) Anterior inferior ilias spine--AIIS
13) Greater sciatic notch
14) Lesser sciatic notch
15) Ischial spine
16) Ischial tuberosity
17) Greater trochanter
18) Gluteal tuberosity
19) Linea aspera
20) Intertrochanteric crest
21) Trochanteric fossa
22) Lesser trochanter

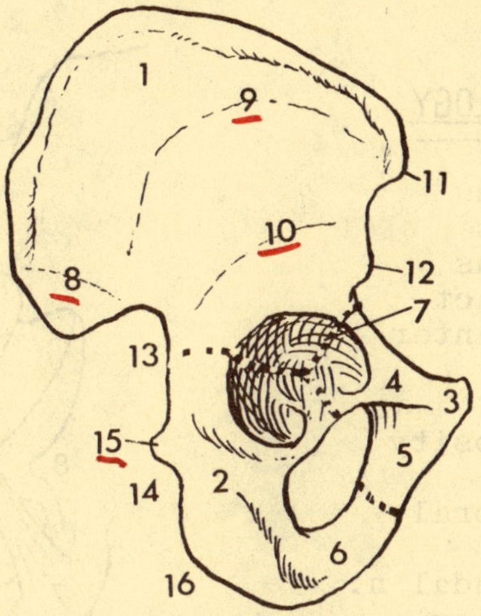

LATERAL

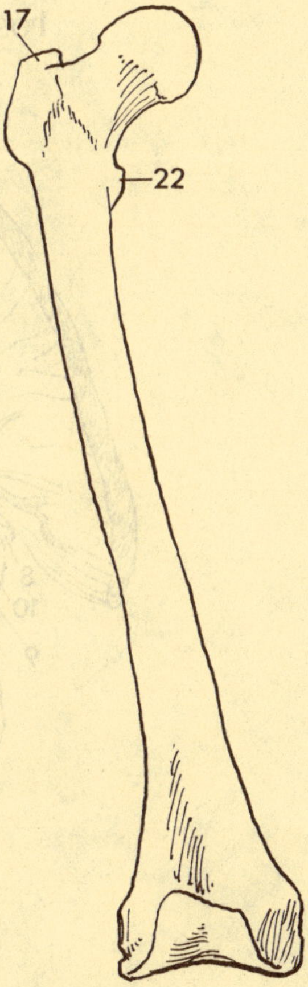

ANTERIOR

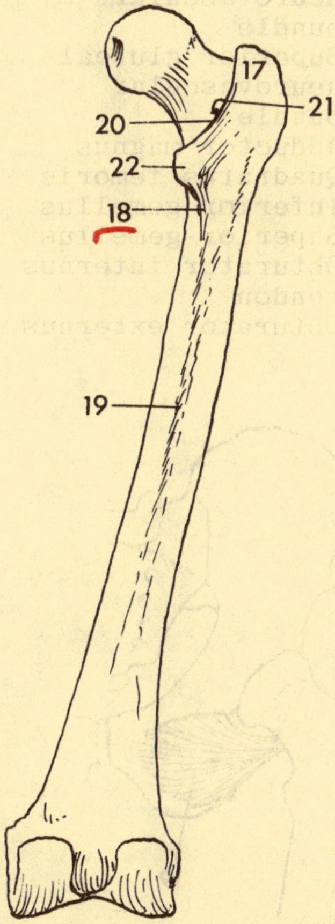

POSTERIOR

MYOLOGY AND ANGIOLOGY

1) Gluteus maximus
2) Gluteus medius
3) Gluteus minimus
4) Iliotibial tract
5) Greater trochanter
6) Sacrotuberous ligament
7) Ischial tuberosity
8) Ischial spine
9) Posterior femoral cutaneous n.
10) Internal pudendal n.
11) Sciatic n.--lies halfway between (7) and (5)
12) Inferior gluteal neurovascular bundle
13) Superior gluteal neurovascular bundle
14) Adductor magnus
15) Quadratus femoris
16) Inferior gemellus
17) Superior gemellus
18) Obturator internus tendon
19) Obturator externus

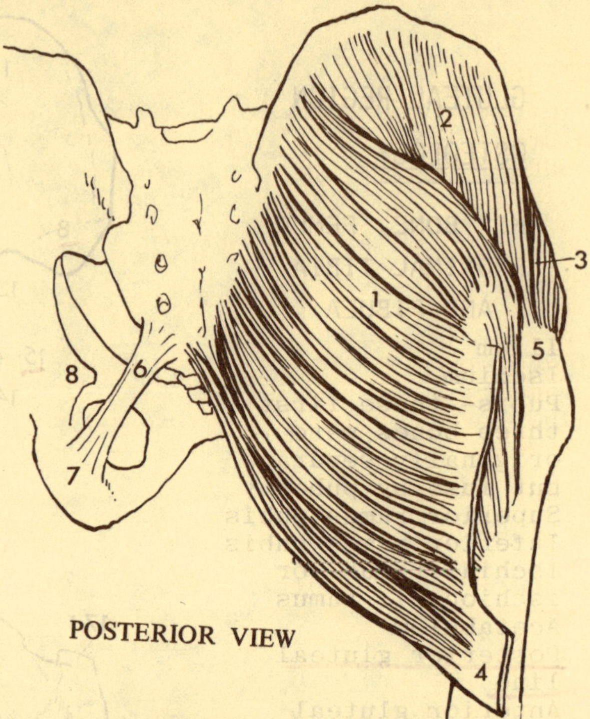

POSTERIOR VIEW

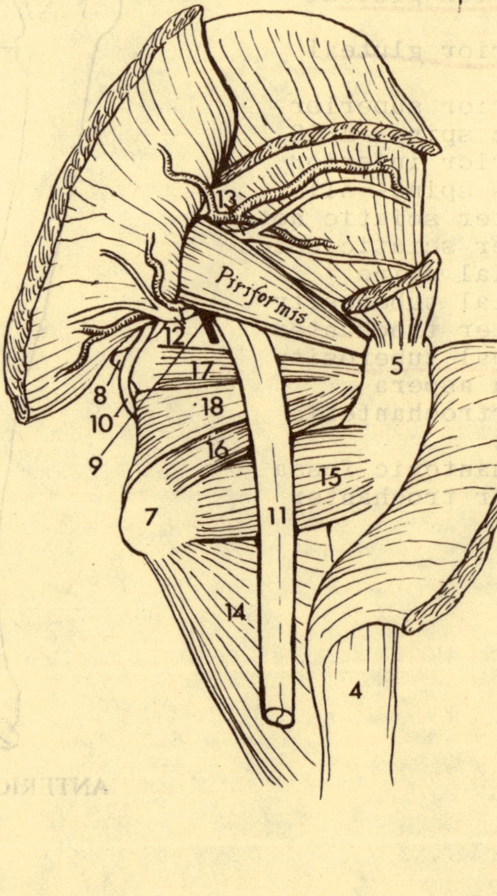

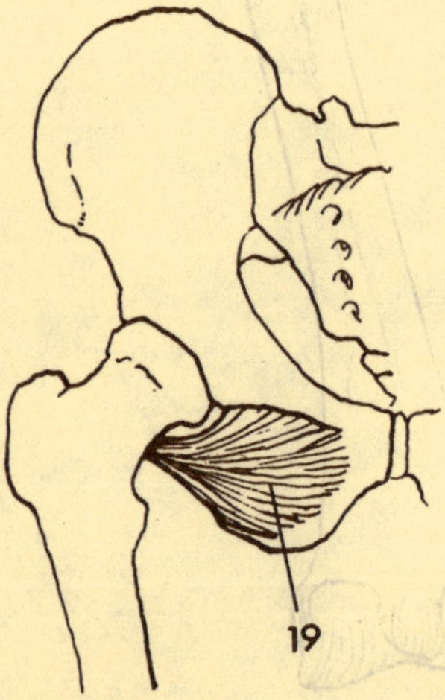

ANTERIOR VIEW

GLUTEUS MAXIMUS

O-Sacrum, coccyx, sacrotuberous ligament, posterior gluteal line
I-Gluteal tuberosity, iliotibial tract, fascia lata
A-Abduction, lateral rotation and extension of thigh
N-Inferior gluteal nerve
B-Inferior and superior gluteal vessels

GLUTEUS MEDIUS

O-Between anterior and posterior gluteal lines
I-Greater trochanter
A-Abduction and medial rotation of thigh
N-Superior gluteal nerve
B-Superior gluteal vessels and anastomoses with inferior gluteal vessels

GLUTEUS MINIMUS

O-Between inferior and anterior gluteal lines
I-Greater trochanter
A-Abduction and medial rotation of thigh
N-Superior gluteal n.
B-Superior gluteal vessels and anastomoses with inferior gluteal vessels

PIRIFORMIS

O-Anterior aspect of sacrum
I-Greater trochanter
A-Abduction and lateral rotation of thigh
N-Branches from spinal nerves L5 to S2
B-Inferior gluteal vessels

SUPERIOR GEMELLUS

O-Ischial spine
I-Greater trochanter
A-Abduction and lateral rotation of thigh
N-Nerve to obturator internus
B-Inferior gluteal vessels

OBTURATOR INTERNUS

O-Bone surrounding obturator foramen, obturator membrane
I-Greater trochanter
A-Abduction and extension of thigh
N-Nerve to the obturator internus
B-Obturator, gluteal arteries through cruciate anastomoses

INFERIOR GEMELLUS

O-Ischial tuberosity
I-Greater trochanter
A-Abduction and lateral rotation of thigh
N-Nerve to quadratus femoris
B-Inferior gluteal artery, superior gluteal artery through cruciate anastomoses

QUADRATUS FEMORIS

O-Ischial tuberosity
I-Greater trochanter
A-Lateral rotation of thigh
N-Nerve to quadratus femoris
B-Inferior and superior gluteal arteries through cruciate anastomosis

OBTURATOR EXTERNUS

O-Bone surrounding obturator foramen, obturator membrane
I-Trochanteric fossa
A-Lateral rotation of thigh
N-Posterior division of obturator
B-Obturator artery, cruciate anastomosis

THE CRUCIATE ANASTOMOSIS is formed about the posterior surface of the hip by anastomoses of the medial and lateral circumflex, inferior gluteal and first perforating branch of the profunda femoris (deep femoral) arteries.

LATERAL VIEW OF GLUTEAL REGION AND THIGH

1) Gluteus maximus
2) Gluteus medius
3) Gluteus minimus
4) Tensor fascia lata

    O-ASIS, iliac crest
    I-Iliotibial tract
    A-Tense fascia lata, extension and lateral rotation of thigh
    N-Superior gluteal
    B-Gluteal vessels and cruciate anastomosis

5) Fascia lata
6) Iliotibial tract--the deep or muscular fascia of the thigh is quite thick and is called the fascia lata, the iliotibial tract is a denser portion of the lateral part of the fascia lata

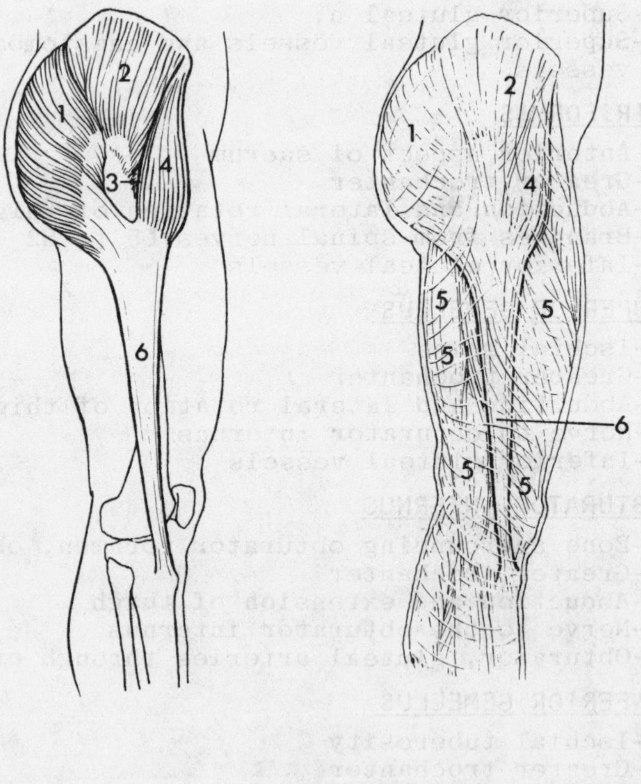

THE HIP JOINT

This is a synovial ball and socket joint of great mobility and stability. The stability is required to support the weight and movement of the body. The head of the femur fits into the acetabulum; the acetabulum is made deeper by a rim of cartilage called the acetabular labrum and the acetabular notch is bridged by the transverse ligament. Blood supply of the hip is of great importance, due to the large number of fractures of the neck of the femur.

LIGAMENTS:

1) Iliofemoral or Y ligament
2) Pubofemoral
3) Ischiofemoral

These ligaments surround the head of the femur in such a way that the head is 'screwed' into the acetabulum during extension.

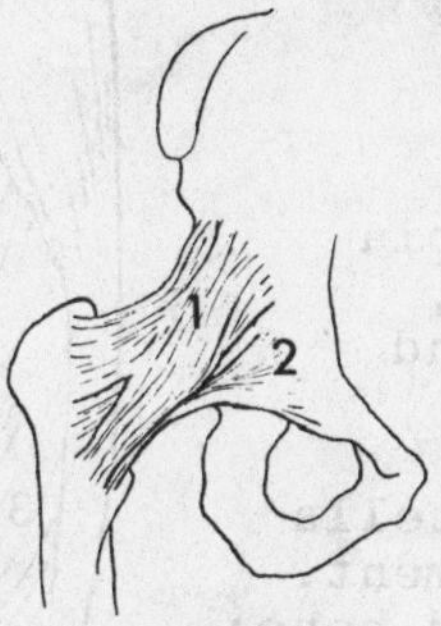

ANTERIOR

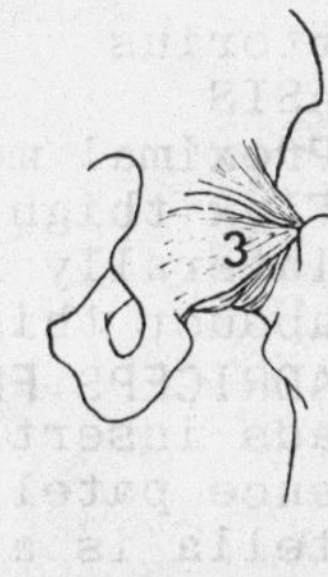

POSTERIOR

BLOOD SUPPLY:

4) Acetabulum
5) Transverse ligament
6) Ligamentum teres
7) Artery of the ligamentum teres--from obturator a.
8) Branch of medial femoral circumflex a.
9) Branch of lateral femoral circumflex a.
10) Femoral a.
11) Obturator a.

The artery of the ligamentum teres is too small to supply the head and neck; these structures will necrose if other aa. are destroyed.

ANTERIOR

III. THE THIGH

ANTERIOR OR QUADRICEPS COMPARTMENT

All muscles innervated by femoral nerve, vascularized by femoral a. or lateral branches of the femoral a., most extend the leg, other actions are indicated.

1) Sartorius
   O-ASIS
   I-Proximal medial tibia
   A-Flex thigh and leg, laterally rotate and abduct thigh

THE QUADRICEPS FEMORIS:
All heads insert into patella and thence patellar ligament. The patella is a sesamoid bone; therefore the patellar ligament is actually a tendon.

2) Rectus femoris
   O-AIIS
   A-Extend leg, flex thigh
3) Vastus lateralis
   O-Greater trochanter, linea aspera, intertrochanteric line
   A-Extend leg
4) Vastus medialis
   O-linea alba, medial intermuscular septum
   A-Extend leg
5) Vastus intermedius
   O-Lateral intermuscular septum and femoral shaft
   A-Extends leg

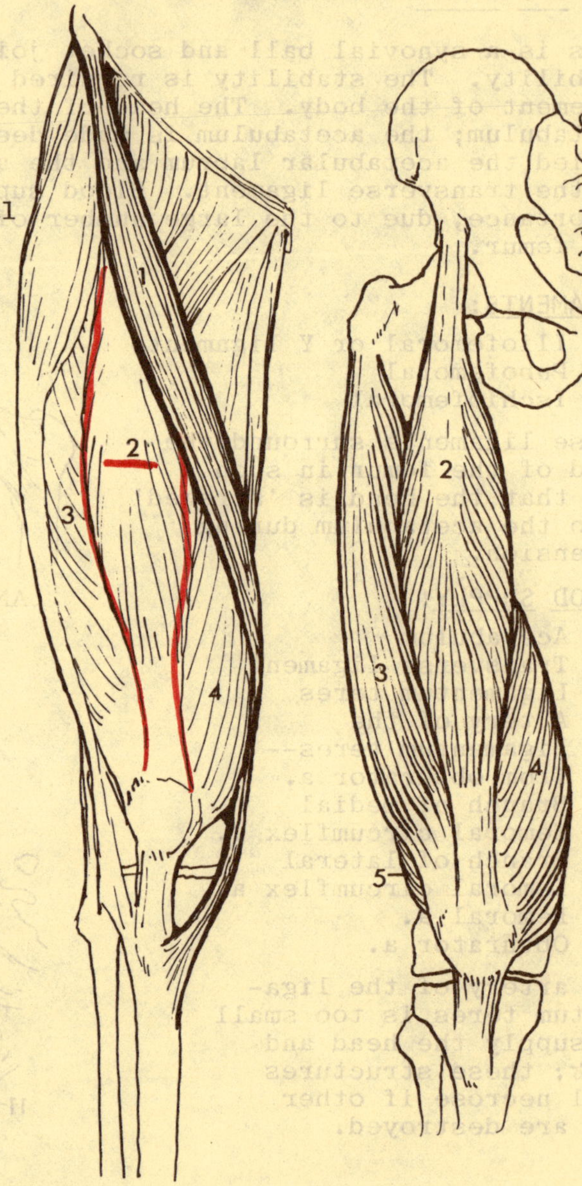

MEDIAL OR ADDUCTOR COMPARTMENT

Contains muscles which originate from the pecten or body of the pubis; all adduct the thigh but other actions are noted; most are innervated by the obturator nerve. The obturator nerve divides into anterior and posterior divisions at the superior border of the adductor brevis, the adductor brevis lies between the divisions.

1) Iliopsoas--actually the combined tendon of I of the iliacus and psoas muscles.
   O-Iliac fossa and lumbar transverse processes
   I-Lesser trochanter
   A-Mainly a flexor of the thigh
   This muscle is an exception to the generalization above.
2) Adductor longus
   I-Femur
   N-Anterior division obturator n.
3) Adductor brevis
   I-Femur
   N-Obturator n.
4) Adductor magnus
   --has a 'hamstring' portion associated with posterior muscles N by sciatic n.
   O-Pubis and ischial tuberosity
   I-Femur and adductor tubercle of femur
5) Pectineus
   I-Lesser trochanter
   N-Femoral n.
6) Gracilis
   I-Proximal medial tibia
   N-Obturator

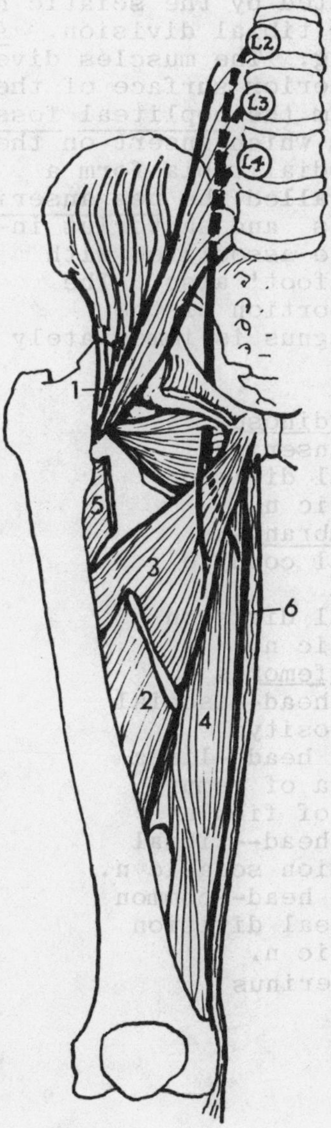

POSTERIOR OR HAMSTRING COMPARTMENT

Contains muscles which originate from the ischial tuberosity. All are innervated by the sciatic n., most by its tibial division. All flex the leg. The muscles diverge at the posterior surface of the knee to form the popliteal fossa. The muscles which insert on the proximal medial tibia form a structure called the pes anserinus. The gracilis and sartorius insertions are associated with the 'goose foot' also. The hamstring portion of the adductor magnus is immediately superior.

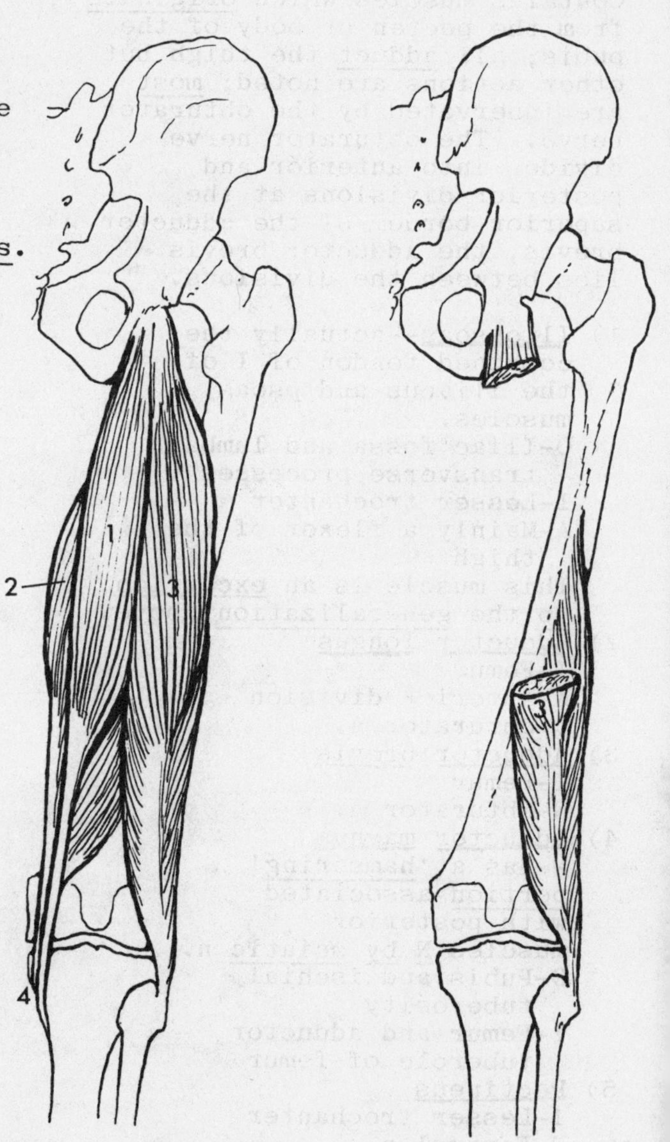

1) Semitendinosus
   I-Pes anserinus
   N-Tibial division sciatic n.
2) Semimembranosus
   I-Medial condyle tibia
   N-Tibial division sciatic n.
3) Biceps femoris
   O-Long head--ischial tuberosity
   Short head--linea aspera of femur
   I-Head of fibula
   N-Long head--tibial division sciatic n.
   Short head--common peroneal division sciatic n.
4) Pes anserinus

FEATURES OF THE THIGH

FEMORAL TRIANGLE

BOUNDARIES: Inguinal ligament, sartorius, adductor longus
FLOOR: Iliopsoas and pectineus muscles
CONTENTS: Femoral nerve, lateral femoral cutaneous nerve, Femoral Sheath containing the femoral vessels, and the Femoral Canal between the sheath and lacunar ligament, the canal serves to transmit lymphatics and for expansion of vessels in sheath. Femoral hernias pass through the canal.
MNEMONIC: NAVEL--for structures lateral to medial--femoral Nerve, femoral Artery, femoral Vein, Empty space or canal and Lymphatics

1) Lateral femoral cutaneous n.
2) Inguinal ligament
3) Sartorius m.
4) Iliopsoas m.
5) Femoral nerve
6) Femoral a.
7) Femoral v.
8) Femoral canal
9) Pectineus
10) Adductor longus

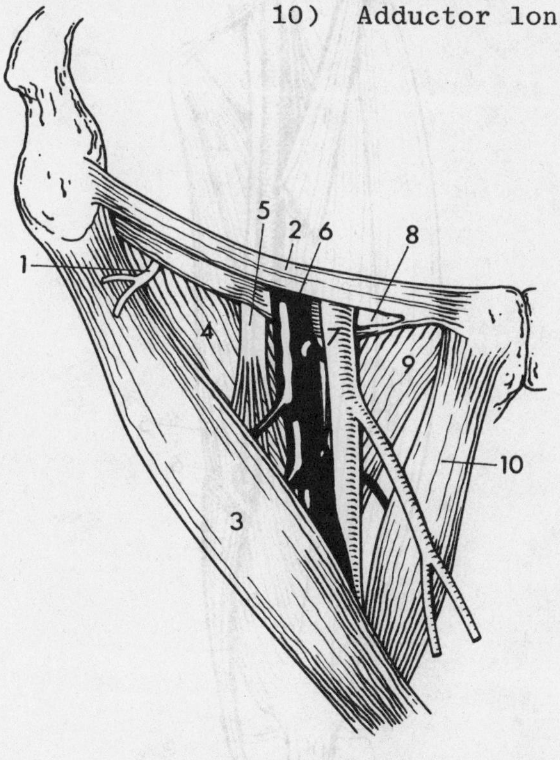

ADDUCTOR CANAL

THE ADDUCTOR CANAL, HUNTER'S CANAL OR SUBSARTORIAL CANAL is formed by a groove between the adductors, quadriceps and sartorius. It begins at the apex of the Femoral Triangle and terminates at the Adductor hiatus.

BOUNDARIES: Sartorius, adductors magnus and longus and vastus medialis

CONTENTS: Nerve to vastus medialis, saphenous nerve, and femoral vessels. The femoral vessels pass through the adductor hiatus to become the popliteal vessels; the nerve to vastus medialis and saphenous nerve <u>do not</u> pass through the hiatus.

1) Sartorius
2) Vastus medialis
3) Femoral nerve, artery and vein
4) Adductor longus
5) Adductor magnus
6) Adductor hiatus

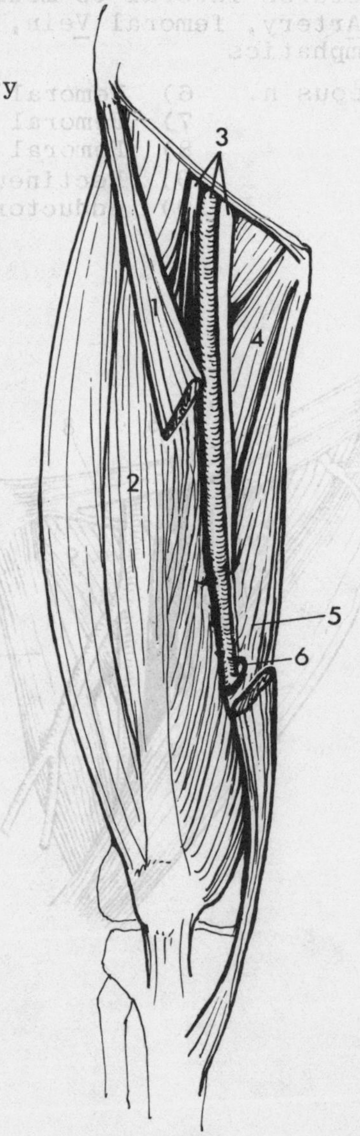

IV. THE KNEE AND POPLITEAL FOSSA

THE KNEE

One would judge the knee joint to be one of the weakest joints simply because of its architecture; in reality it is one of the most structurally sound. It is a synovial condylar joint made secure by surrounding ligaments and muscles.

1) Patella
2) Patellar ligament
3) Medial (tibial) collateral ligament
4) Lateral (fibular) collateral ligament
5) Lateral meniscus
6) Medial meniscus
7) Semimembranosus tendon
8) Oblique ligament
9) Arcuate ligament

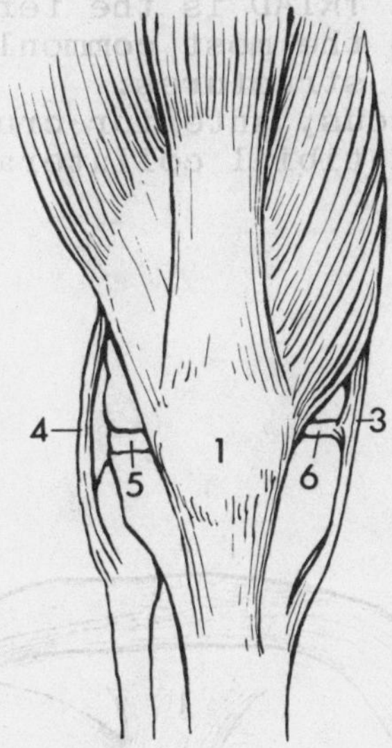

ANTERIOR

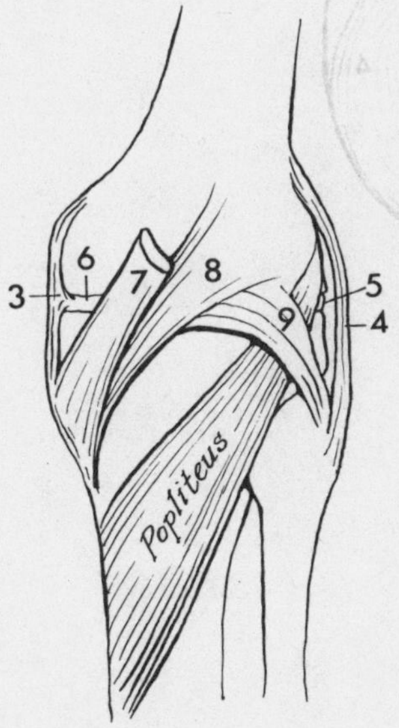

POSTERIOR

MENISCI AND CRUCIATE LIGAMENTS:

1) Anterior cruciate ligament
2) Posterior cruciate ligament
3) Medial meniscus--the tibial collateral ligament is attached to the medial meniscus
4) Lateral meniscus

THE TERRIBLE TRIAD is the term referring to the most commonly damaged knee structures: Medial meniscus, anterior cruciate ligament and tibial collateral ligament.

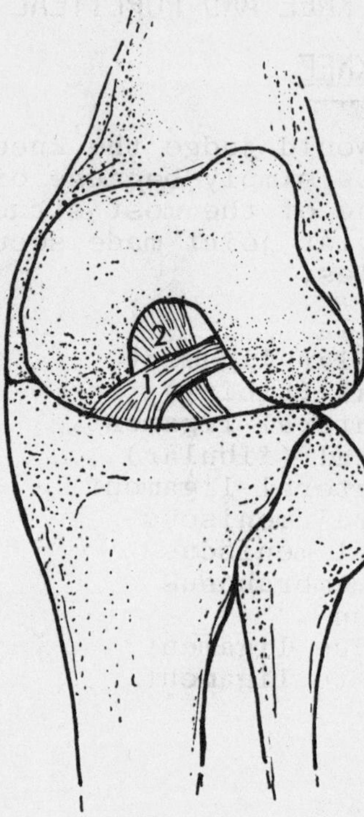

ANTERIOR

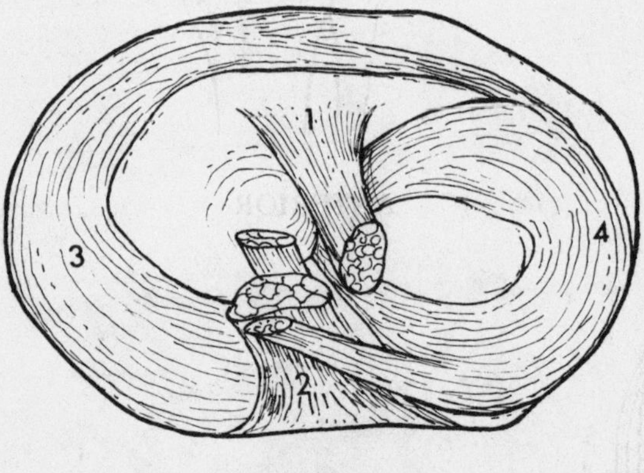

SUPERIOR

THE POPLITEAL FOSSA

BOUNDARIES: Semimembranosus, biceps femoris, lateral and medial head of the gastrocnemius.

FLOOR: Distal femur, proximal tibia, popliteus, and a portion of the head of the fibula.

CONTENTS:

1) Sciatic nerve
2) Common peroneal n.
3) Tibial n.--usually the sciatic n. has already divided into common peroneal and tibial before the fossa
4) Popliteal a.
5) Popliteal v.
6) Lesser saphenous v.
7) Anterior tibial a.
8) Posterior tibial a.

The anterior and posterior tibial veins are not shown

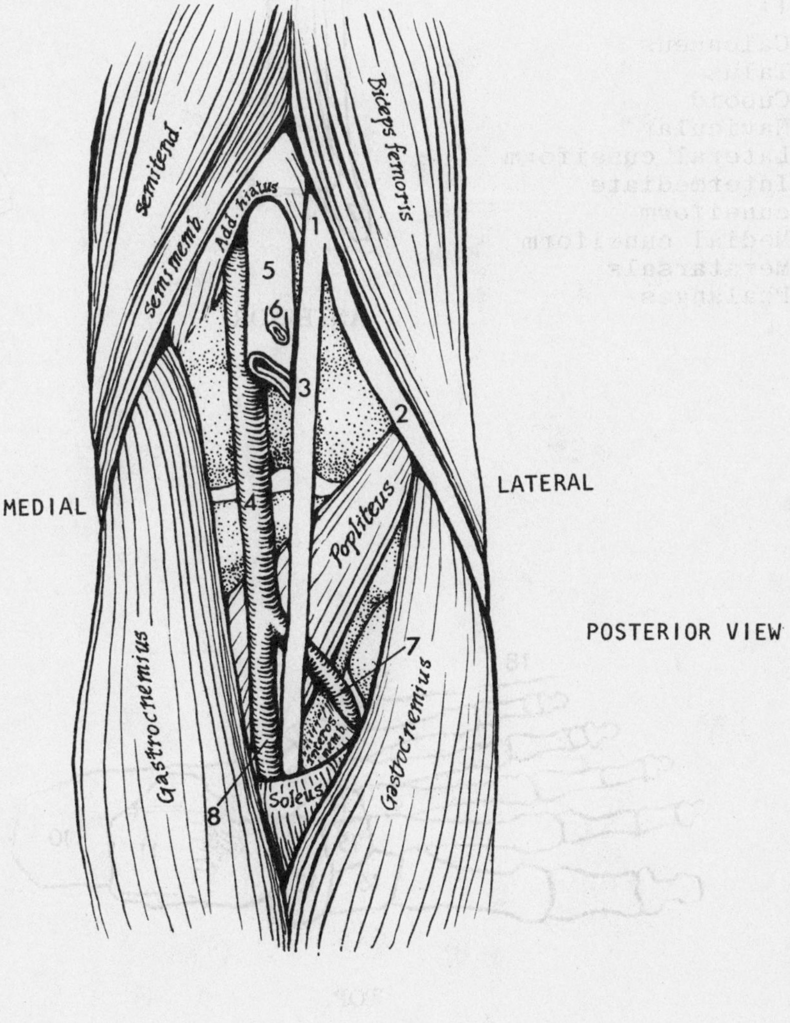

V. LEG, ANKLE AND FOOT

OSTEOLOGY

TIBIA:

1) Condyles
2) Soleal line
3) Tuberosity
4) Anterior border
5) Medial malleolus
6) Interosseous crest

FIBULA:

7) Lateral malleolus
8) Interosseous crest
9) Head

FOOT:

10) Calcaneus
11) Talus
12) Cuboid
13) Navicular
14) Lateral cuneiform
15) Intermediate cuneiform
16) Medial cuneiform
17) Metatarsals
18) Phalanges

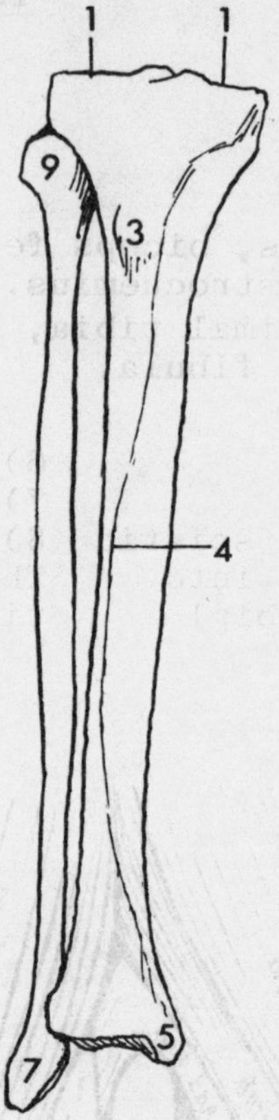

ANTERIOR

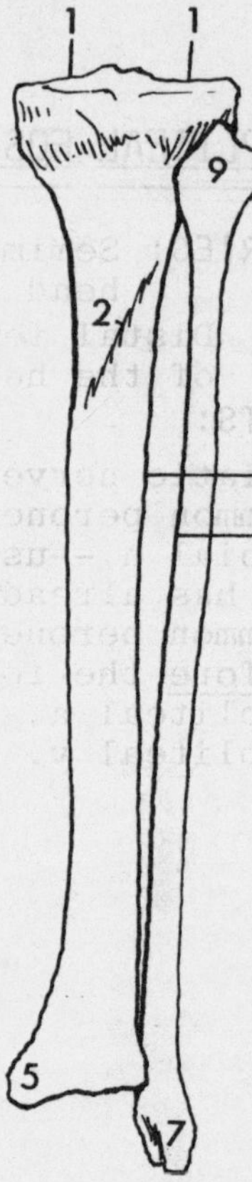

POSTERIOR

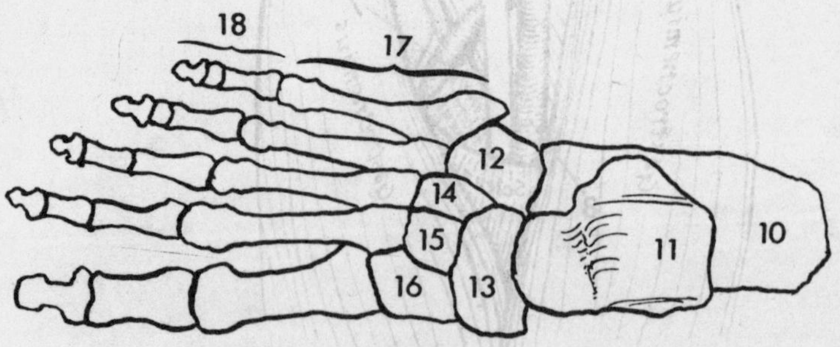

TOP

THE ANTERIOR COMPARTMENT

CONTAINS muscles which dorsiflex the foot. Anatomically these are extensors--compare them to extensors of the hand--but functionally they are included in the flexion reflex.

1) TIBIALIS ANTERIOR
   - O-Tibia, interosseous membrane, intermuscular septum
   - I-Talus, 1st metatarsal medial cuneiform
   - A-Inversion and dorsiflexion of foot
   - N-Deep peroneal
   - B-Anterior tibial a.

2) EXTENSOR HALLUCIS LONGUS
   - O-Fibula and interosseous membrane
   - I-Base distal phalanx great toe
   - A-Extension great toe, dorsiflexion foot
   - N-Deep peroneal
   - B-Anterior tibial a.

3) EXTENSOR DIGITORUM LONGUS
   - O-Tibia, fibula, interosseous membrane
   - I-Dorsal digital expansion
   - A-Dorsiflexion foot, extension lateral toes
   - N-Deep peroneal
   - B-Anterior tibial a.

4) PERONEUS TERTIUS
   - O-Distal fibula
   - I-Base 5th metatarsal
   - A-Dorsiflexion foot
   - N-Deep peroneal
   - B-Anterior tibial a.

ALSO:

5) Anterior tibial a.
6) Dorsalis pedis
7) Common peroneal n.
8) Deep peroneal n.
9) Superficial peroneal n.
10) Gastrocnemius
11) Soleus
12) Interosseous membrane
13) Peroneus longus
14) Peroneus brevis

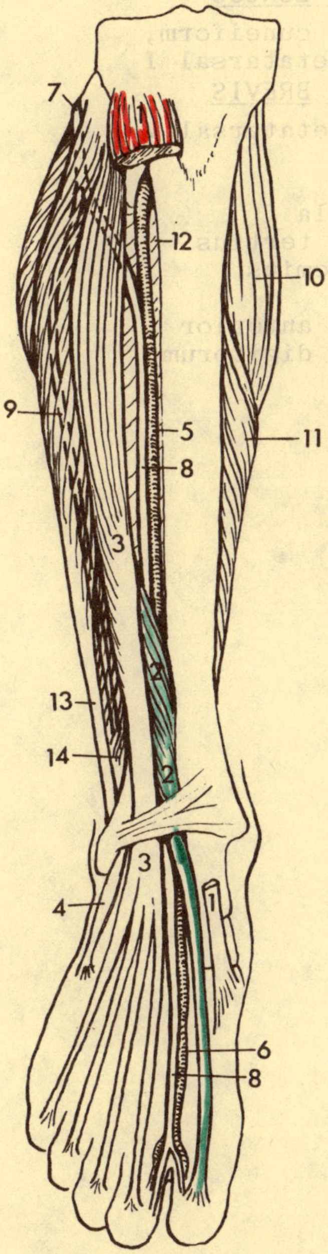

LATERAL OR EVERTOR COMPARTMENT

BOTH MUSCLES originate from the fibula, plantar flex and evert the foot and are innervated by the superficial peroneal nerve.

CONTENTS:

1) PERONEUS LONGUS

   I-Medial cuneiform, base metatarsal I

2) PERONEUS BREVIS

   I-Base metatarsal V

ALSO:

3) Retinacula
4) Peroneus tertius
5) Gastrocnemius
6) Soleus
7) Tibialis anterior
8) Extensor digitorum longus

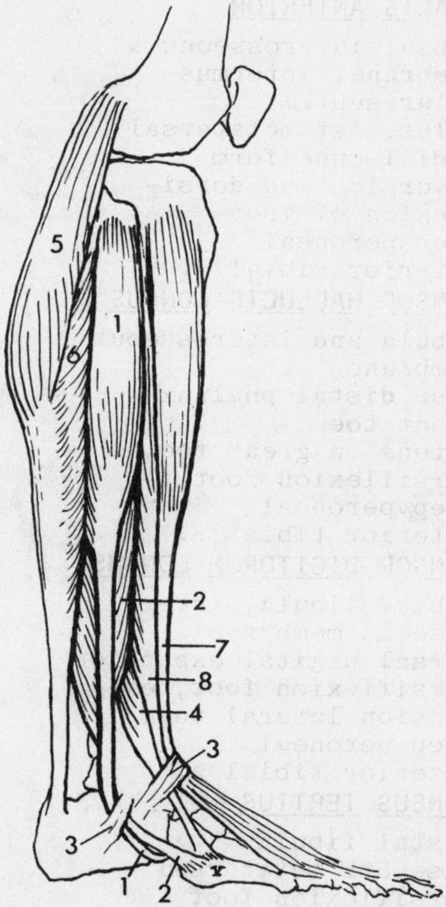

POSTERIOR COMPARTMENT

SUPERFICIAL COMPARTMENT contains muscles innervated by the tibial nerve. They insert into the calcaneal tendon--'Achilles' Tendon--and plantar flex the foot. Other actions are indicated. They are vascularized by the posterior tibial a., popliteal a. and peroneal a.

1) PLANTARIS

   O-Posterior distal femur, superior to gastrocnemius

THE TRICEPS SURAE OR CALF MUSCLES

2) GASTROCNEMIUS
   O-Medial head--medial femoral condyle
   Lateral head--lateral femoral condyle
   A-Plantar flex foot, flex leg at knee
3) SOLEUS

   O-Proximal tibia and fibula

ALSO:

4) Calcaneal tendon
5) Plantaris tendon--also known as Freshman's Nerve

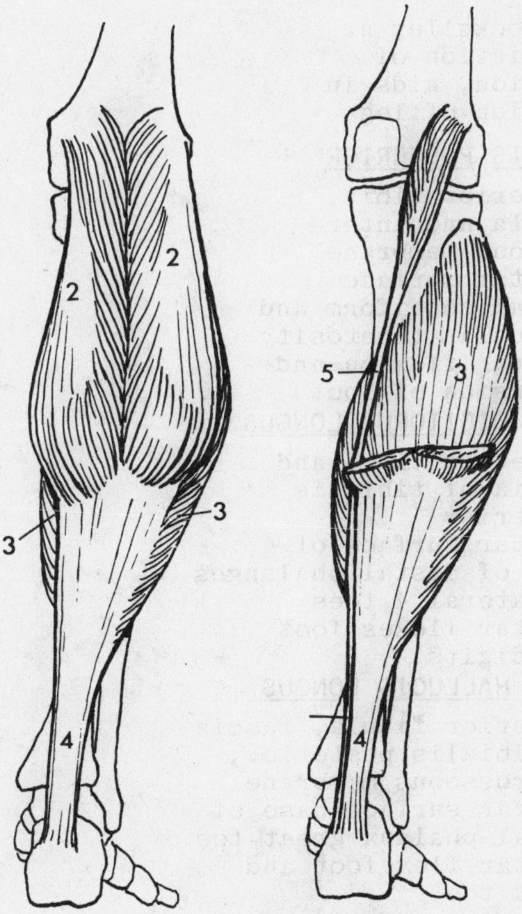

DEEP COMPARTMENT contains muscles innervated by the tibial n. and vascularized by the popliteal, posterior tibial and peroneal arteries.

1) POPLITEUS
   - O-Lateral condyle of femur and lateral meniscus
   - I-Proximal posterior tibia
   - A-'Unlocks'leg at initiation of flexion, aids in flexion of leg

2) TIBIALIS POSTERIOR
   - O-Posterior tibia, fibula and interosseous membrane
   - I-Plantar surface medial cuneiform and navicular tuberosity
   - A-Plantar flexion and inversion of foot

3) FLEXOR DIGITORUM LONGUS
   - O-Posterior tibia and fascia of tibialis posterior
   - I-Plantar surface of base of distal phalanges of lateral 4 toes
   - A-Plantar flexes foot and digits

4) FLEXOR HALLUCIS LONGUS
   - O-Posterior fibula, fascia of tibialis posterior, interosseous membrane
   - I-Plantar surface base of distal phalanx great toe
   - A-Plantar flex foot and great toe

ALSO:

5) Common peroneal n.
6) Tibial n.
7) Popliteal a.
8) Anterior tibial a.
9) Posterior tibial a.
10) Peroneal a.
11) Peroneus longus
12) Soleus

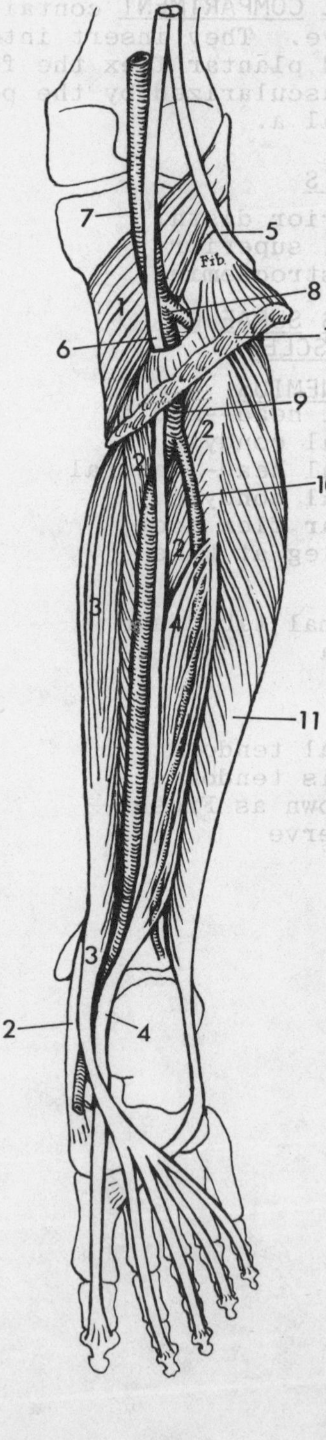

ANKLE

THE TRUE ANKLE JOINT is formed by the articulation of the tibia and fibula with the talus. It allows plantar and dorsiflexion with little inversion and eversion; it is a synovial joint with medial and lateral collateral ligaments. Most talar joints are unimportant and allow only a little gliding motion, an exception is the SUBTALAR JOINT which allows for inversion and eversion. The Subtalar Joint is made up of the posterior talocalcaneal joint and the talocalcaneonavicular joint.

1) Medial collateral ligament--the Deltoid Ligament
2) Plantar calcaneonavicular ligament--the Spring ligament--of the talocalcaneonavicular joint

Lateral collateral ligaments:

3) Anterior talofibular
4) Posterior talofibular
5) Calcaneofibular
6) Bifurcate ligament--of the talocalcaneonavicular joint
7) Interosscous talocalcaneal ligament--of the posterior talocalcaneal joint

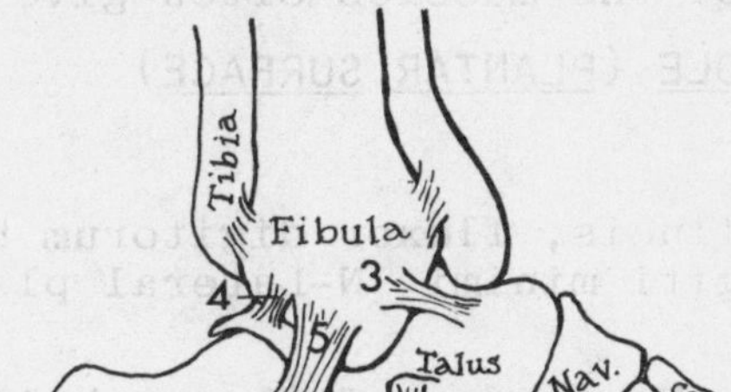

LATERAL ASPECT

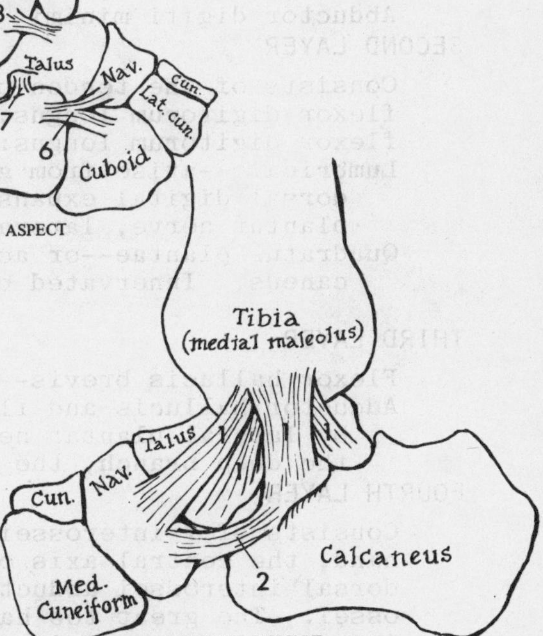

MEDIAL ASPECT

STRUCTURES AT THE ANKLE:

8) Tendon tibialis posterior
9) Tendon flexor digitorum longus
10) Posterior tibial artery
11) Tibial nerve
12) Flexor hallucis longus

ALSO:

13) Gastrocnemius
14) Calcaneal tendon

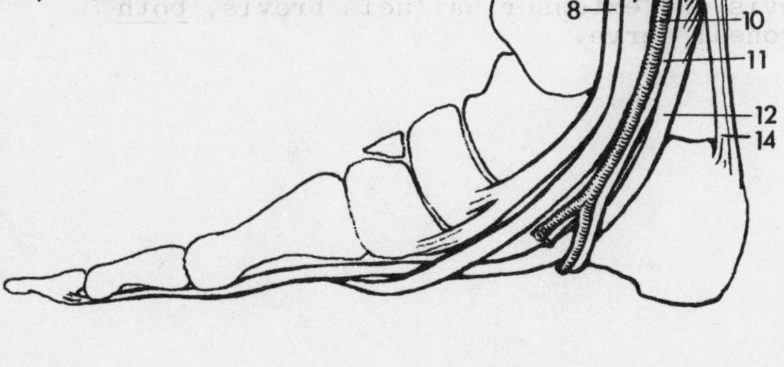

FOOT

Although the foot contains muscles remarkably structured like those of the hand, individual actions of the muscles of the foot are not as important. The foot has been modified for support of the body during movement. Muscles of the foot tend to act more as living springs than as independent muscles. The sole is classically described as consisting of 4 layers. While these are presented here, origins, insertions and actions have been dropped--names of the muscles often give their anatomical action.

LAYERS OF THE SOLE (PLANTAR SURFACE)

SUPERFICIAL

Abductor hallucis, flexor digitorum brevis N-Medial plantar n.
Abductor digiti minimi N-Lateral plantar n.

SECOND LAYER

Consists of the tendon of flexor hallucis longus and the flexor digitorum longus. Two sets of muscles attach to the flexor digitorum longus:

Lumbricals--arise from great toe side of tendon, insert into dorsal digital expansions. Medial 1 innervated by medial plantar nerve, lateral 3 by lateral plantar nerve.

Quadratus plantae--or accessory flexor inserts into the calcaneus. Innervated by the Lateral plantar nerve.

THIRD LAYER

Flexor hallucis brevis--innervated by medial plantar nerve

Adductor hallucis and flexor digiti minimi brevis--innervated by lateral plantar nerve. The adductor is innervated by the deep branch, the flexor by the superficial branch.

FOURTH LAYER

Consists of 7 interossei--3 plantar and 4 dorsal. Unlike the hand, the central axis of the foot is the 2nd digit. Since dorsal interossei abduct, the 2nd digit has 2 dorsal interossei. The great toe has a separate adductor, so all 3 plantar interossei are associated with the lateral 3 toes. All are innervated by the deep branch of the lateral plantar nerve.

THE DORSUM OF THE FOOT contains two muscles of little consequence, the extensor digitorum brevis and extensor hallucis brevis, both innervated by the deep peroneal nerve.

ARCHES OF THE FOOT

Under normal conditions the plantar surface of the foot should resemble half of a bell, with the medial portion off the ground and the lateral portion, heel and distal metatarsals touching the ground. The foot has two axes, longitudinal and transverse, strengthened by ligaments and long tendons.

LONGITUDINAL

1) Medial portion--supported by flexors hallucis longus and digitorum longus, also tibiales posterior and anterior, spring ligament and plantar aponeurosis
2) Lateral portion--supported by peroneus longus, flexor digitorum longus, plantar aponeurosis, long and short plantar ligaments

TRANSVERSE

Supported by adductor hallucis, peroneus longus, and interosseous ligaments

ILLUSTRATED:

3) Spring ligament
4) Tendon tibialis posterior
5) Tendon of peroneus longus
6) Long plantar ligament
7) Short plantar ligament

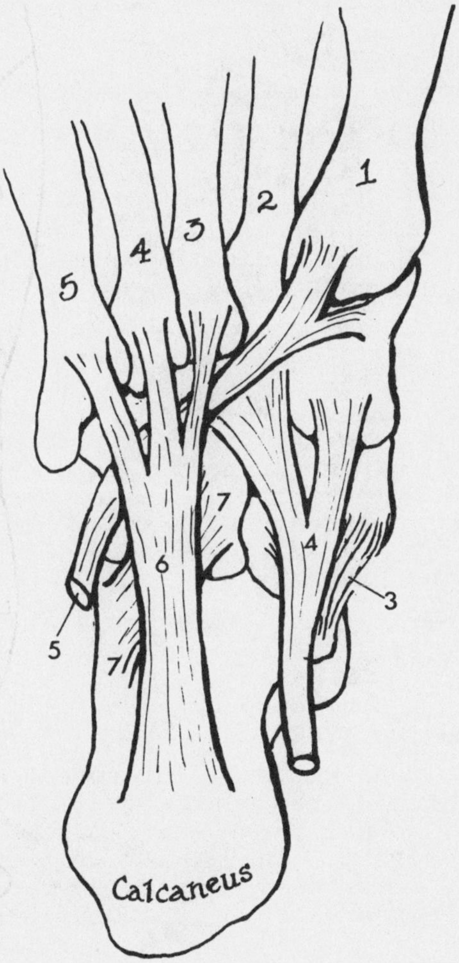

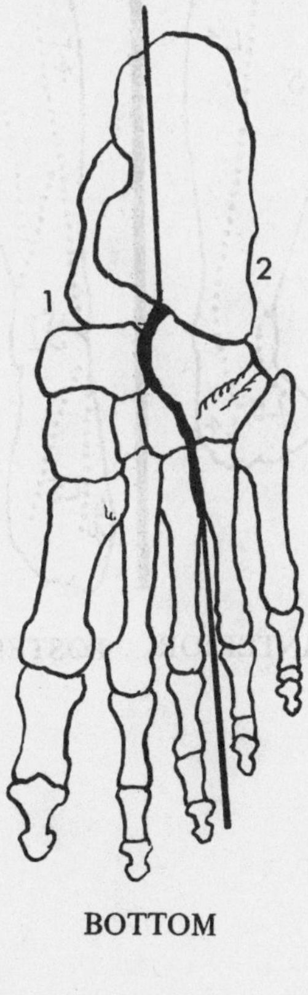

BOTTOM

VI. CUTANEOUS STRUCTURES

DERMATOMES

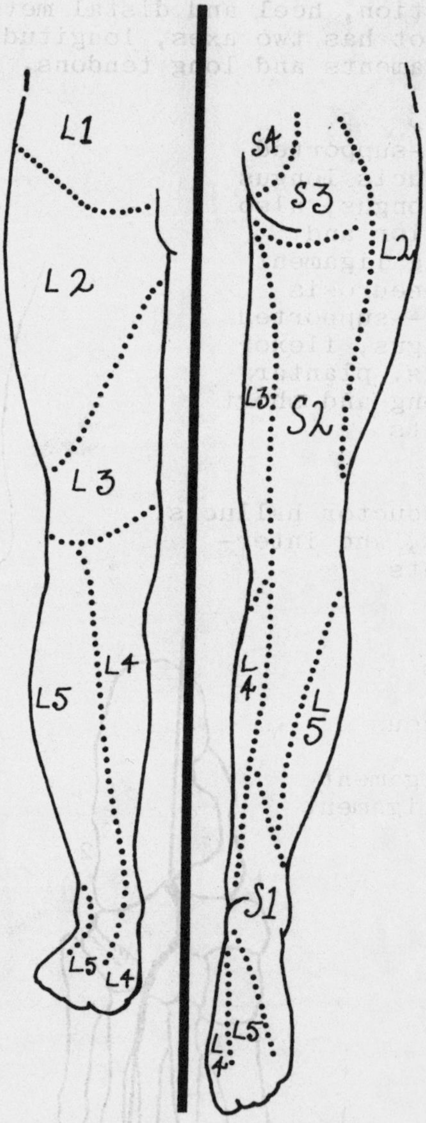

ALSO:

1) Saphenous n.
2) Greater saphenous v.
3) Dorsal venous arch
4) Sural n.
5) Lesser saphenous v.

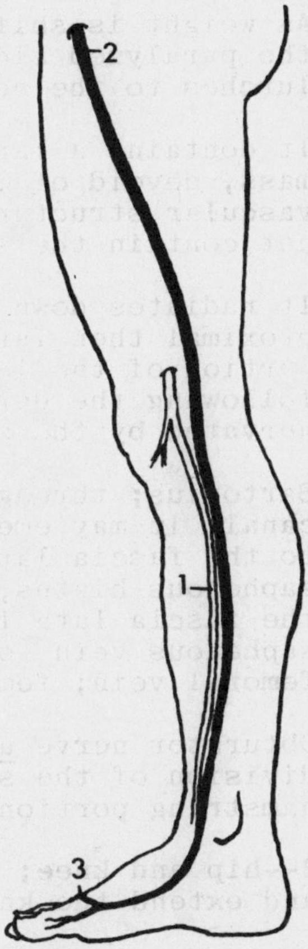

MEDIAL ASPECT OF RIGHT LEG

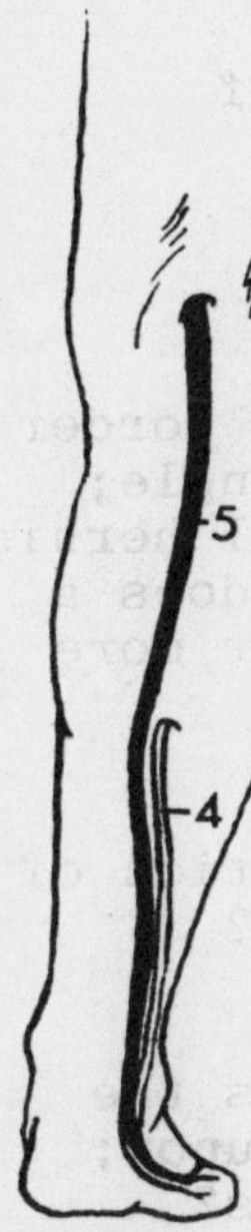

POSTERIOR ASPECT OF RIGHT LEG

STUDY QUESTIONS

1. Which muscles are medial rotators of the hip; which is the most effective?

   Gluteus medius, gluteus minimus, iliopsoas and gracilis; iliopsoas

2. Where does the cruciate anastomosis occur?

   Deep in the gluteal region on the posterior surface of the hip joint.

3. What is the gait of a person with a paralyzed gluteus medius?

   As weight is shifted onto the paralyzed side, the body lurches to the normal side.

4. Why is the superolateral quadrant of the buttock used for intramuscular injections?

   It contains a large muscle mass, devoid of major neurovascular structures; it does not contain the sciatic n.

5. Where is the pain of sciatica felt?

   It radiates down the posterior proximal then lateral distal portion of the lower extremity, following the dermatomes innervated by the sciatic n.

6. What is the lateral border of the femoral triangle; where does a femoral hernia occur; in what sex does a femoral hernia occur more often?

   Sartorius; through the femoral canal, it may emerge from deep to the fascia lata through the saphenous hiatus, a defect in the fascia lata for the greater saphenous vein to enter the femoral vein; female.

7. What is the innervation of the adductor magnus?

   Obturator nerve <u>and</u> tibial division of the sciatic for the hamstring portion.

8. How many joints does the rectus femoris act upon; what are the actions?

   2--hip and knee; flex the hip and extend the knee.

9. The profunda femoris gives rise to perforating branches which supply the posterior compartment and participate in the cruciate anastomosis. T OR F

   T

10. What is the Terrible Triad; is the fibular collateral ligament of the knee attached to the lateral meniscus?

    Anterior cruciate ligament, medial meniscus and medial (tibial) collateral ligament; no.

11.	How are the cruciate ligaments named?	By their attachments to the tibia.
12.	The artery of the round ligament of the head of the femur carries all blood required by the head and neck. T OR F	F.
13.	What happens to the foot after a 'bumper injury'; why?	It is incapable of dorsiflexion and eversion, therefore it tends to hang plantar flexed and inverted; the common peroneal n. is damaged.
14.	What is the soleus pump?	Muscle action of the calf muscles helps propel venous blood towards the heart.
15.	Why is the popliteus important?	It helps 'unlock' the knee during the initiation of movement by producing a twisting action of the femur upon the tibia.
16.	What vessels of the lower extremity are commonly used as transplant vessels during coronary bypass surgery?	The saphenous veins, preferably the greater saphenous vein.
17.	What muscles help support the longitudinal arch of the foot?	Flexor hallucis longus, flexor digitorum longus, tibiales anterior and posterior, peroneus longus, abductor digiti minimi, flexor digitorum brevis, and abductor hallucis.
18.	The peroneus tertius is innervated by the superficial peroneal n. T OR F	F
19.	The extensor hallucis brevis is innervated by the lateral plantar n. T OR F	F
20.	The third digit of the foot has 2 palmar interossei and no dorsal one. T OR F	F
21.	Which collateral ligament of the ankle is torn in an inversion sprain; which is the weakest?	Fibular; fibular.

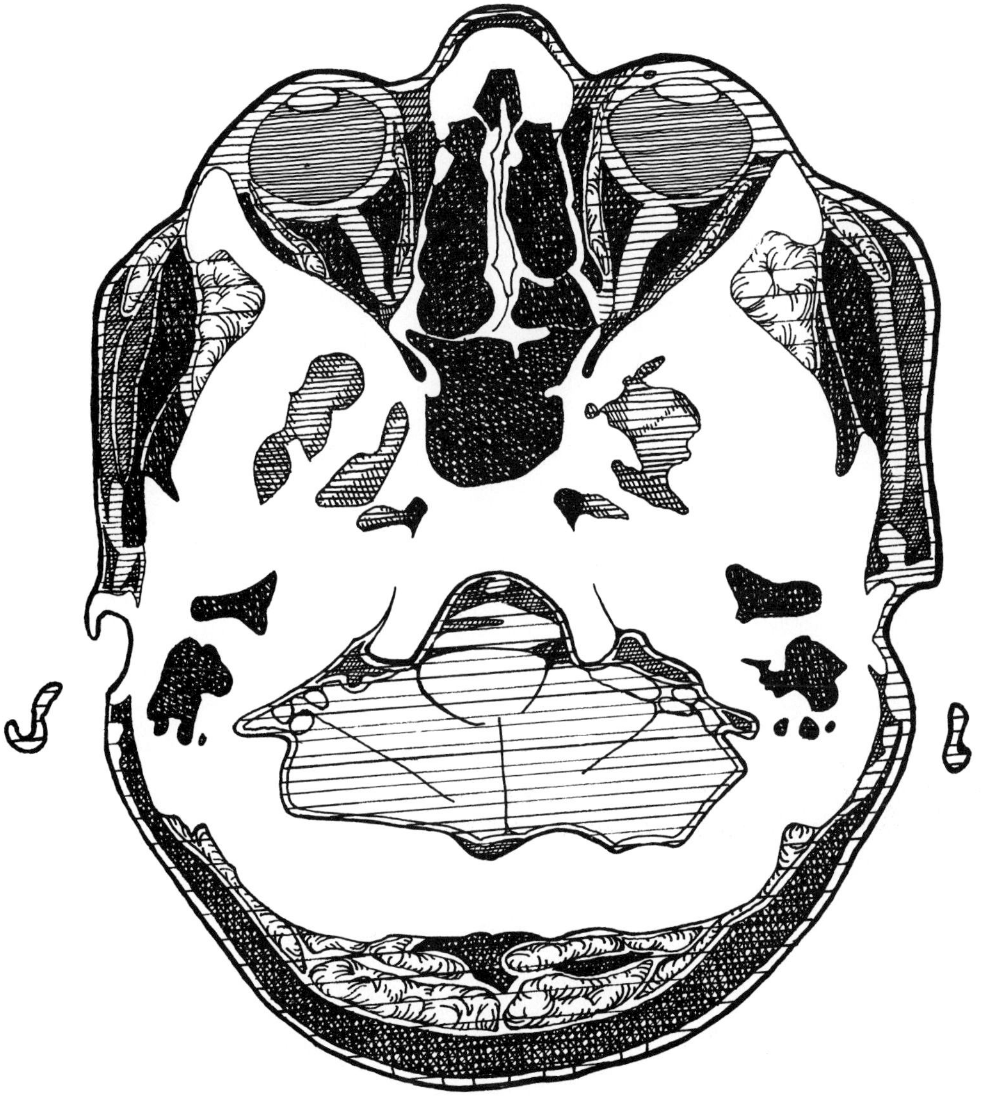

TOMOGRAMS

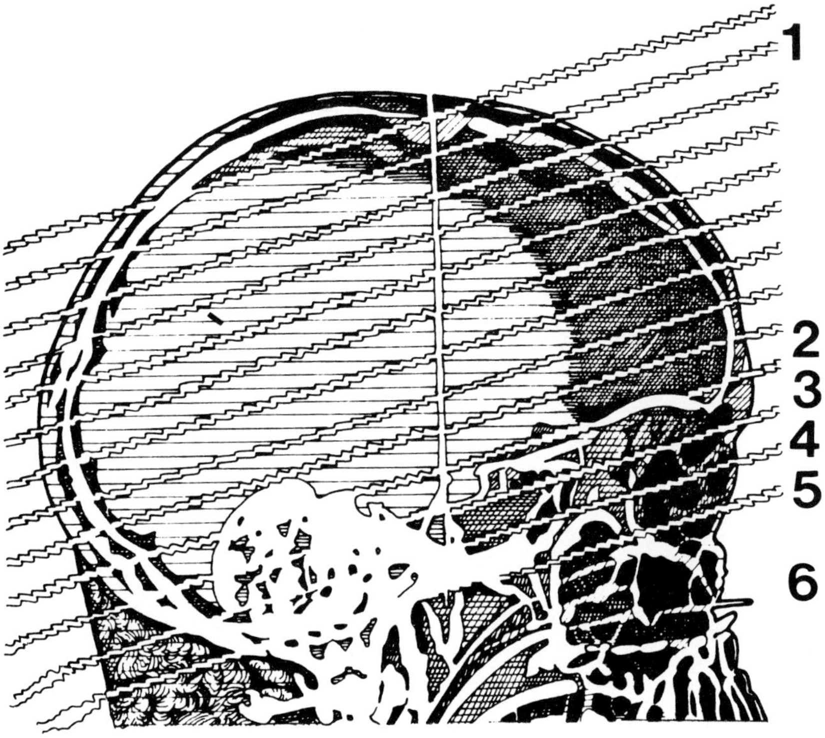

THE NUMBER INDICATES THE
LEVEL OF THE CORRESPONDING
COMPUTERIZED TOMOGRAM

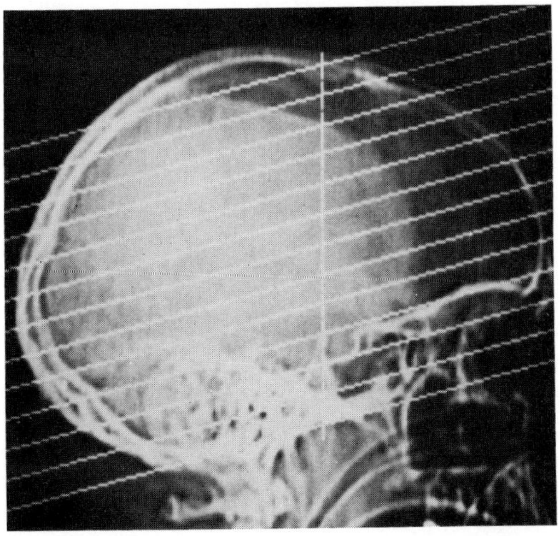

GUIDE

REGION	TOMOGRAM NUMBERS
HEAD	1 - 6
NECK	7 - 11
THORACIC INLET	12 - 13
THORAX	14 - 19
ABDOMEN	20 - 25
MALE PELVIS	26 - 31
FEMALE PELVIS	32 - 34
KNEE	35 - 39

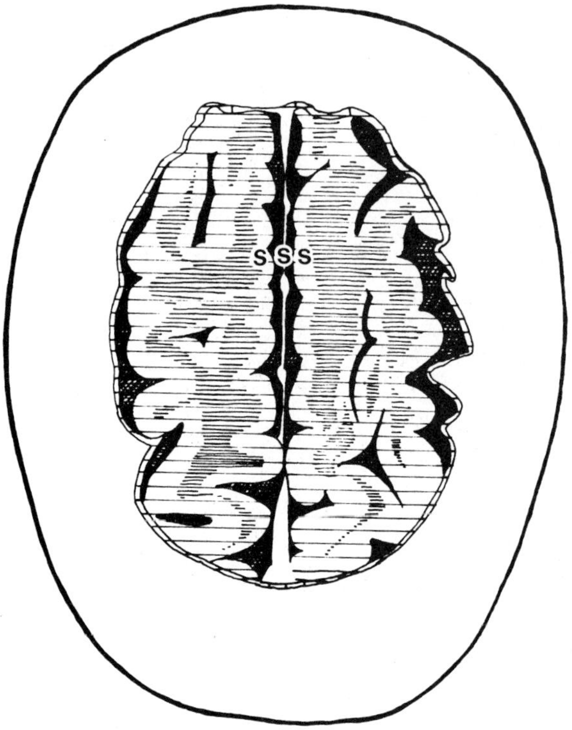

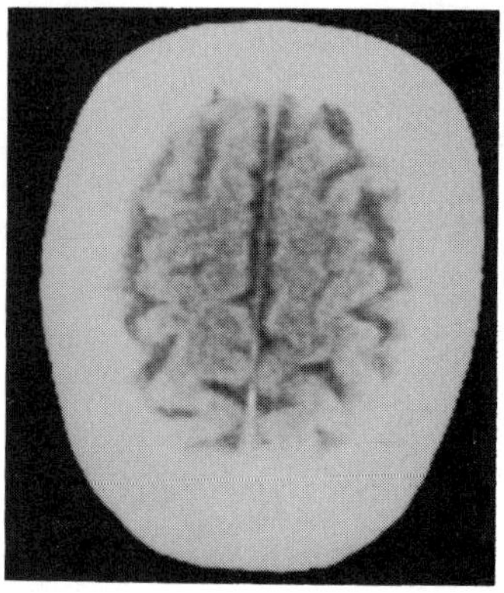

1

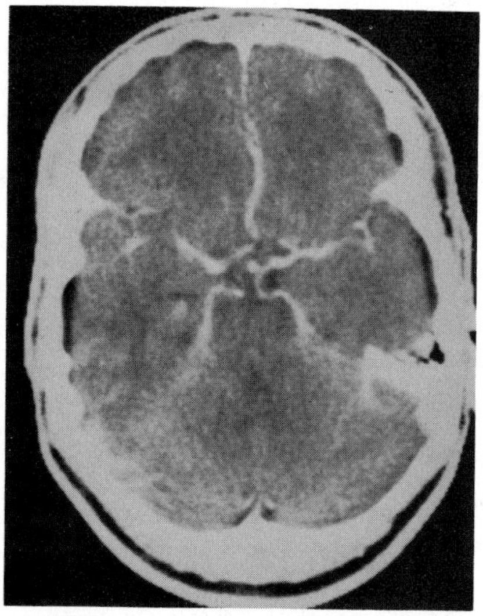

2

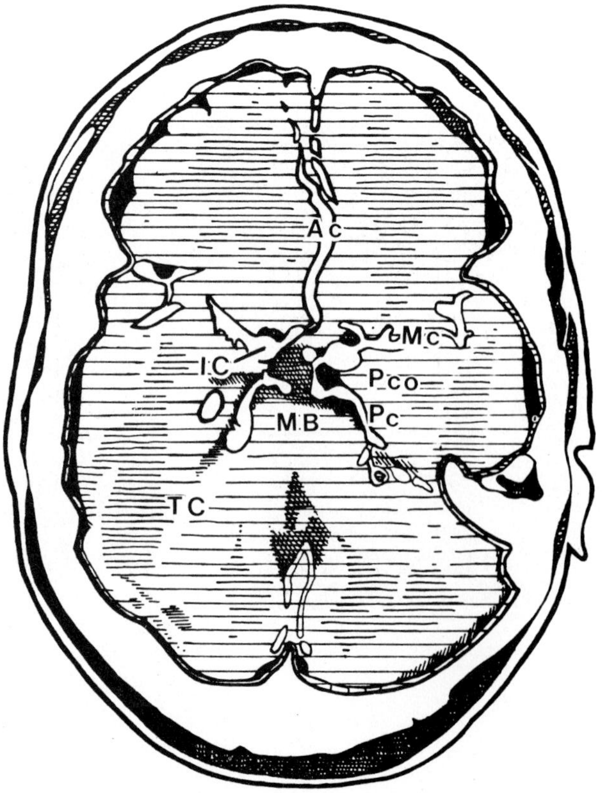

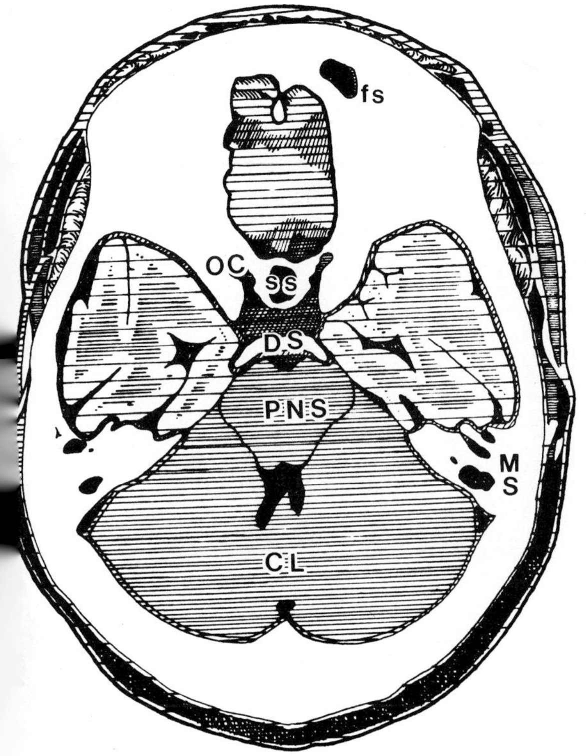

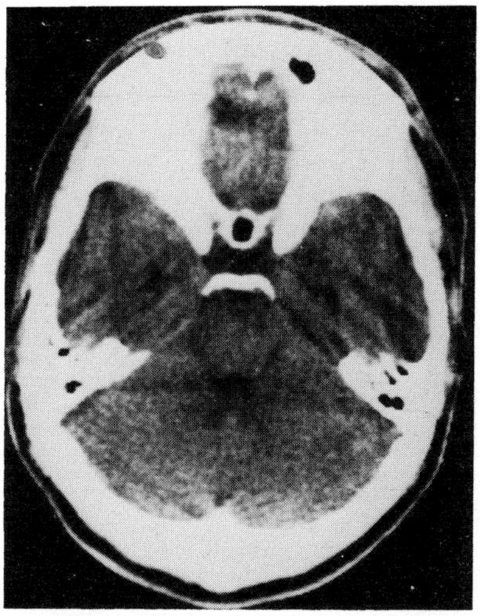

3

4

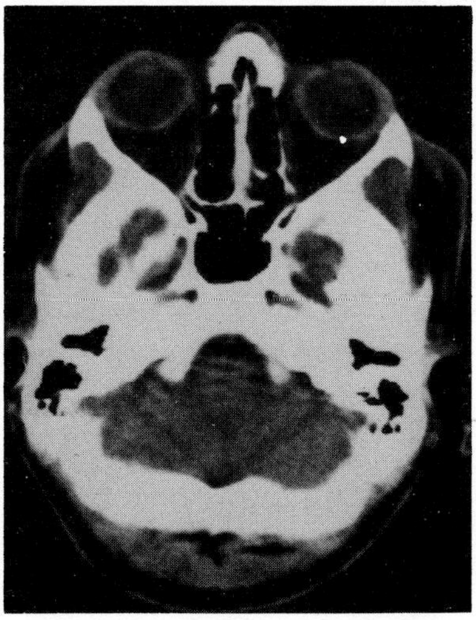

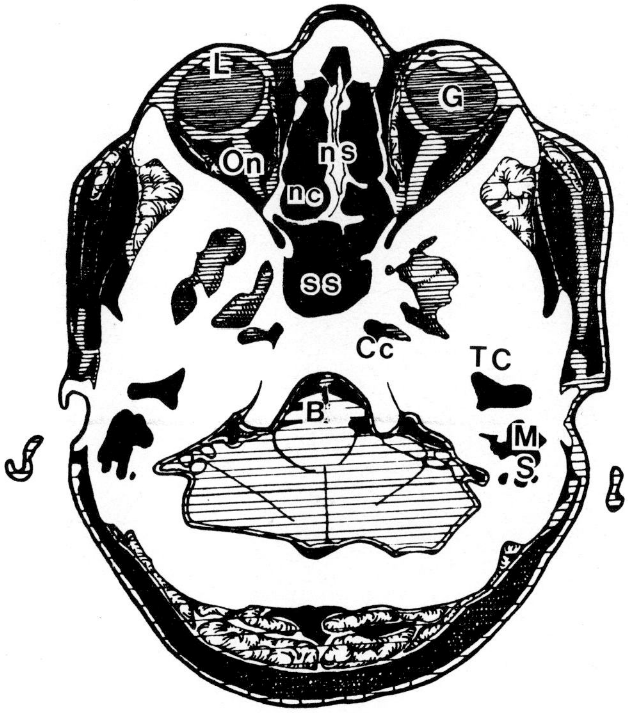

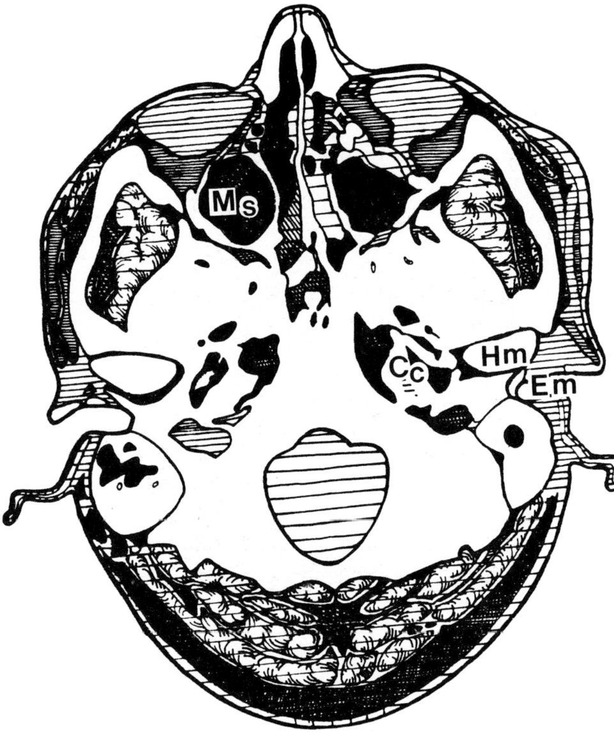

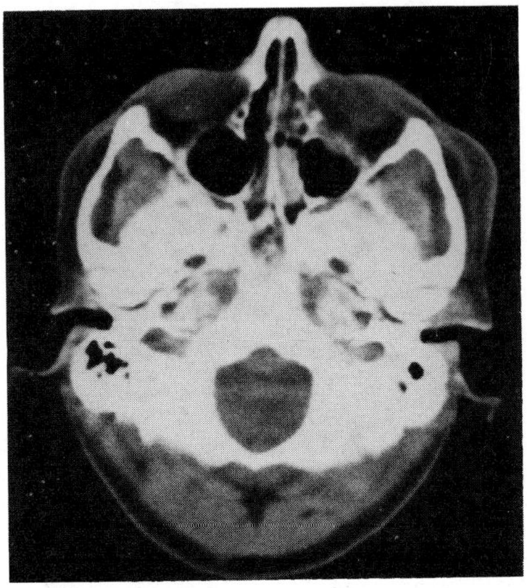

5

6

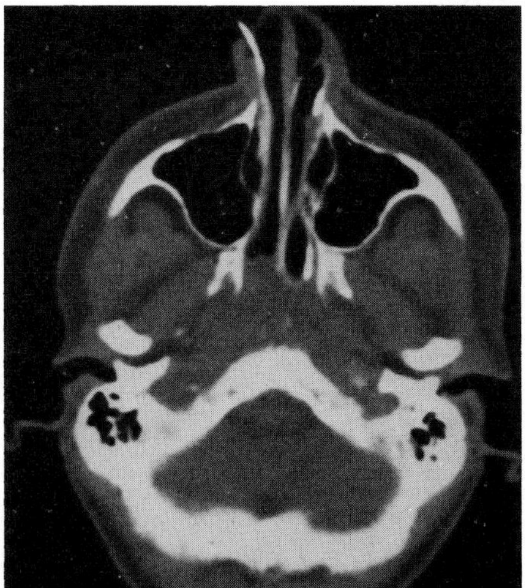

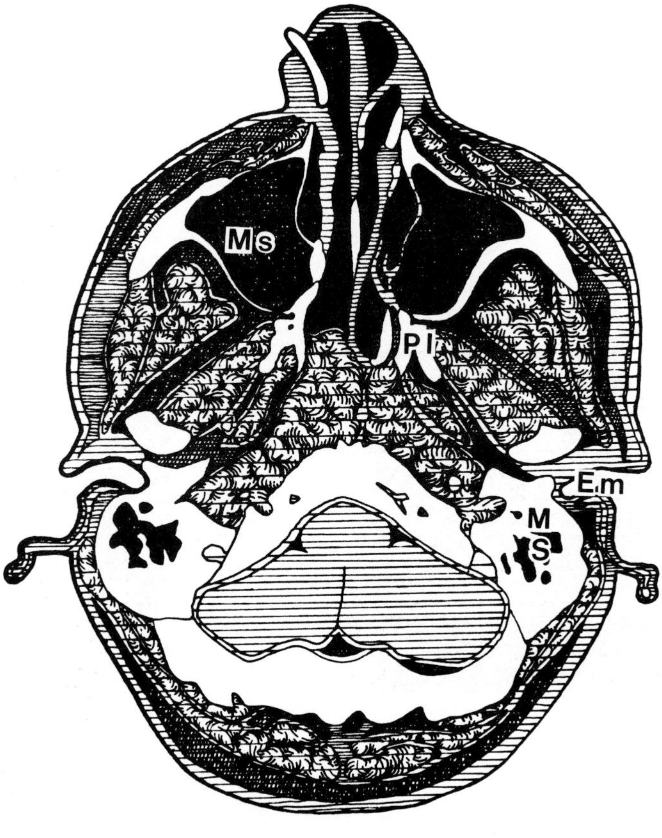

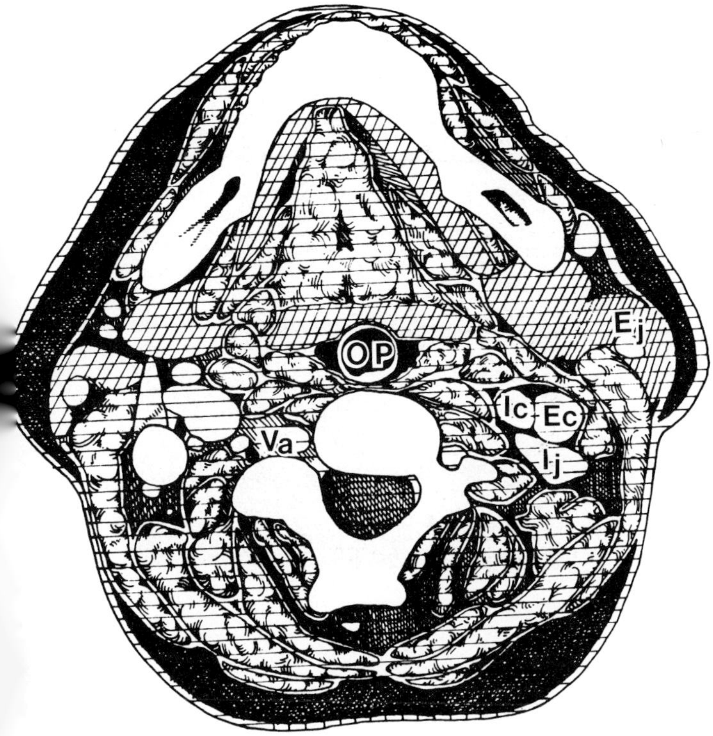

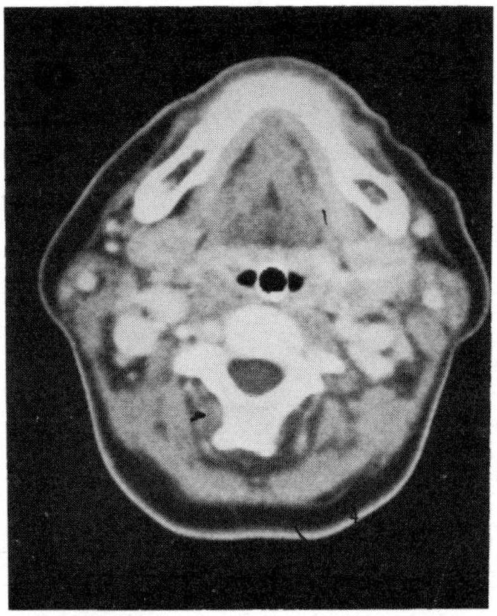

7

8

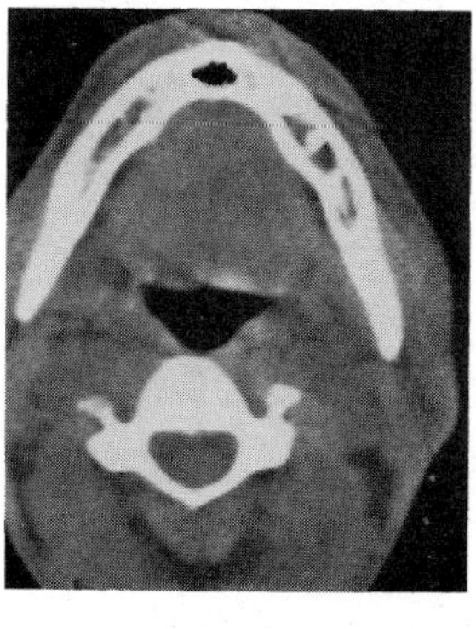

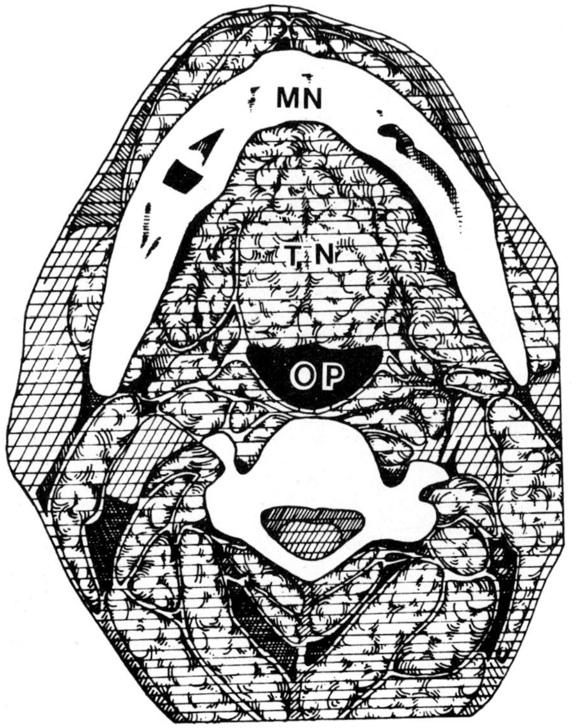

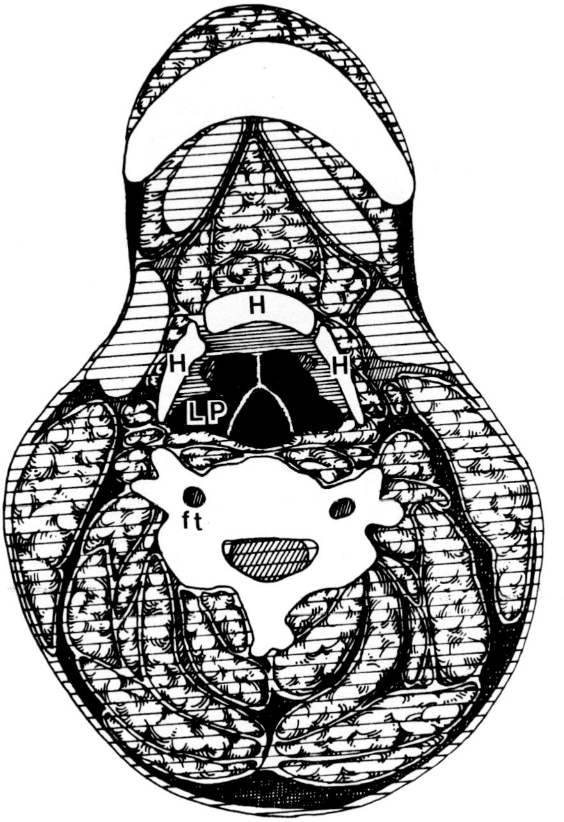

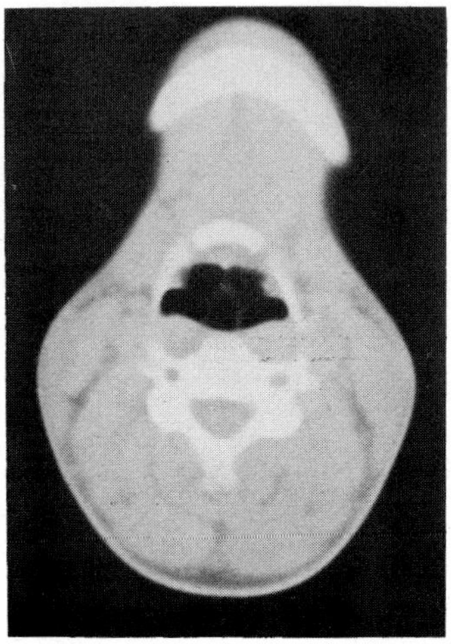

9

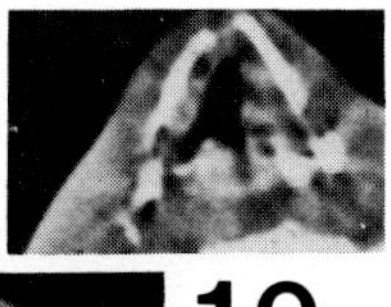

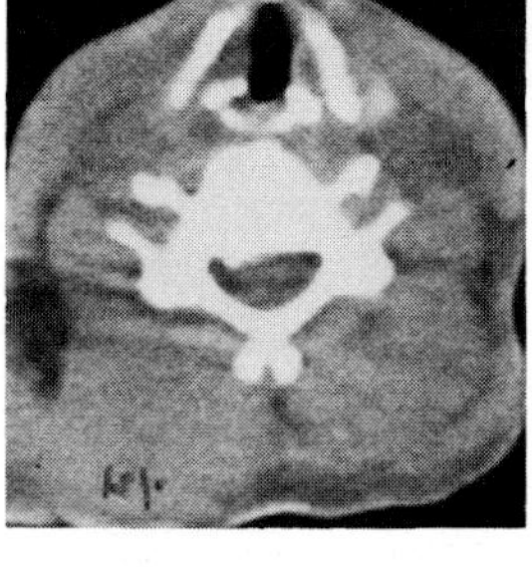

10

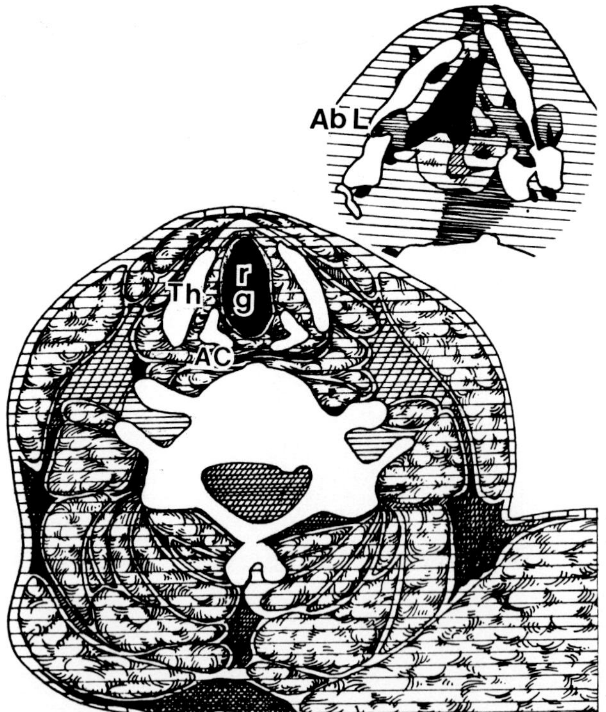

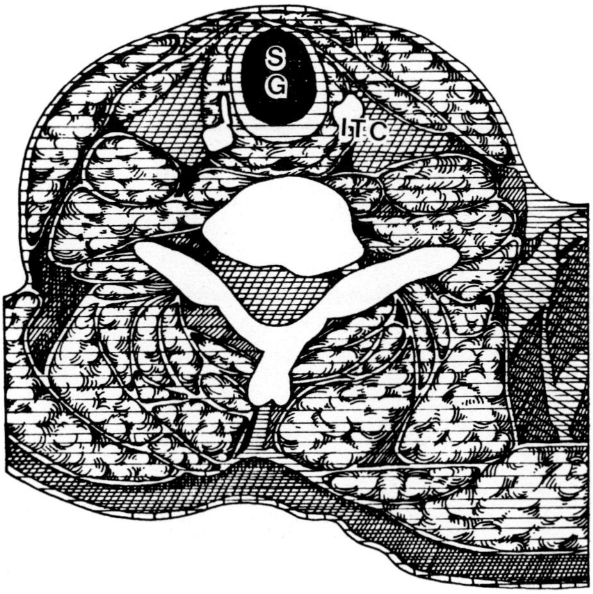

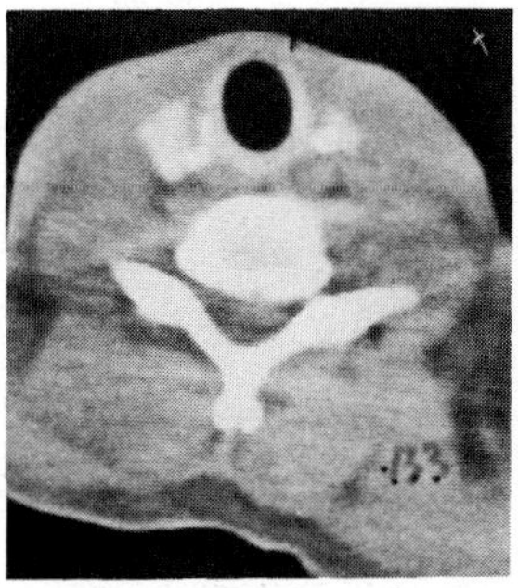

11

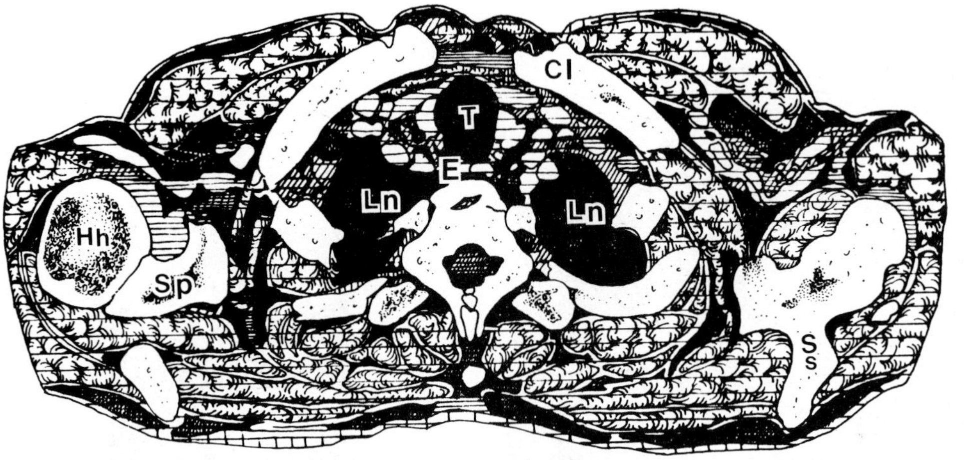

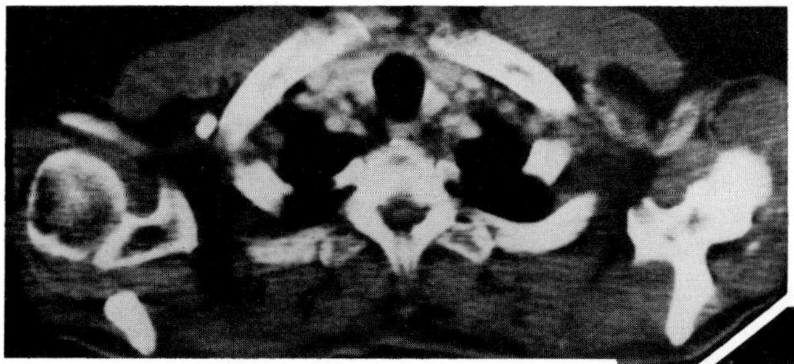

12

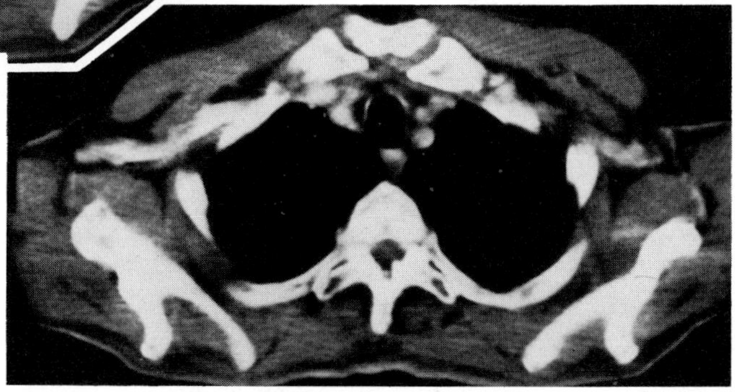

13

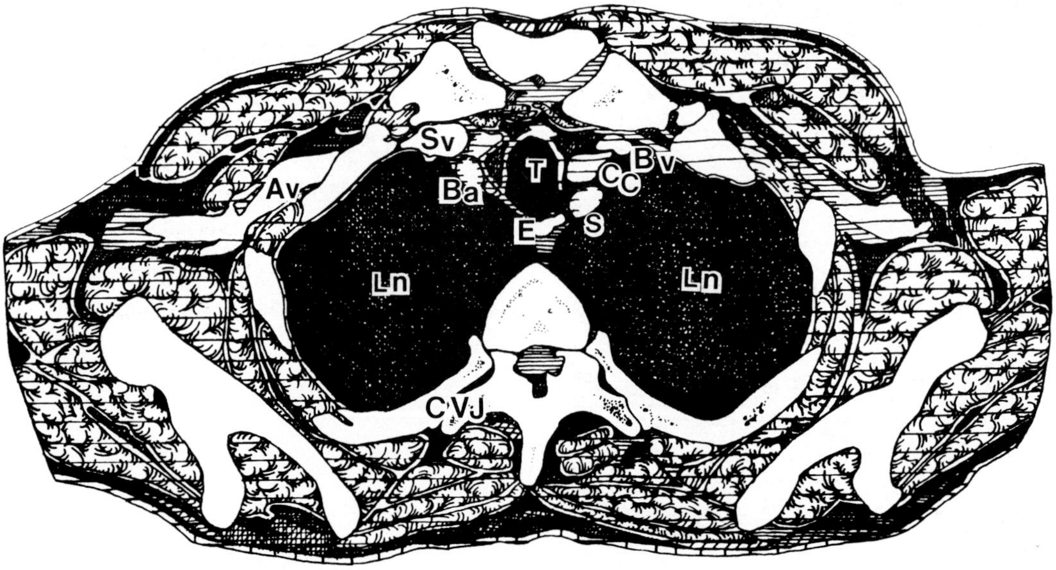

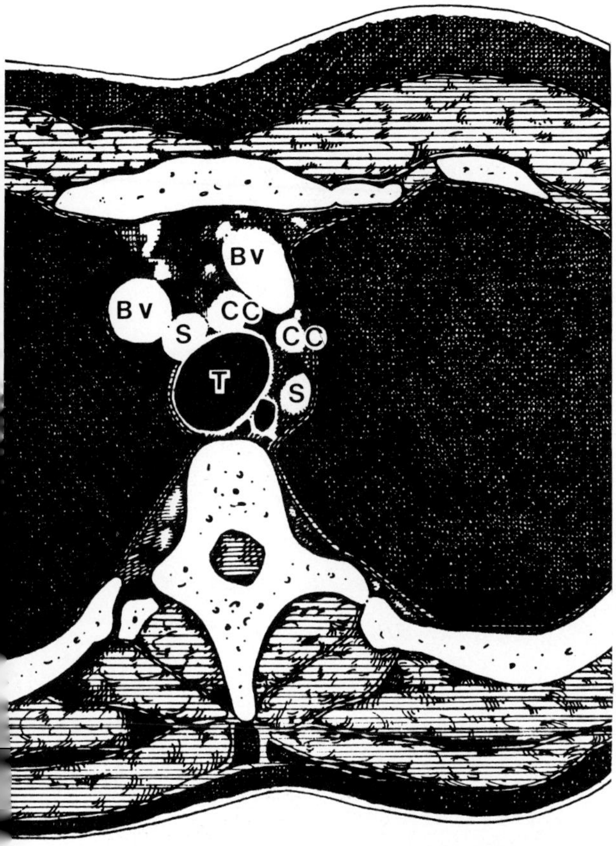

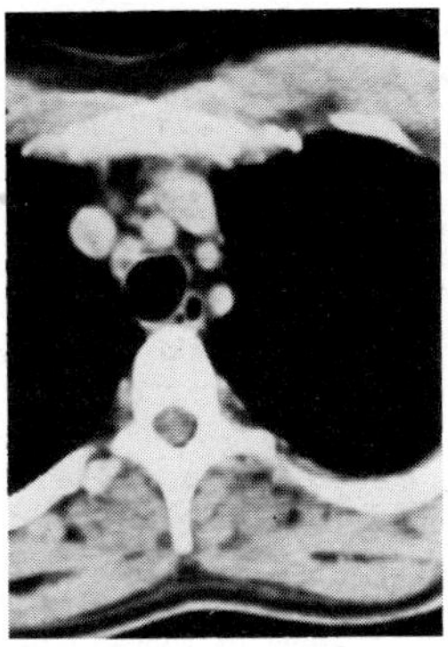

14

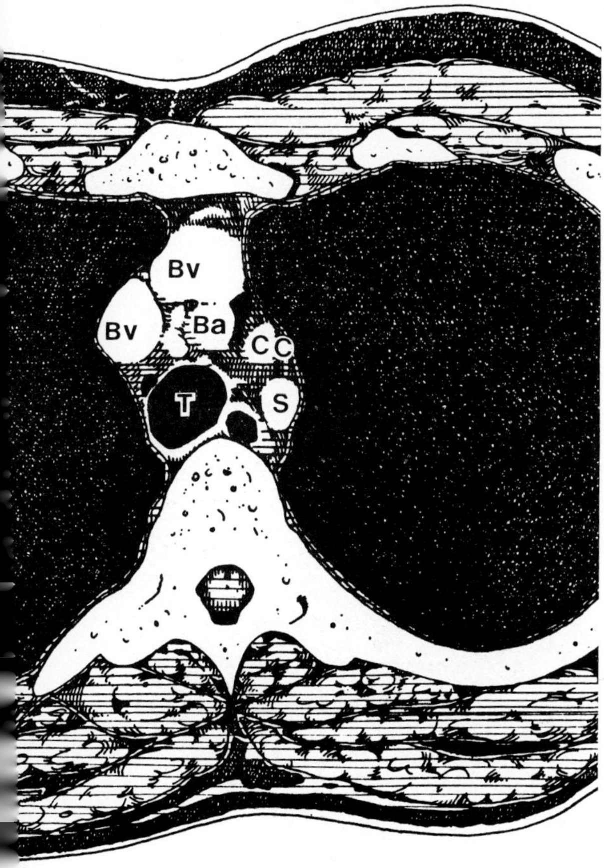

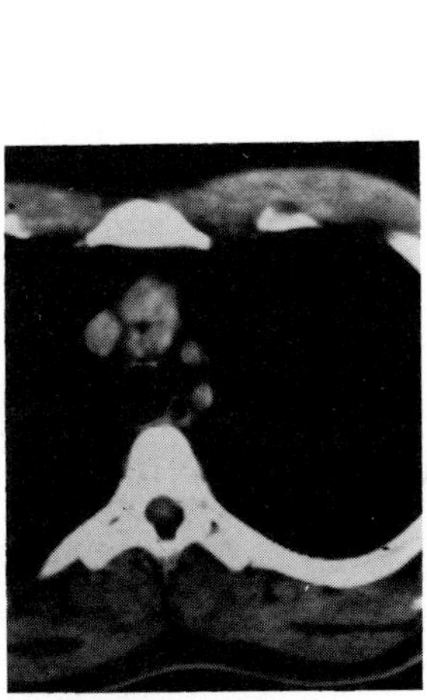

15

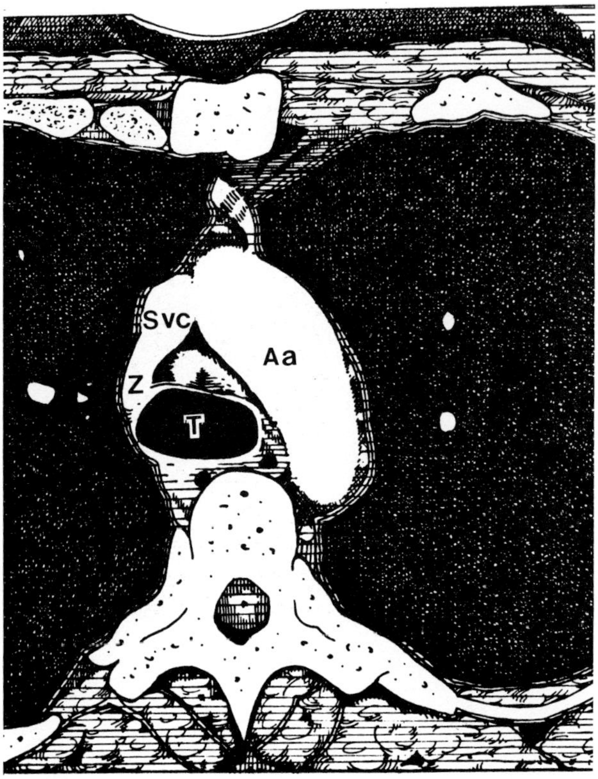

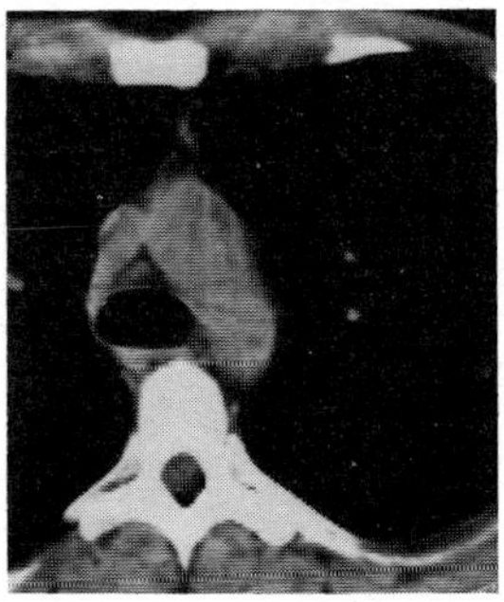

16

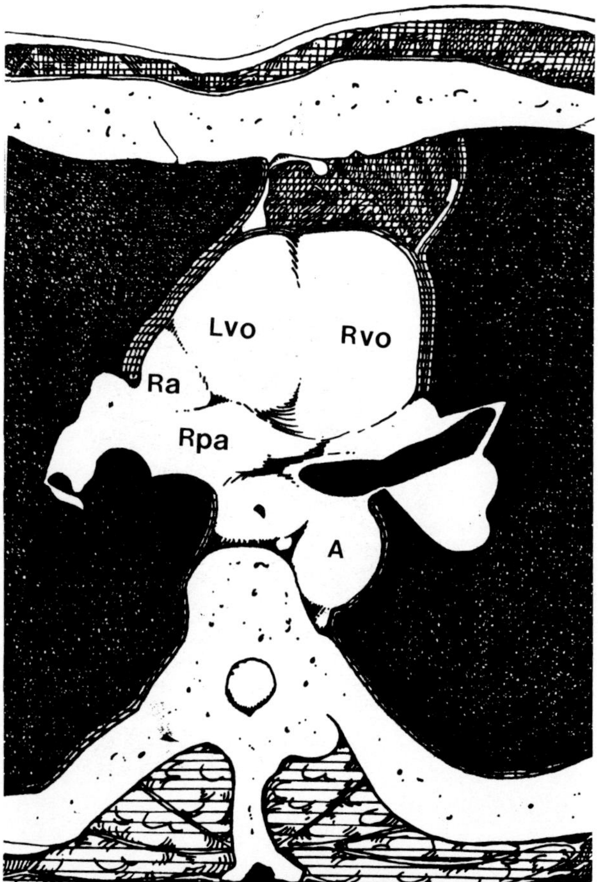

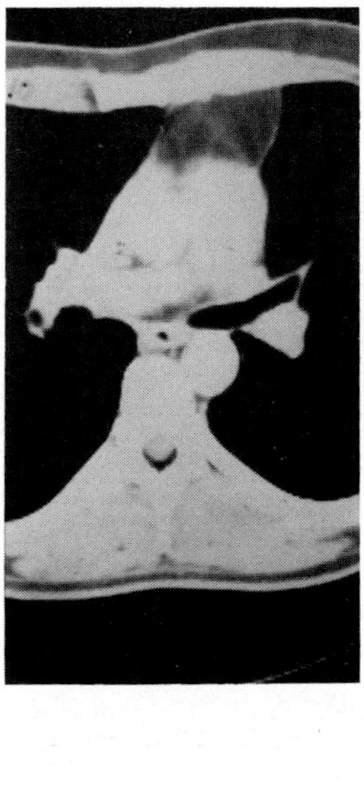

17

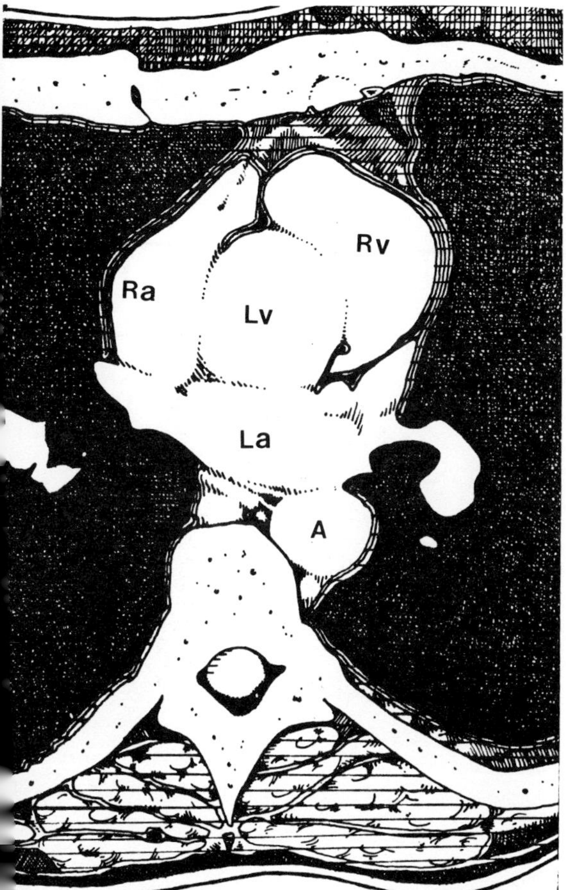

18

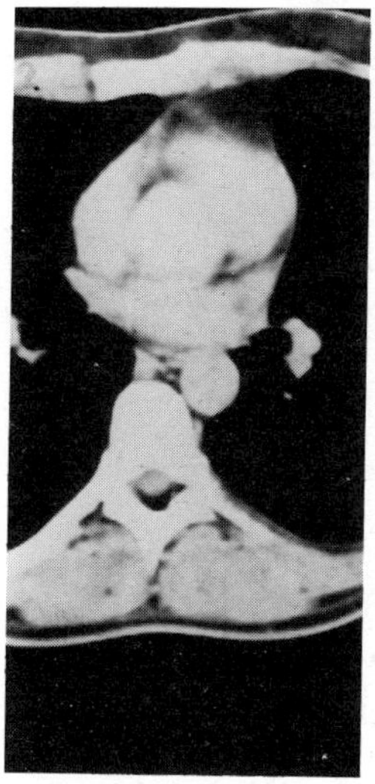

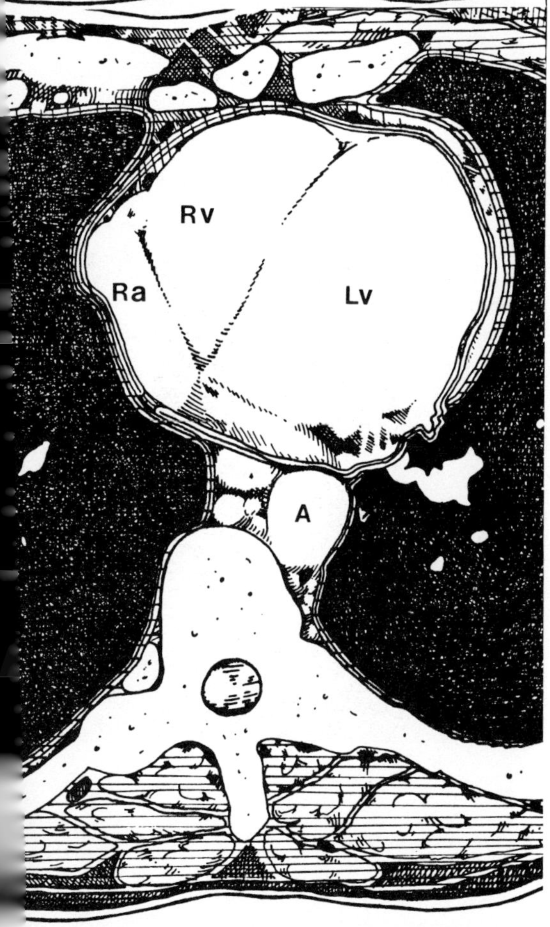

19

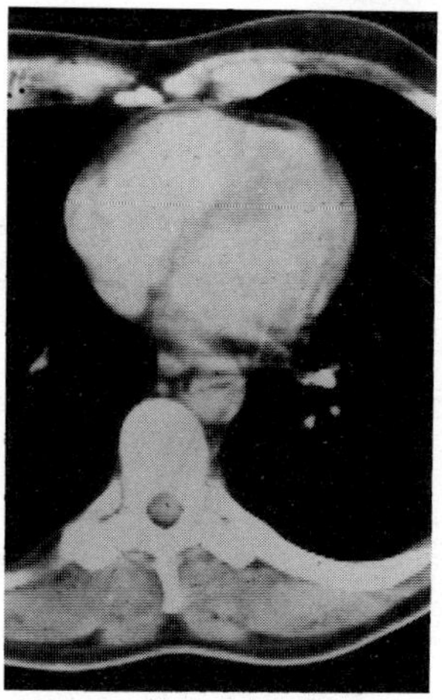

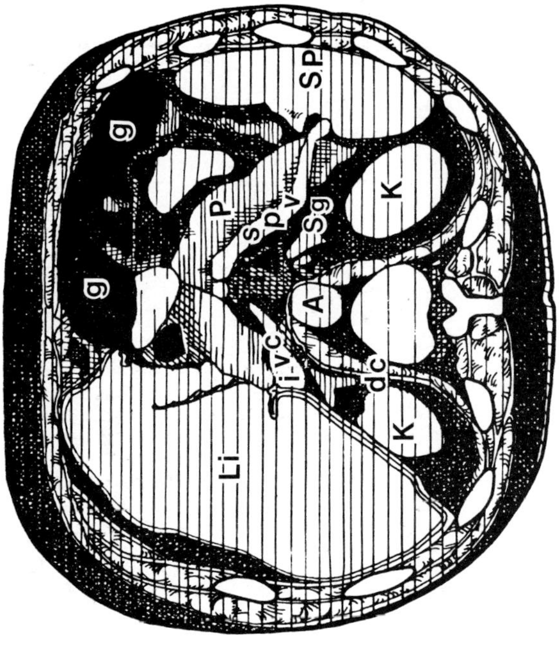

21

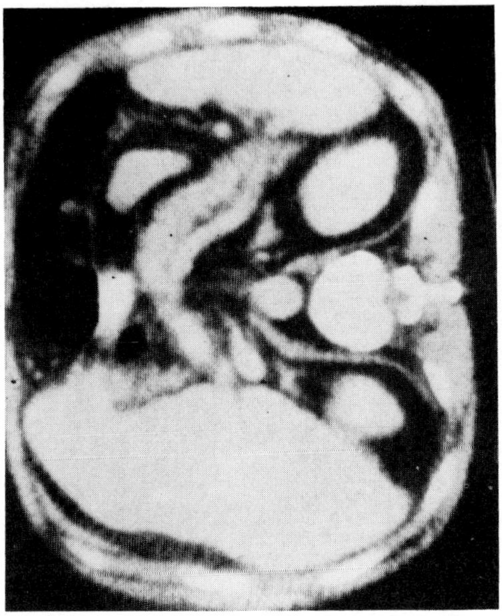

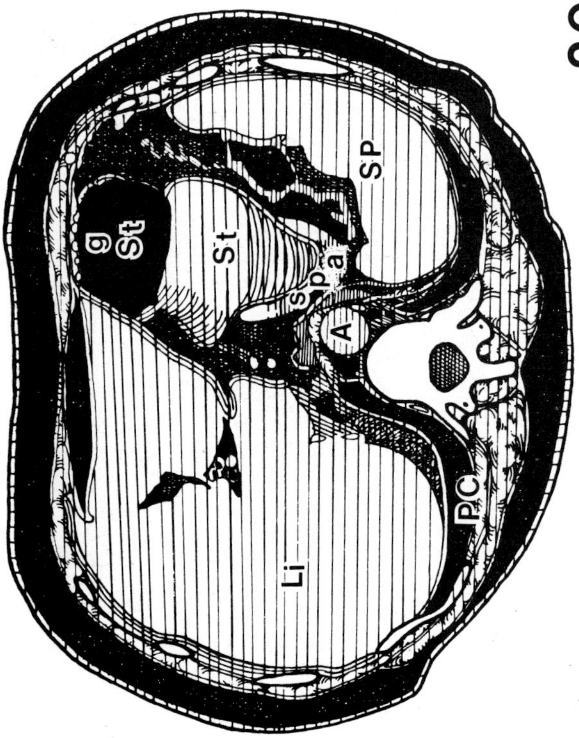

20

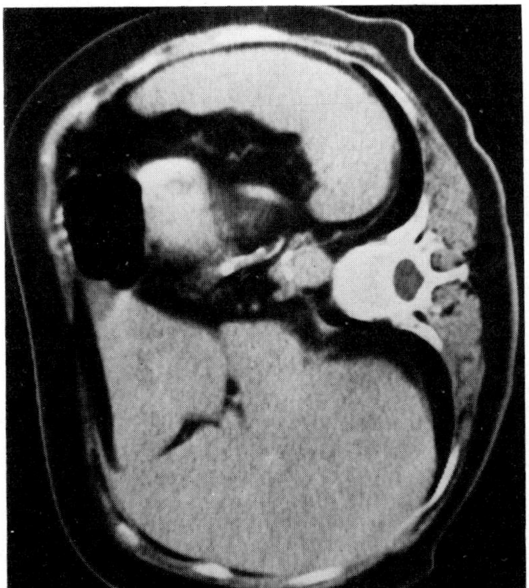

23

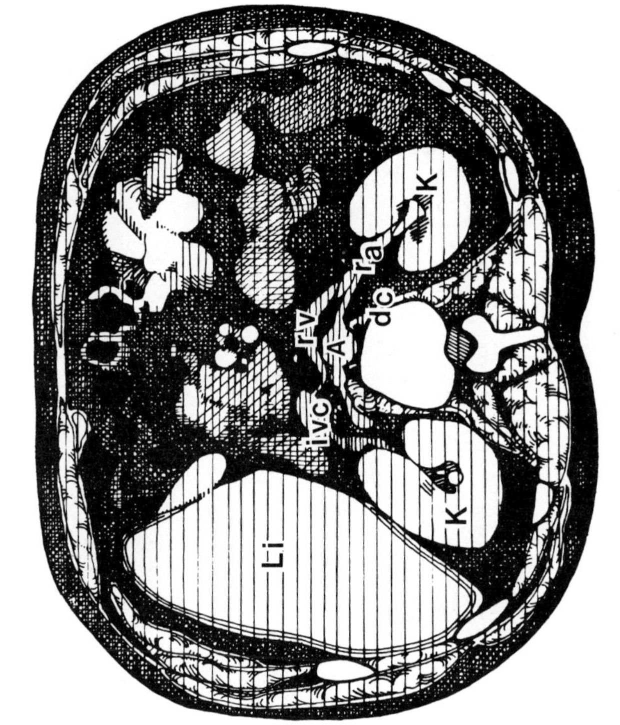

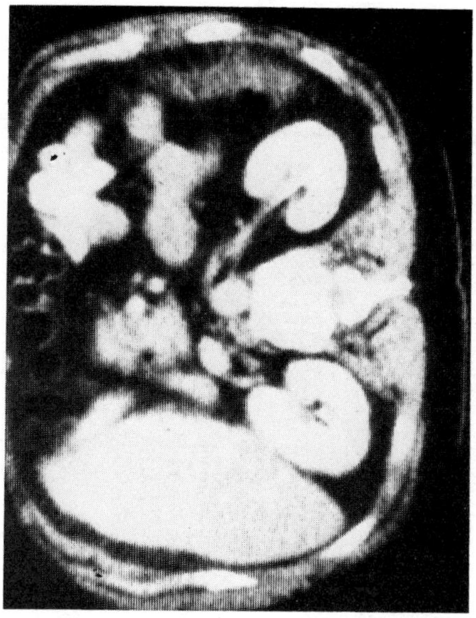

22

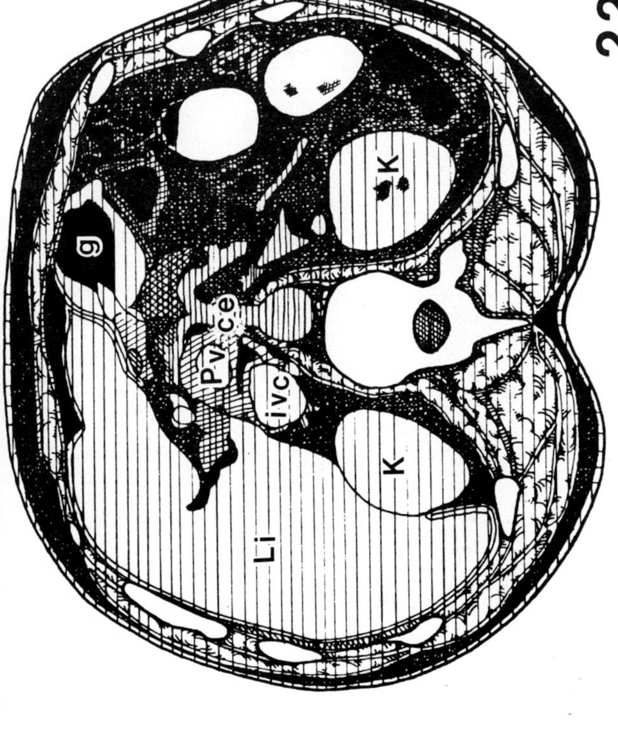

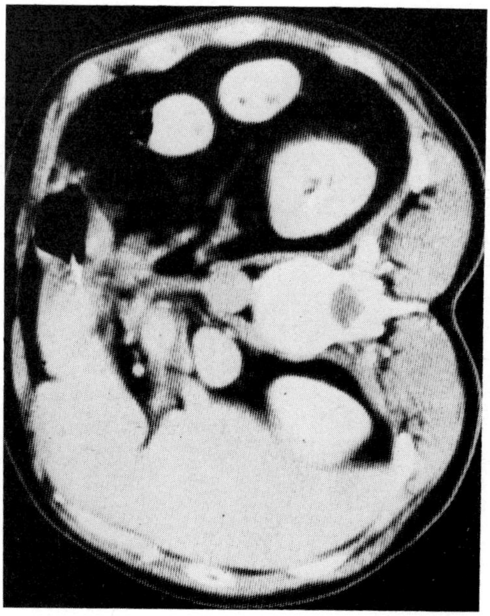

25

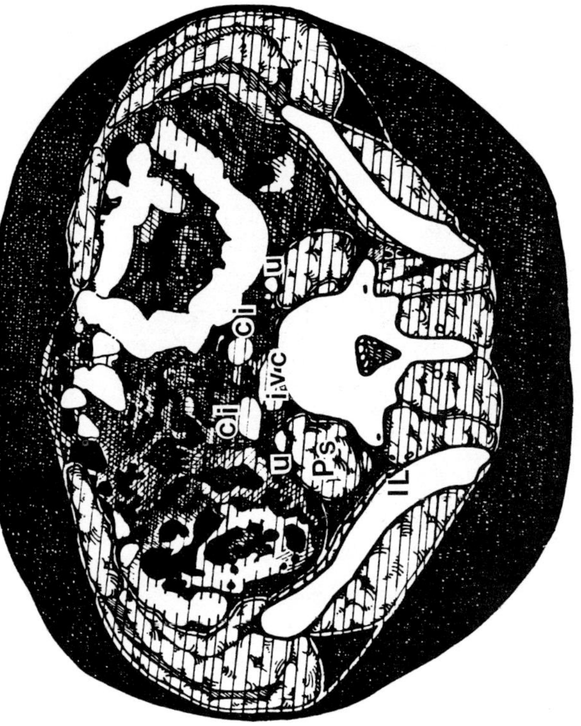

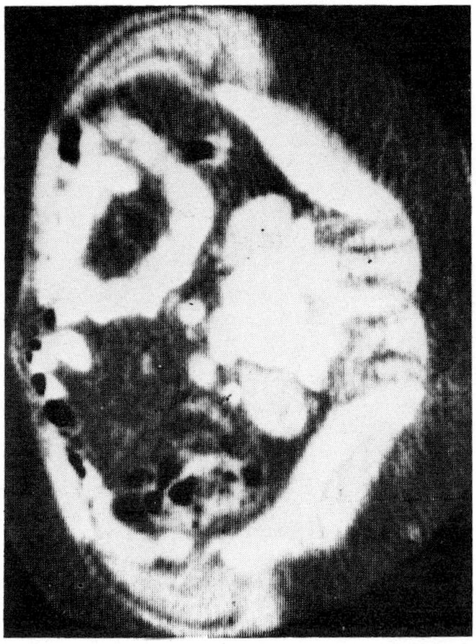

24

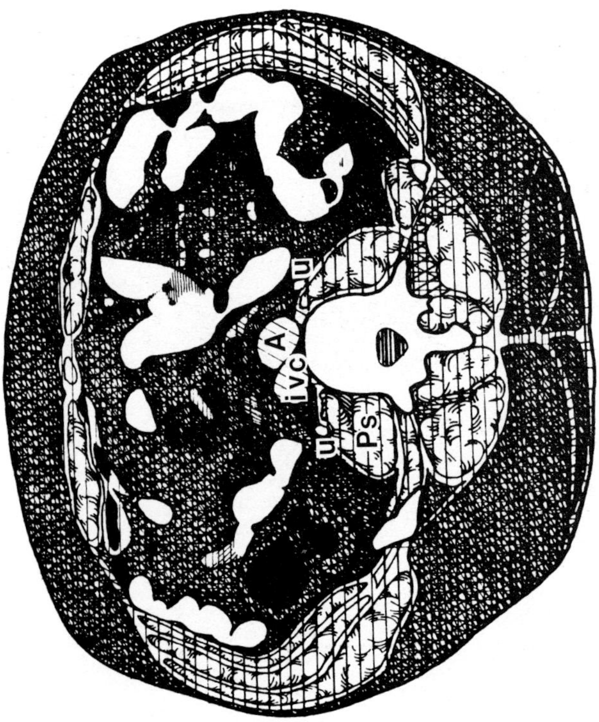

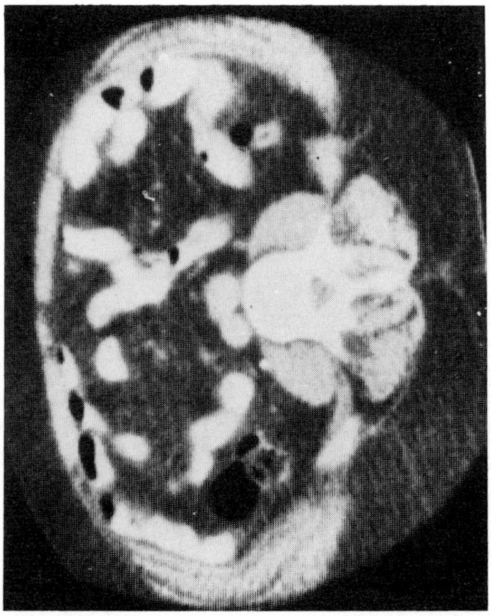

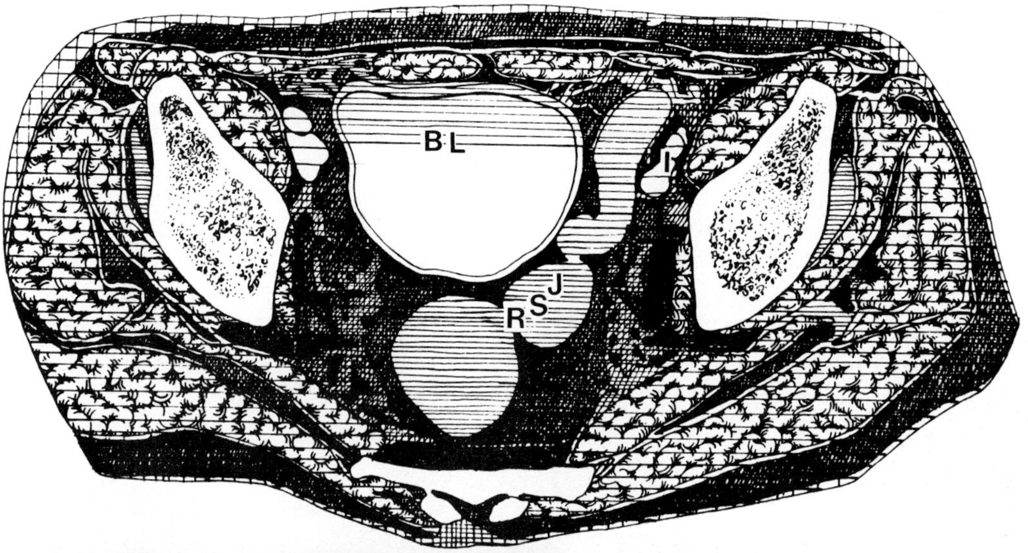

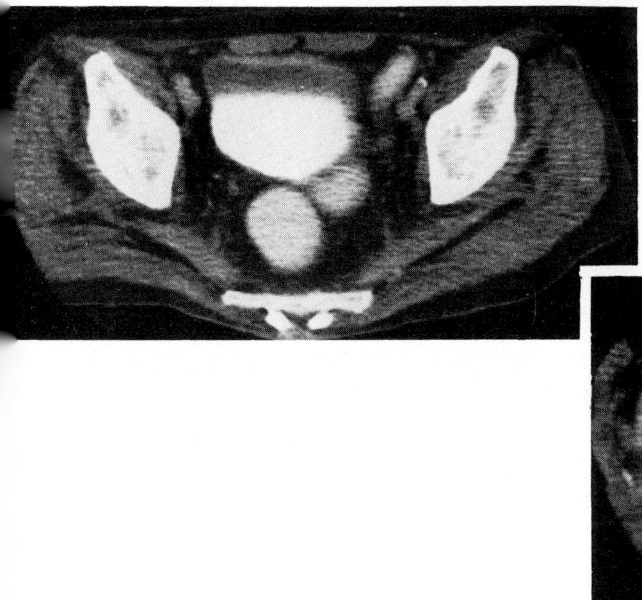

26

27

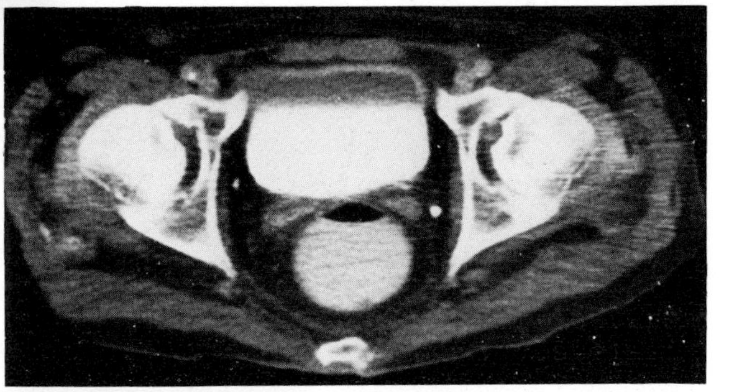

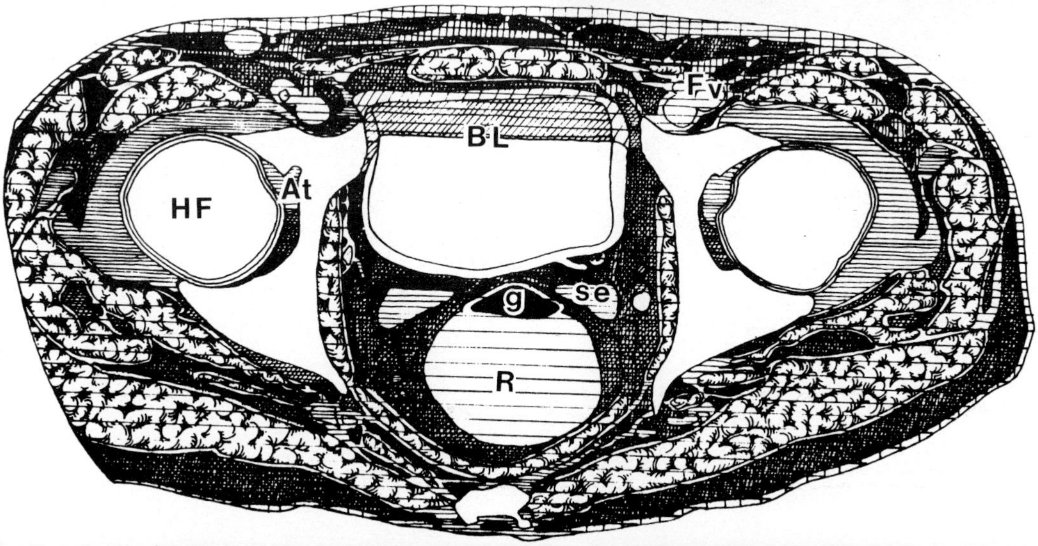

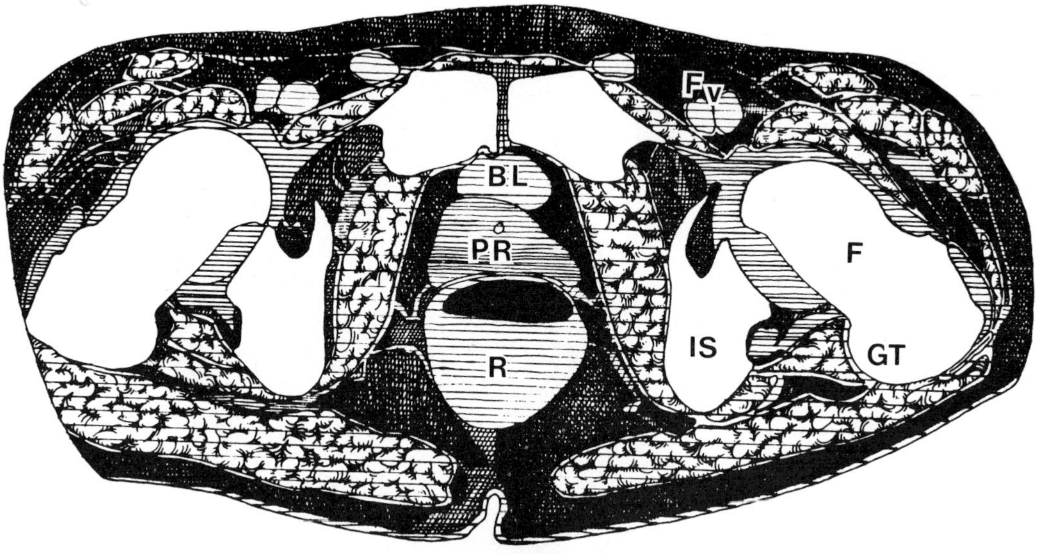

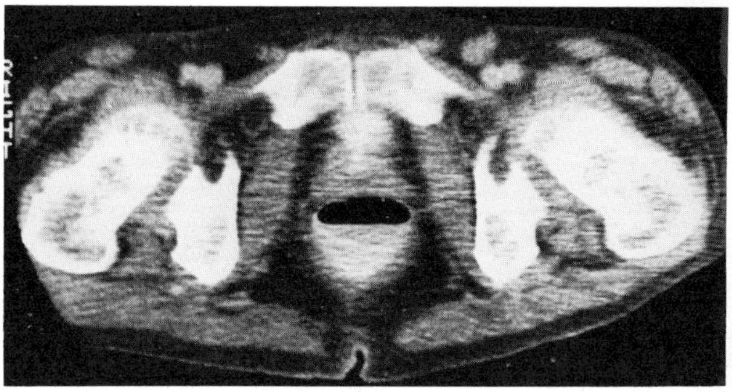

28

29

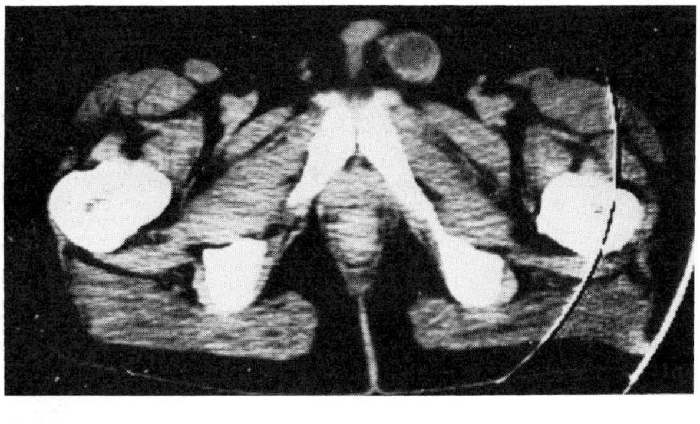

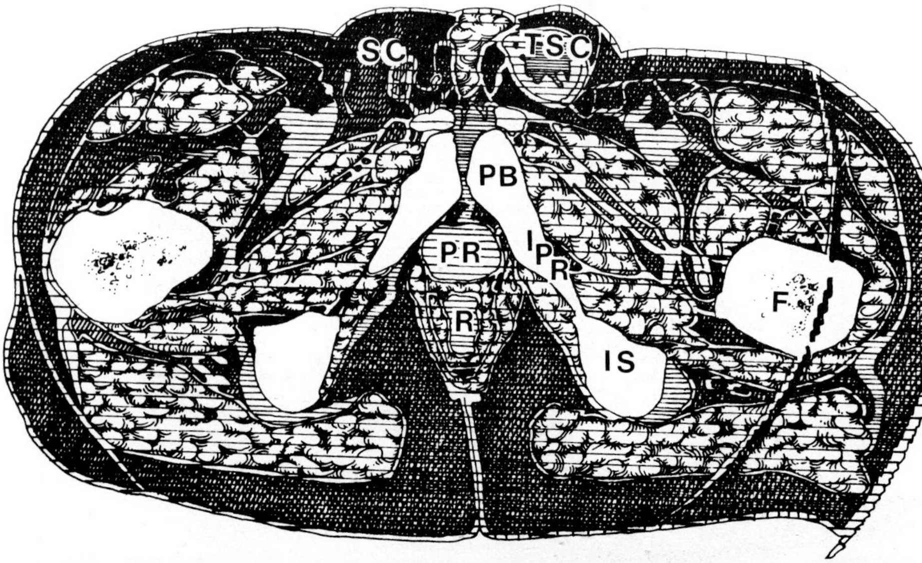

30

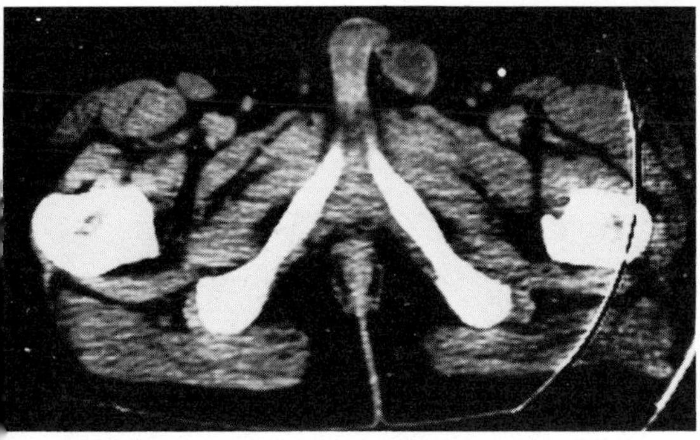

31

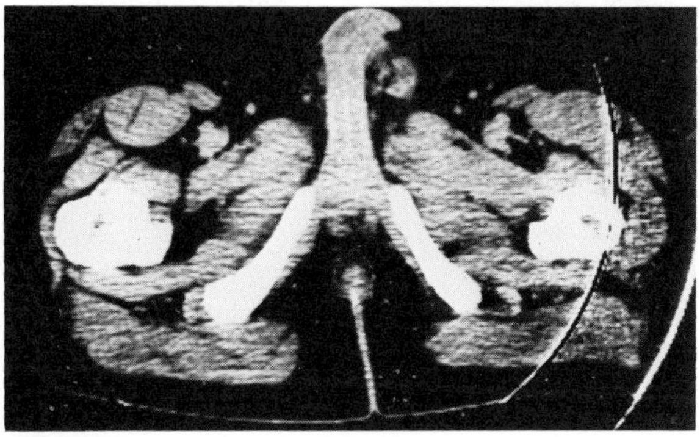

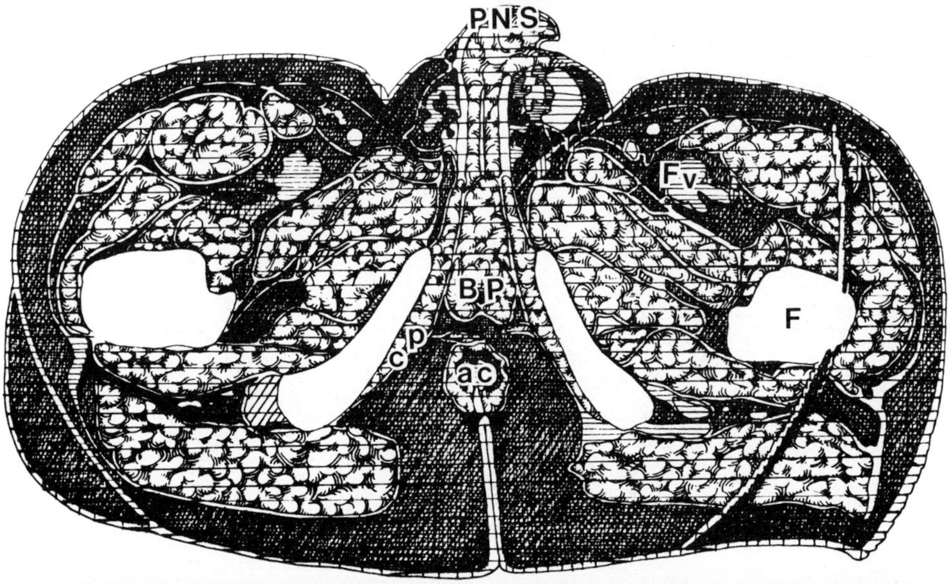

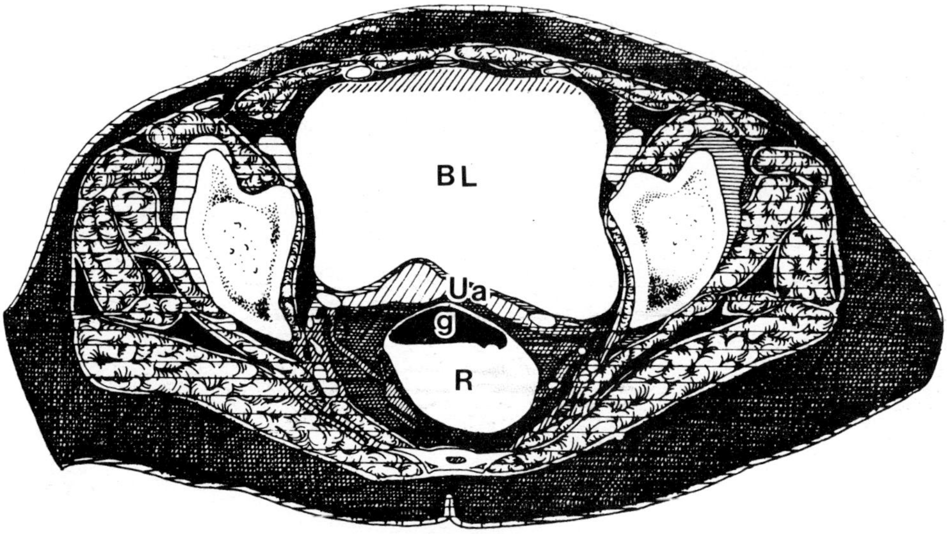

32

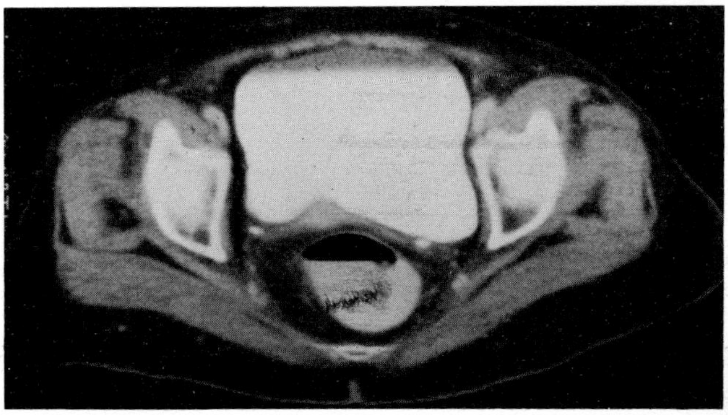

33

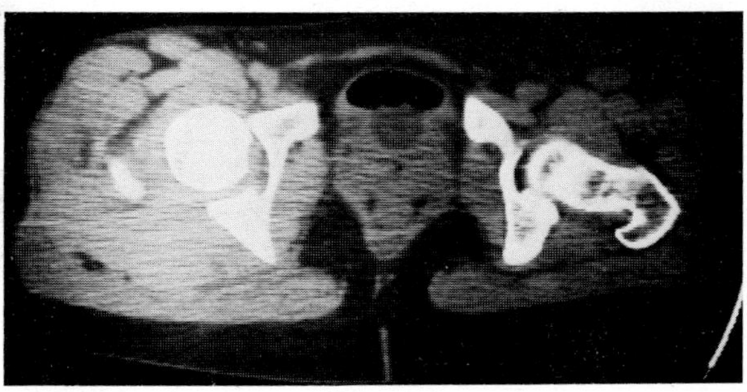

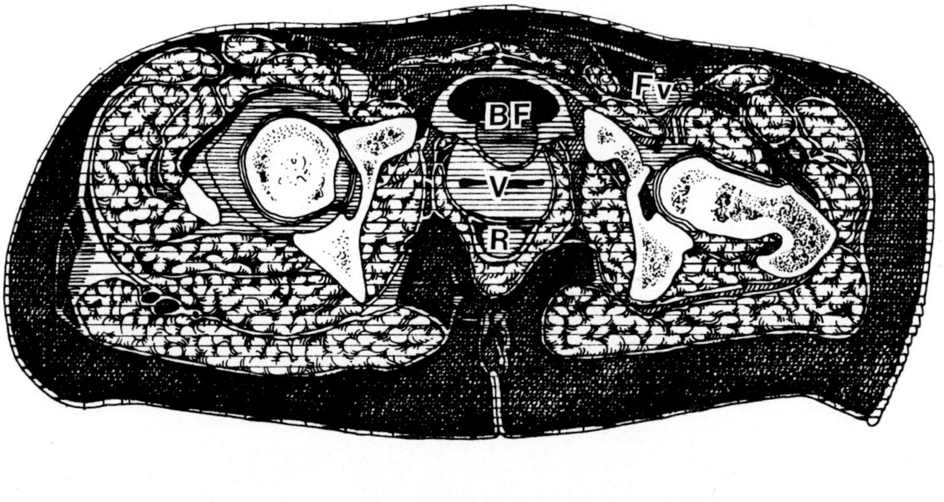

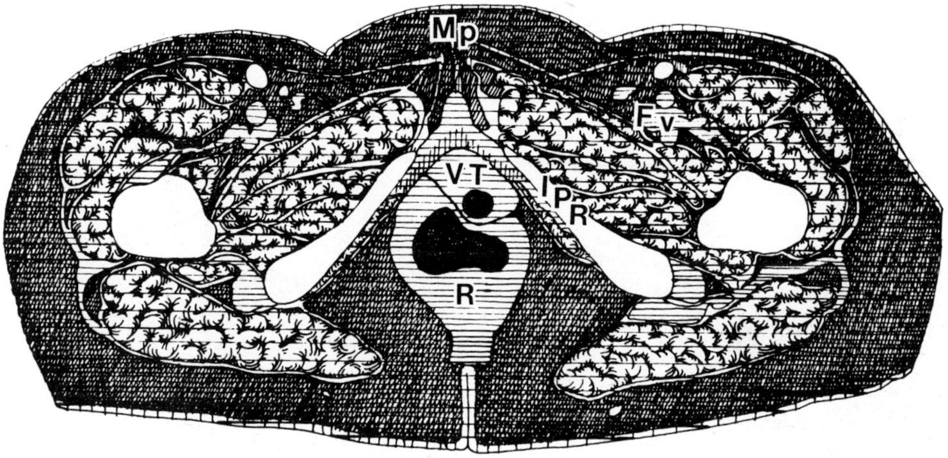

34

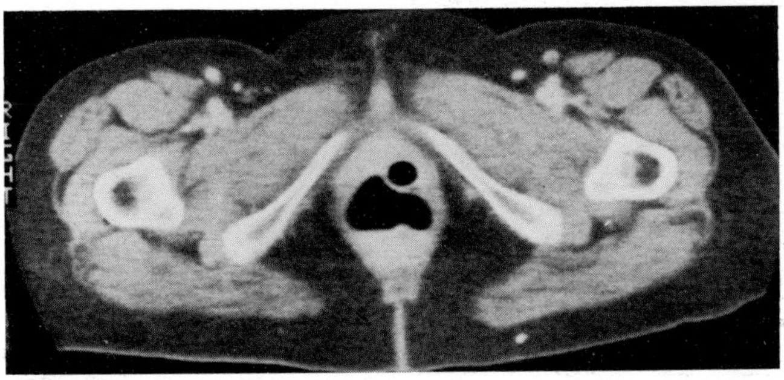

35

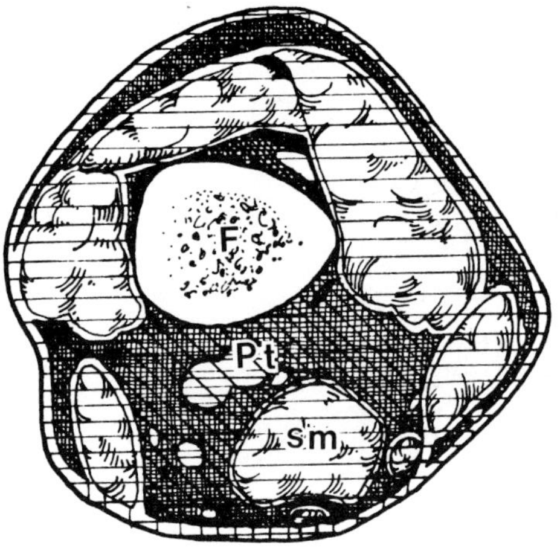

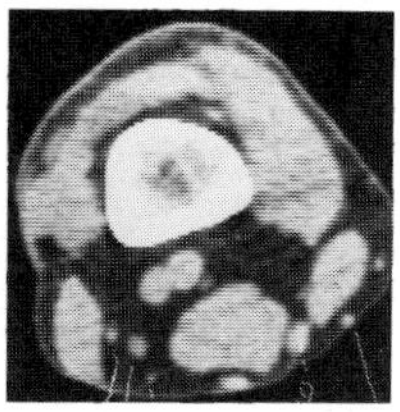

36

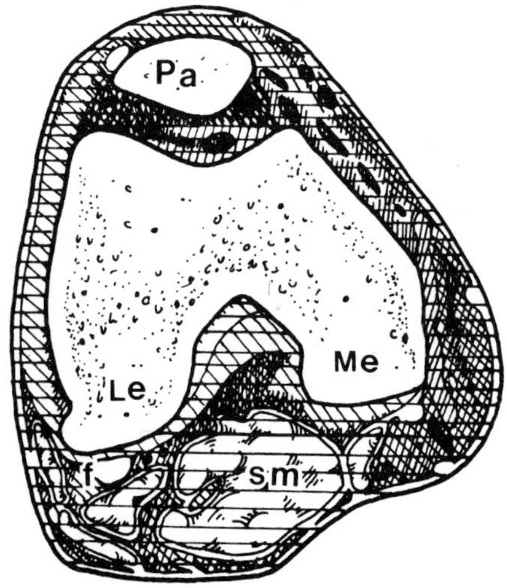

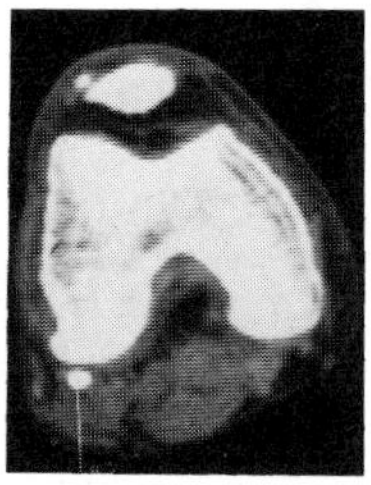

37

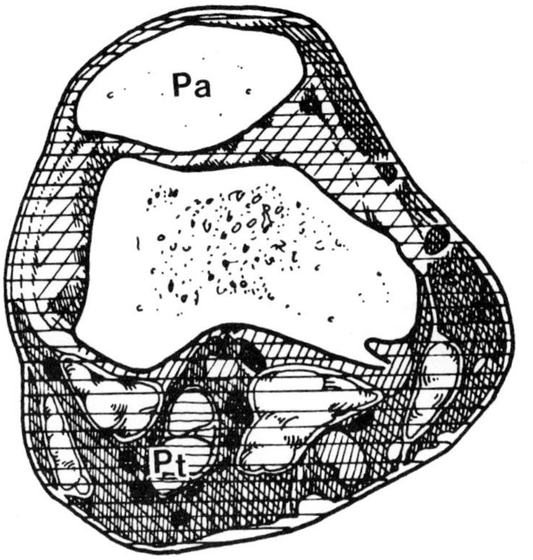

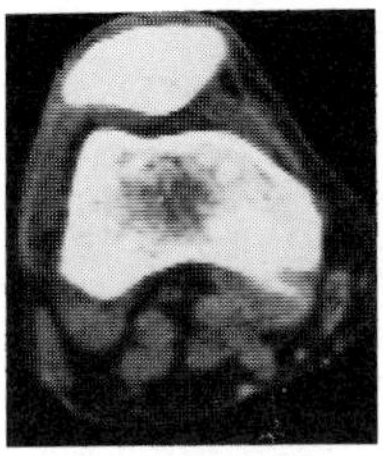

38

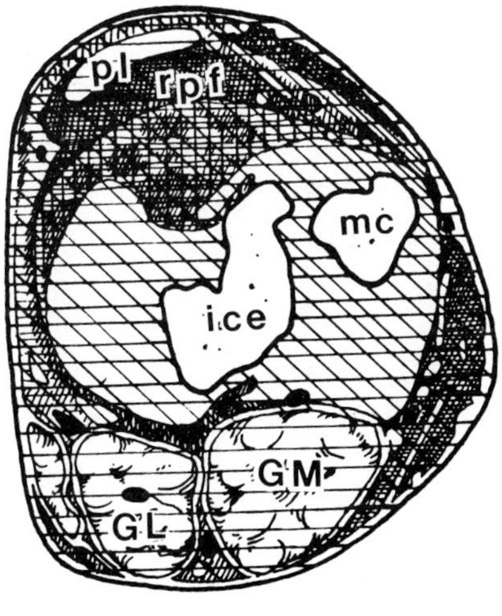

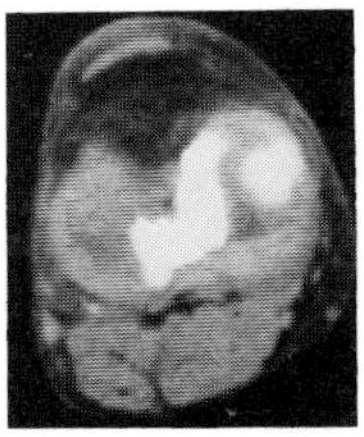

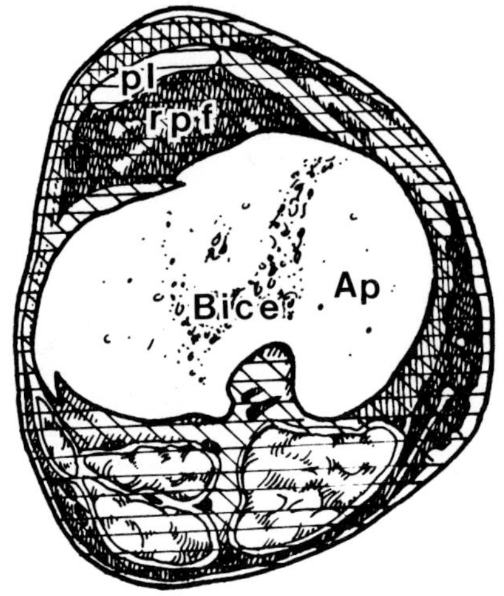

39

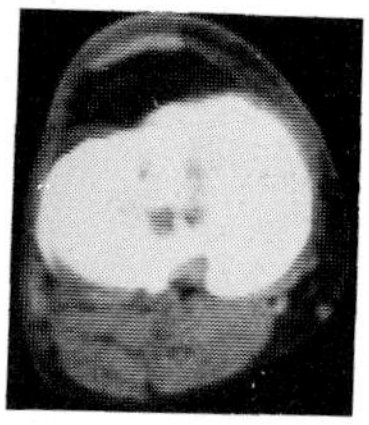

KEY TO COMPUTED TOMOGRAMS

A Aorta
Aa Aortic arch
Abl Abnormal larynx
AC Arytenoid cartilage
ac Anal canal
Ac Anterior cerebral a.
Ap Articular plateau
At Acetabulum
Av Axillary vein

B Basilar a.
Ba Brachiocephalic a.
Bice Base intercondylar eminence
Bl Bladder
BP Bulb of penis
Bv Brachiocephalic v.

CC Common carotid a.
Cc Carotid canal
ce Coeliac a.
ci Common iliac a.
CL Cerebellum
Cl Clavicle
cp Crus of penis
CVj Costovertebral joint

DS Dorsum sellae
dc crus of diaphragm

E Esophagus
Ec External carotid a.
Ej External jugular v.
Em External acoustic meatus

F Femur
fs Frontal sinus
ft Foramen transversarium
Fv Femoral vessels

G Globe
g Gas
GL Gastrocnemius, lateral head
GM Gastrocnemius, medial head
GT Greater trochanter

H Hyoid
HF Head of femur
Hh Head of humerus
Hm Head of mandible

IC Internal carotid a.
ice intercondylar eminence
Ij Internal jugular v.
IL Ilium
IPR Ischiopubic ramus
IS Ischium
ITC Inferior horn of the thyroid cartilage
ivc Inferior vena cava

K Kidney

L Lens
La Left atrium
Le Lateral epicondyle
Li Liver
Ln Lung
LP Laryngopharynx
Lv Left ventricle
Lvo Left ventricular outflow

MB Midbrain
Mc Middle cerebral a.
mc Medial condyle
Me Medial epicondyle
MN Mandible
Mp Mons pubis
MS Mastoid cells
Ms Maxillary sinus

nc Nasal cavity
ns Nasal septum

OC Optic canal
On Optic nerve
OP Oropharynx

P Pancreas
Pa Patella
PB Pubis
PC Pleural cavity
Pc Posterior cerebral a.
Pco Posterior communicating a.
Pl Pterygoid lamina
pl Patellar ligament

PNS Pons
PR Prostate
Ps Psoas
Pv Portal v.
Pt Popliteal neurovascular bundle

R Rectum
Ra Right atrium
ra renal artery
rg rima glottidis
Rpa Right pulmonary a.
rpf Retropatellar fat
Rv Right ventricle
RSJ Rectosigmoid juncture
rv Renal vein
rvo Right ventricular outflow

S Subclavian a.
SC Spermatic cord
se Seminal vesicle
SG Subglottic space
Sg Suprarenal gland
Sm Semimembranosus
sma Superior mesenteric a.
SP Spleen
Sp Scapula
spv Splenic v.
Ss Spine of scapula
ss Sphenoid sinus
sss Superior sagittal sinus
St Stomach
Sv Subclavian vein
Svc Superior vena cava

T Trachea
TC Tentorium cerebelli
Th Thyroid cartilage
TN Tongue
TSC Tumor of spermatic cord

u Ureter
Ua Uterus and adnexa

V Vagina
Va Vertebral a.
VT Vagina with tampon

Z Azygous vein

EPONYMS

BACK

Cruveilhier, Joint of--median atlanto-axial joint.
Cruveilhier, Plexus of--communication of C1-3, deep to semispinalis capitis.
Huber, Ganglion of--a sensory ganglion occasionally found on C1.
Lauth, Ligament of--transverse atlantal ligament.

SCALP, CRANIAL VAULT, MENINGES AND BRAIN

Andersch, Ganglion of--inferior ganglion IX.
Arnold, Canal of--canal for lesser petrosal nerve in temporal bone.
Breschet, Sinus of--sphenoparietal dural venous sinus.
Dorello, Canal of--occasionally found at tip of temporal bone, surrounding inferior petrosal sinus and VI.
Ehrenritter, Ganglion of--superior ganglion of IX.
Englisch, Sinus of--inferior petrosal dural venous sinus.
Fallopius, or Torini, Canal and Foramen--facial canal and foramen.
Galen, Vein of--great cerebral vein.
Gasserian Ganglion--trigeminal ganglion.
Meckel, Cave of--depression in sphenoid bone and the subarachnoid space around the ganglion of V.
Pacchonian Bodies--arachnoidal granulations.
Pacchonian Foramen--tentorial incisure.
Portio major--sensory root V.
Portio minor--motor root V.
Ribe, Ganglion of--a sympathetic ganglion associated with anterior communicating artery of Circle of Willis.
Ridley, Sinus of--circular sinus, the 2 cavernous and 2 intercavernous dural venous sinuses which form a ring around the pituitary.
Walther, Plexus of--plexus of sympathetic nerves in the cavernous sinus from the carotid nerve.
Wormian Bones--sutural bones.

FACE

Mimetic Muscles

Albinius Muscle--risorius.
Canine Muscle--levator anguli oris.
Coiter, Muscle of--corrugator supercilii.
Santorini, Muscle of--risorius.

Parotid Region

Stensen, Duct of--parotid duct.
Stensen, Plexus of--venous network around parotid duct.

Sinuses

Highmore, Antrum of--maxillary sinus.

Temporomandibular Joint

Ferrein, Ligament of--lateral ligament of TMJ.
Sappey, Ligament of--posterior portion of TMJ capsule.

Vascular

External Maxillary Artery--facial artery.
Internal Maxillary Artery--maxillary artery.

ORBIT

Orbit, Lacrimal Apparatus

Beraud, or Krause, Valve of--fold at the junction of the lacrimal sac and duct.
Bianchi, Valve of--at lower end of the nasolacrimal duct.
Bochdalek, Valve of--in lacrimal canaliculus at the punctum.
Bruch, Gland of--lymph nodes found in the palbebral conjunctiva.
Ciaccio, Glands of--accessory lacrimal tissue.
Galen, Innominate Gland of--palbebral portion of the lacrimal gland.
Hasner, or Rosemuller, Valve of--lacrimal fold or plica.
Krause, Gland of--accessory lacrimal glands.
Meibomian Glands--tarsal glands.
Moll, Glands of--sudoriferous ciliary glands.
Muller, Muscle of--usually superior tarsus, can also refer to other smooth muscle in the orbit.
Soemmering, Ligament of--suspensory ligament of the lacrimal gland.
Terson, Glands of--conjunctival glands.
Waldeyer, or Wolfring, Glands of--coiled gland near the margins of the lids.
Zeis, Glands of--sebaceous ciliary glands.

Cross-Section of the Eye

Bowman, Muscle of--ciliary muscle.
Bozzi, Foramen of--macula retinae.
Fontana, space of--interval or gap at the iridocorneal angle.
Hesselbach, Fascia of--cribiform fascia.
Soemmering, Foramen of--fovea centralis.

Optic Nerve

Schwalbe, Space of--deep to sheath of optic nerve.
Tiedemann, Nerve of--nerve traveling with the central artery of the retina.
Zinn, Artery of--central artery of the retina.

Miscellaneous

Horner, Muscle of--lacrimal portion of the orbicularis oculi.
Papyracea, Lamina of--orbital plate of the ethmoid.
Mauchart, Ligaments of--check ligaments of the globe.
Tenon, Space of--episcleral space.
Schacher, Ganglion of--ciliary ganglion.
Lockwood, Ligament of--suspensory ligament of the globe.
Zinn, Ring of--common tendinous ring.

NOSE

Bochdalek, Ganglion of--in plexus of alveolar nerve about the root of maxillary canine teeth.
Bock, Nerve of--pharyngeal ramus of pterygopalatine ganglion.
Dental arteries--alveolar arteries.
Dental nerves--alveolar nerves.
Kiesselbach, Plexus of--arterial anastomoses in base of nasal septum.
Meckel, Ganglion of--pterygopalatine ganglion.
Scarpa, Nerve of--nasopalatine nerve.
Scarpa, Foramen of--openings in intermaxillary suture for nasopalatine nerves.
Stensen, Foramen of--incisive foramen.
Stensen, Canal of--incisive canal.
Valentin, Ganglion of--ganglion of superior alveolar nerve.
Valentin, Nerve of--connects pterygopalatine ganglion with the 6th cranial nerve.
Vidian, Artery of--artery of the pterygoid canal.
Vidian, Nerve of--nerve of the pterygoid canal.

MOUTH

Bartholin, Duct of--sublingual duct.
Langley, Ganglion of--collection of cells in the submandibular gland.
Morgagni, Foramen of--foramen cecum of tongue.
Rivinus, Ducts of--lesser sublingual ducts.
Wharton, Duct of--submandibular duct.

PHARYNX

Gillette, Suspensory Ligament of--cricoesophageal ligament.
Haller, Plexus of--branches of external laryngeal and sympathetic nerves on the inferior pharyngeal constrictor.
Henke, Spaces of--areas filled with connective tissue between the pharynx and esophagus, and the vertebral column.
Morgagni, Sinuses of--1) pharyngeal recess, 2) gap between occipital bone and superior constrictor, 3) laryngeal ventricle.
Passavant, Fold of--a ridge that may develop on posterior pharyngeal wall.
Petik, Diverticulum of--excessive pharyngeal recess.
Rosenmuller, Fossa of--pharyngeal recess.
Santorini, Cartilage of--corniculate.
Tourtual, Membrane of--quadrangular membrane.
Tourtual, Sinus of--supratonsillar fossa.
Waldeyer, Ring of--pharyngeal lymphatic ring.
Wrisberg, Cartilage of--cuneiform.

EAR

External

Arnold, Nerve of--auricular branch X.
Arnold, Ganglion of--otic ganglion.
Glaserian Fissure--petrotympanic fissure.
Meyer, Sinus of--a small depression in the floor of the auditory canal, near the ear drum.

Inner

Bottcher, or Cotunnius, Space of--endolymphatic space.
Bottcher, Canal of--connects saccule and utricle.
Bottcher, Ganglion of--on cochlear nerve in the internal acoustic meatus.
Cotunnius, Canal of--vestibular aqueduct.
Hensen, Canal of--ductus reuniens.
Scarpa, Ganglion of--vestibular ganglion.

Middle

Andersch, Nerve of--tympanic branch of IX.
Glaserian Artery--anterior tympanic artery.
Helmholtz, Axial Ligament of--axial ligament of malleus.
Huguier, Sinus of--fossa on medial tympanic wall between oval and round window.
Jacobson, Nerve of--tympanic branch IX.
Jacobson, Plexus of--tympanic plexus.
Prussak, Pouch of--tympanic recess.
Prussak, Space of--between head of malleus and pars flaccida.
Reissner, Membrane of--vestibular membrane.
Scarpa, Membrane of--secondary tympanic membrane, at round window.
Toynbee, Muscle of--tensor tympani.

Tympanic Membrane

Wilde or Woelde, Triangle of--cone of light.
Shrapnell or Rivinus, Membrane of--pars flaccida.

NECK

Beclard, Triangle of--bounded by greater cornu hyoid, posterior belly of the digastric, posterior border of the hypoglossus.
Chassaignac, Tubercle of--carotid tubercle of 6th cervical vertebra.
Farabeuf, Triangle of--bounded by internal jugular vein, hypoglossal nerve and facial nerve.
Gley, Glands of, or Sandstorm, Bodies of--parathyroid glands.
Heister, Diverticulum of--superior bulb of internal jugular vein.
Hering, or DeCastro, Nerve of--nerve to the carotid sinus from IX.
Lesser, Triangle of--between XII and two bellies of the digastric muscle.
Malgaigne, Triangle of--carotid triangle.
Neubauer, Artery of--thyroidea ima artery.
Pirogoff, Triangle of--formed by intermediate tendon of the digastric, posterior border of the mylohyoid, and XII.
Wolfler, Gland of--accessory thyroid gland(s).

UPPER EXTREMITY

Anterior Thoracic Nerve--pectoral nerves.
Bell, Nerve of--long thoracic.
Caldani, Ligament of--coracoclavicular.
Campbell, or Gerdy, Ligament of--suspensory ligament of the axilla.
Casser, Perforated Muscle of--coracobrachialis.
Dupuytren, Fascia of--palmar aponeurosis.

Erb, Point of--at the side of the neck, where the brachial plexus may be compressed, results in Erb's Paralysis.
Haver, Glands of--a misnomer for synovial membrane-covered depositions of adipose tissue in synovial joints.
Lacertus fibrosus--bicipital aponeurosis.
Lister, Tubercle of--dorsal radial tubercle.
Parono, Spaces of--between deep muscles of the forearm.
Pozzi, Muscle of--extensor digitorum brevis.
Weitbrecht, Cartilage of--articular wedge of the acromioclavicular joint.

THORAX

Breast

Chaissaignac, Space of--retromammary space.
Cooper, Ligament of--suspensory ligaments of the breast.
Marcacci, Muscle of--smooth muscle deep to the areola and nipple.
Montgomery, Glands of--large sebaceous gland in mammary areola.
Sappey, Plexus of--lymphatics in the areola.
Spence, Tail of--axillary extension of breast tissue, a normal phenomenon.

Posterior and Superior Mediastinum

Bock, or Laumonier, Ganglion of--carotid ganglion.
Cyon, or Ludwig, Nerve of--aortic nerve.
Innominate Vessels--brachiocephalic vessels.
Lobstein, Ganglion of--occasional enlargement of greater splanchnic nerve superior to diaphragm.
Teutleben, Ligament of--pulmonary ligament.
Van Hoorne, Canal of--thoracic duct.
Vieussens, anulus of--ansa subclavia.

Middle Mediastinum

Albinius, Nodules of--nodules on margins of the A-V valves.
Arantius, Morgagni, or Bianchi, Nodules of--nodules on cardiac semilunar valves.
Bezold, Ganglion of--nerve cells in the interatrial septum.
Botallo, Duct and Ligament of--duct and ligamentum arteriosum.
Eustachius, Valve of--valve of the IVC.
Keith and Flack, or Koch, Node of--sinoatrial node.
Lannelongue, Foramina of--openings of the thebesian veins.
Ludwig, Ganglion of--collection of parasympathetic nerve cells in the interatrial septum.
Luschka, Ligament of--sternopericardial ligament.
Marshall, Vein of--oblique vein of the left atrium.
Oehl, Muscle of--muscle fibers in the chordae of the left A-V valve.
Rindfleisch, Fold of--semilunar shaped folds of serous pericardium at the beginning of the aorta.
Tawara, Node of--A-V node.
Thebesian valve--valve of the coronary sinus.
Thebesian veins--least cardiac veins.
Valsalva, or Petit, Sinus of--aortic sinuses.
Vieussens, Ring of--limbus of fossa ovalis.

Westberg, Space of--around origin of aorta, where it is invested with pericardium.
Wrisberg, Ganglion of--cardiac ganglion.

Miscellaneous

Burns, Space of--suprasternal.
Godman, Fascia of--continuation of pretracheal fascia into the thorax and onto pericardium.
Guenau de Mussy, Point of--from sternal line to bony termination of 10th rib, painful in diaphragmatic pleurisy.
Internal Mammary Artery--internal thoracic artery.
Lisfranc, Tubercle of--scalene tubercle of 1st rib.
Sibson, Fascia of--suprapleural membrane.
Sibson, Muscle of--scalenus minimus.
Troisier, Ganglion of--enlarged supraclavicular node, a "signal node."

ABDOMEN

Colon

Gehrlach, Valve of--fold or 'valve' at ostium of vermiform appendix.
Hartmann, Line of--where lowest sigmoid artery anastomoses with the superior rectal artery.
Hensing, Ligament of--left superior colic ligament.
Riolan, Arc of--anastomoses of left and middle colic arteries.
Sudek, Point of--where lowest sigmoid artery anastomoses with the superior rectal artery.
Treves, Bloodless fold of--ileocaecal fold.

Diaphragm

Bochdalek, Foramen of--pleuroperitoneal hiatus, incomplete partitioning of pleural and abdominal cavities by the diaphragm; occurs posterior lateral portions of diaphragm; most common on left side.
Morgagni, Foramen of--failure of sternal and costal elements of diaphragm to fuse; may be bilateral at parasternal line.

Kidney

Brodel, Line of--bloodless division between posterior and anterior branching of the renal artery.
Gerota, Fascia of--renal fascia.
Toldt, Membrane of--anterior layer of the renal fascia.

Stomach

Braune, Valve of--fold of mucous membrane at the cardioesophageal junction.
Gavard, Muscle of--obliquely oriented fibers of the stomach wall.
Remak, Ganglion of--autonomic ganglia associated with nerves of the stomach.
Winslow, or Duverney, Foramen of--epiploic foramen.

Liver and Gallbladder

Arantius, Ligament of--ligamentum venosum.
Boyden, Sphincter of--sphincter of the cystic duct.
Calot, Triangle of--cystohepatic triangle, bounded by cystic and hepatic ducts of the gallbladder, and the liver.
Glisson, Capsule of--capsule of the liver.
Hartmann, Pouch of--pelvis of the gallbladder.
Heister, or Amussat, Spiral valve of--spiral valve of the cystic duct.
Labbe, Triangle of--where the stomach touches anterior abdominal wall, bounded by the inferior margin of the liver, a horizontal line at the left 9th rib, and the line of the false ribs.
Morison, Pouch of--hepatorenal pouch.
Retzius or Ruysch, Veins of--retroperitoneal porta-caval anastomoses.
Spigelius, Lobe of--caudate lobe of the liver.

Nerves

Latarjet, Nerve of--superior hypogastric nerve.
Zuckerkandl, Organ of--para-aortic catecholamine-producing tissue bodies near origin of inferior mesenteric artery.

Pancreas and Spleen

Malpighi, Capsule of--splenic capsule.
Oddi or Glisson, Sphincter of--sphincter of the hepatopancreatic ampulla.
Santorini or Bernard, Duct of--accessory (lesser) pancreatic duct.
Toldt, Fascia of--fascia posterior to the body of the pancreas.
Treitz, Fascia of--fascia posterior to the head of the pancreas.
Vater, Ampulla of--hepatopancreatic ampulla.
Vater, Fold of--a fold of the mucous membrane proximal to the hepatopancreatic ampulla.
Wirsung, Duct of--main pancreatic duct.

Small Intestine

Wilkie, Artery of--supraduodenal artery.
Pecquet, Cistern of--cisterna chyli.
Aselli, Gland of--mesenteric lymph nodes.
Vieussens, Ganglion of--celiac ganglia, associated nodes and fibers.
Auerbach, Plexus of--myenteric division of gastro-intestinal nerve plexus.
Bauhin, Varolius or Tulp, Valve of--ileocaecal valve.
Mayo, Vein of--prepyloric vein.
Brunner, Gland of--duodenal glands.
Heller, Plexus of--small arteries of the intestinal wall.
Meissner or Remak, Plexus of--submucosal division of gastro-intestinal nerve plexus.
Varolius, Sphincter of--sphincter of terminal ileum.
Meckel, Diverticulum of--diverticulum of ileum, persistent part of the proximal yolk sac.

Topography

Bochdalek, Gap of--lumbocostal triangle.
Douglas, Line of--arcuate line of the rectus sheath.
Gimbernat, Ligament of--lacunar ligament.

Grynfeldt, or Lesshaft, Triangle of--triangle bounded by the quadratus lumborum, internal abdominal obliques and serratus posterior superior.
Henle, Ligament of--portion of the conjoint tendon, with transversalis fascia and transversus abdominus.
Hesselbach, Ligament of--fascia associated with inferior epigastric vessels.
Hunter, Line of--linea alba.
Imlach, Ring of--the portion of the inguinal canal in the female which contains the round ligament of the uterus.
Lian, Point of--2/3 distance from umbilicus to ASIS where a trocar may be inserted in paracentesis.
Mayo-Robson, Point of--superior and right of the umbilicus, painful in pancreatic disease.
McBurney, Munro or Clado, Point of--junction arcuate and interspinous planes, rebound point of pain in appendicitis, overlies position of appendix.
Petit, Triangle of--lumbar triangle.
Poupart, Vesalius or Fallopius, Ligament of--inguinal ligament.
Poupart, Line of--mid-inguinal line.
Sappey, Veins of--veins of falciform ligament.
Spigelius, Line of--semilunar line.
Velpeau, Canal of--inguinal canal.

Vasculature, Posterior Abdominal Wall

Abernathy, Fascia of--areolar tissue anterior to external iliac artery.
Adamkiewicz, Artery of--great spinal radicular artery, usually occurs on left side, major blood supply to spinal cord below thoracic level.
Elaut, Triangle of--bounded by iliac arteries and sacral promontory.
Hypogastric artery--internal iliac.

PELVIS AND PERINEUM

Male Pelvis

Henle, Ampulla of--ductus deferens
Proust, Space of--rectovesical space
Weber, Organ of--prostatic utricle
Tyrrell, or Denonvillier, Fascia of--rectovesical septum

Female Pelvis

Clado, Ligament of--mesenteric fold from the broad ligament to the appendix on the right side.
Gartner, Duct of--mesonephric remnant in the mesometrium.
Douglas, Pouch of--rectouterine.
Jacque, Plexus of--nerve plexus in the muscle of the Fallopian tubes.
Jarjavay, or Petit, Ligament of--uterosacral ligament.
Hunter, Ligament of--round ligament of the uterus.
Mackenrodt, Ligament of--cardinal ligaments of the uterus.
Poirer, Gland of--node on the uterine artery at the juncture with the ureter.
Petit, Fold of--uterosacral fold.
Lee or Frankenhauser, Ganglion of--uterovaginal nerve plexus.
Nuck, Canal of--persistent processus vaginalis.

Rectum

Braune, Muscle of--puborectalis.
Houston, of Kohlrausch, Valves of--transverse rectal folds.
Hyrtl, or Nelaton, Sphincter of--inconstant circular bands in rectal wall about 3/4 inch above anus.
Obeirne, Sphincter of--circular muscle in upper rectum.

Bony Pelvis

Bichat, Ligament of--lower portion posterior sacroiliac joint.
Capuron, Points of--iliopubic eminences and sacroiliac joints, 4 fixed points of the pelvic inlet.
Walther, Ganglion of--ganglion impar.
Weber, Point of--center of gravity of body, on sacral promontory.

Bladder

Bell, Muscle of--muscle underlying lateral edges of the urinary trigone, running from the uvula to ureteric orifices.
Guthrie, Muscle of--sphincter of urethrae.
Lietaud, Trigone of-urinary trigone.
Mercier, Bar of--interureteric ridge.
Mercier, Valve of--name given an occasional membranous fold in the bladder which may restrict urine outflow at urethra.
Retzius, Space of--retropubic space.
Voillemier, Point of--mid-sagittal point 2 1/2" below interspinous plane, where bladder may be punctured when distended.

Anal Triangle

Alcock, Canal of--pudendal canal.
Ball, Valve of--anal valves.
Cruveilhier, Fascia of--superficial perineal fascia.
Gay, Glands of--circumanal glands.
Hilton, or Stroud, Line of--pecten of anal canal.
Krause, Ligament of--transverse perineal ligament.
Luschka, Body of--coccygeal body, an arteriovenous anastomoses on the anterior surface of the coccyx.
Morgagni, Columns of Sinuses of--anal columns and anal sinuses.
Quenu, Hemorrhoidal Plexus of--lymphatic plexus around anus.
Schaumann, Body of--central tendon of perineum.
Symington, Body of--anococcygeal ligament.
Theile, Muscle of--superficial transverse perineal muscle.

Female Perineum

Bartholin, or Tiedemann or Duverney, Gland of--major vestibular glands.
Guerin, Glands of--urethral glands of the female.
Skene, Glands of and Ducts of--paraurethral glands and ducts in the female.

Male Perineum

Cowper, or Mery, Glands of--bulbourethral glands.
External Spermatic Artery--cremasteric artery.
Internal Spermatic Artery--testicular artery.

Guerin, Fold of--valve of the fossa navicularis.
Highmore, Body of--mediastinum testis.
Littre, Glands of--urethral glands.
Morgagni, Appendix of--appendix of the testis or epoophoron.
Riolan, Muscle of--cremaster muscle.
Sebileau, Muscle of--deep fibers of Darto's tunic.
Veslingius, Line of--scrotal raphe.

LOWER EXTREMITY

Astragalus--talus.
Bertin, or Bigelow, 'Y' Ligament of--iliofemoral ligament.
Chopart, Joint of--transverse tarsal joints.
Cloquet, or Rosenmuller, Node of--nodes at the femoral ring.
Cruveilhier, Ligament of--glenoid or plantar ligaments of metatarso-phalangeal articulations.
Hunter, Subsartorial Canal of--adductor canal.
Lisfranc, Joint of--tarso-metatarsal joints.
Maissat, Tract of--iliotibial tract.
Retzius, Ligament of--fundiform ligament of the foot.
Scarpa, Triangle of--femoral triangle.
Weber, Triangle of--from sole of the foot and center of the plantar surface to the heads of the 1st and 5th metatarsals.
Wrisberg, or Robert, Ligament of--posterior meniscofemoral ligament.

SOURCE MATERIAL

A Manual of Human Anatomy, J. Aitken, G. Causey, J. Joseph, J. Young, Churchill Livingstone, Edinburgh, 1972.

An Introduction to Functional Anatomy, D. Sinclair, Blackwell Science Publications, Oxford, 1970.

Anatomy, E. Gardner, D. Gray, R. O'Rahilly, W.B. Saunders Co., Philadelphia, 1969.

Anatomy for Anaesthetists, H. Ellis, S. Feldman, Blackwell Science Publications, Oxford, 1977.

Anatomy of the Human Body, R. Lockhart, G. Hamilton, F. Fyfe, J.B. Lippincott Co., Philadelphia, 1969.

The CIBA Collection of Medical Illustrations, F. Netter, CIBA Publications, New Jersey, 1962.

Clinical Anatomy, O. Brantigan, McGraw-Hill Book Co., Inc., New York, 1963.

Clinical Anatomy, H. Ellis, Blackwell Science Publications, Oxford, 1971.

Clinical Anatomy for Medical Students, R. Snell, Little, Brown & Co., Boston, 1981.

Clinically Oriented Anatomy, K. L. Moore, Williams & Wilkins, Baltimore, 1980.

Cunningham's Manual of Practical Anatomy, J. Brash and E. Jameson, Editors, Oxford University Press, New York, 1940.

Cunningham's Text-Book of Anatomy, A. Robinson, Editor, William Wood & Co., New York, 1914.

Essentials of Human Anatomy, R. T. Woodburne, Oxford University Press, New York, 1978.

Gray's Anatomy, 28th American Edition, C. M. Goss, Editor, Lea and Febiger, Philadelphia, 1966.

Gray's Anatomy, 35th British Edition, R. Warwick and P. Williams, Editors, W. B. Saunders Co., Philadelphia, 1973.

Human Anatomy A Synoptic Approach, N. Mizeres, Elsevier, New York, 1981.

Morris' Human Anatomy, C. Jackson, Blakiston's Son & Co., Philadelphia, 1925.

Primary Anatomy, J. Basmajian, Williams & Wilkins, Baltimore, 1976.

Synopsis of Gross Anatomy, J. Christensen, Harper & Row, Philadelphia, 1982.

Textbook of Anatomy, W. H. Hollinshead, Harper & Row, Hagerstown, 1974

Textbook of Human Anatomy, R. Crafts, John Wiley & Sons, New York, 1979.